Battle, Bards and Brāhmins

PAPERS OF THE 13TH WORLD SANSKRIT CONFERENCE

Edinburg, Scotland (UK) [10-14 July 2006]

General Editors

Peter Bisschop

Paul Dundas

ISBN: 978-81-208-3529-0(Set)

PAPERS OF THE 13TH WORLD SANSKRIT CONFERENCE
Volume II

Battle, Bards and Brāhmins

Edited by
JOHN BROCKINGTON

MOTILAL BANARSIDASS PUBLISHERS
PRIVATE LIMITED • DELHI

First Edition : Delhi, 2012

ISBN: 978-81-208-3604-4

MOTILAL BANARSIDASS
41 U.A. Bungalow Road, Jawahar Nagar, Delhi 110 007
8 Mahalaxmi Chamber, 22 Bhulabhai Desai Road, Mumbai 400 026
203 Royapettah High Road, Mylapore, Chennai 600 004
236, 9th Main III Block, Jayanagar, Bangaluru 560 011
Sanas Plaza, 1302 Baji Rao Road, Pune 411 002
8 Camac Street, Kolkata 700 017
Ashok Rajpath, Patna 800 004
Chowk, Varanasi 221 001

Printed in India
by RP Jain at NAB Printing Unit,
A-44, Naraina Industrial Area, Phase I, New Delhi–110028
and published by JP Jain for Motilal Banarsidass Publishers (P) Ltd,
41 U.A. Bungalow Road, Jawahar Nagar, Delhi-110007

General Preface

The Thirteenth World Sanskrit Conference was held in Edinburgh, Scotland on the 10th–14th July 2006. To the delight of the organisers it was attended by a record number of participants from all parts of the world. At a time when the study of Sanskrit and related Indological subjects is becoming increasingly embattled in certain countries, such regular gatherings assume a particular importance in confirming scholarly solidarity as well as disseminating the most recent fruits of research.

The conference was divided into fourteen subject sections, each chaired by two scholars: 1. Veda; 2. Epics; 3. Purāṇas; 4. Āgamas and Tantra; 5. Vyākaraṇa; 6. Linguistics; 7. Poetry, Drama and Aesthetics; 8. Scientific Literature; 9. Buddhist Literature; 10. Jaina Studies; 11. Philosophy; 12. History, Epigraphy and the Arts; 13. Law and Society; and 14. Culture and Society.

As now customary, arrangements for the publication of the proceedings became the responsibility of the conference organisers. To this end, the various section chairs were requested to edit the proceedings volume of their subject after inviting the submission of papers from participants and then submit a manuscript to the general editors for inspection and final formatting. It should be noted that the chairs of two thematic subpanels of the Philosophy section ('Śāstrārambha-Philosophical Introduction' and 'New Directions in the Study of Yoga'), of History, Epigraphy and the Arts, and of Law and Society, wished to make their own arrangements for publication elsewhere. In addition, it was decided that papers in the Culture and Tradition section which had been invited for publication should be placed in appropriate volumes of the proceedings.

The general editors would like to thank Petteri Koskikallio and Lumi Sammallahti for taking expert control of the final formatting and technical production of the proceedings. It is anticipated that these Proceedings of the Thirteenth World Sanskrit Conference will reveal the impressive breadth and depth of Indological research at the present time and gain the appreciation of the interested scholarly world.

PETER BISSCHOP
PAUL DUNDAS

General Editors,
Proceedings of the 13th World Sanskrit Conference

Introduction

The range and diversity of the Sanskrit epics themselves was reflected in the diversity of the papers delivered in the epics section of the 13th World Sanskrit Conference held at Edinburgh in July 2006, while the number of participants testifies to the strength of epic studies within the discipline of Sanskrit studies as a whole. Papers were presented by 25 scholars in all and 19 of those are included in this volume in revised form, taking account of the active discussion which they prompted during the conference. In addition, the following papers are being published elsewhere: Robert P. Goldman's 'Rules of Engagement: War Crimes, *Rākṣasa* rights and the Political and Military strategies of the Great Sanskrit Epics' (in *Śrutimahatī*, a felicitation volume for Professor R. K. Sharma, the retiring President of the International Association of Sanskrit Studies), John Smith's 'Consistency and character in the Mahābhārata' (in the *Bulletin of the School of Oriental and African Studies*), Przemysław Szczurek's 'Juggling with *ātman*: Remarks on the Bhagavadgītā 6.5–6' (in the felicitation volume for Professor M. K. Byrski), and Laurie L. Patton's '"How do you conduct yourself?": dialogical gender in the Mahābhārata', which she had intended to present at the conference but was unable to give for personal reasons (in *Gender and Narrative in the Mahābhārata*, ed. by Simon Brodbeck & Brian Black, 2007).

Inevitably, there was an overlap with the section on Purāṇas and, while Yaroslav Vassilkov's paper on the boar myth in the epics and Purāṇas was included in the epics section and so appears in this volume, that by André Couture on the grouping of the four Vṛṣṇi heroes in the *Harivaṃśa* was presented in the Purāṇa section (and has since been published in the *Journal of Indian Philosophy* 34 [2006]: 571–585). The majority of the articles in this volume relate primarily to the *Mahābhārata* but there are also five on the *Rāmāyaṇa*, if Vidyut Aklujkar's article relating to the *Ānandarāmāyaṇa* is included under that head, and a single article on the *Harivaṃśa*, which is still the poor relation of epic studies, despite the efforts at the 3rd Dubrovnik International Conference on the Sanskrit Epics and Purāṇas to generate greater interest in it.

Approaches to the epics were indeed extremely varied. While this was only to be expected from the broad group of scholars assembled in Edinburgh from all over the world, it has made the task of arranging the papers into a coherent and meaningful order a problematic one. There are of course certain themes which either form the main subject of certain papers or underlie the discussion in others. Textual issues are addressed most directly in the article by Wendy Phillips-Rodriguez on constructing a stemma for *Mahābhārata* manuscripts making use of techniques drawn from the natural sciences and statistics (phylogenetics and cladistics) but they form an element in several others. Two articles, those by Nick Allen and Simon Brodbeck, examine genealogical issues. Alf Hiltebeitel charts the terminology of friendship in relation to the concept of *bhakti*. Narrative techniques are discussed along rather different lines by James Hegarty in relation to the narrative on *tīrtha*s in the *Mahābhārata* and by Mary Brockington in relation to the plot of the *Rāmāyaṇa*. What is superficially a traditional word study by Sven Sellmer on *hṛd* and related terms is enriched in fact by understandings drawn from modern psychology, while Antonella

Cosi looks at the way that similes are used in speeches in two books of the *Mahābhārata*.

Whereas these articles look at the texts in terms of their structure and their major themes, others concentrate more on content. Individual characters and episodes form the focus for several articles. Thus we have studies of Bhīma's quest for the *saugandhika* flowers by Danielle Feller and of Uttaṅka's quest for the earrings by Paolo Magnone, of Duryodhana's truths by Angelika Malinar, of the story of Viśvāmitra by Adheesh Sathaye, of Nikumbhilā and her grove by Sally Sutherland Goldman and – though they are less individualised – of Kṛṣṇa's many wives by Horst Brinkhaus. Jim Fitzgerald examines one of the pairs of *adhyāya*s devoted to Sāṃkhya and Yoga in the Mokṣadharmaparvan (Mbh 12.289–290). John Brockington surveys again the weapons mentioned in the early *Rāmāyaṇa*, while Urmi Shah documents the relationship of the pronouncements on *rājanīti* found in certain *sarga*s of the *Rāmāyaṇa* with those of the later text, the *Nītiprakāśikā* of Vaiśaṃpāyana.

The articles are grouped according to the text that they are mainly focussing on (in the order *Mahābhārata*, *Harivaṃśa*, *Rāmāyaṇa*) and within that in an order which seeks to bring related topics together, beginning with the genealogical issues that underpin all of the narrative in each of the texts and ending with the study of a retelling of one of them, the *Ānandarāmāyaṇa*. The editor has ensured a degree of uniformity of appearance and of bibliographic reference but had no wish to rein in the diversity of expression and approach to be found in the feast of articles here assembled.

JOHN BROCKINGTON

May 2007

Contents

Vaiśaṃpāyana's *Mahābhārata* Patriline

SIMON BRODBECK

Vyāsa's pupil Vaiśaṃpāyana speaks to King Janamejaya at the latter's *sarpasatra*, an event that has been introduced by Ugraśravas, who is re-narrating Vaiśaṃpāyana's speeches on the subsequent occasion of Śaunaka's *satra* in Naimiṣa Forest.[1] In Ugraśravas's story, Janamejaya asks about the Kuru *vaṃśa* (1.62.2), and Vaiśaṃpāyana begins to tell of Janamejaya's ancestors (Janamejaya is in this royal line). Vaiśaṃpāyana tells the story of the lovely Śakuntalā, with whom King Duḥṣanta dallied once in the woods, later spurned at court, but then took back (1.62–69), and whose son Bharata gave his name to the line and to Vaiśaṃpāyana's (and Ugraśravas's, and thence Sukthankar *et al.*'s) story; and then he begins at the beginning, long before Duḥṣanta's time, and goes through the generations: a list of names, with occasional asides (1.70). He breaks at King Yayāti, telling stories in connection with him for eighteen *adhyāya*s (pivoting on 1.79–80, in which Pūru becomes Yayāti's heir) before continuing in 1.89 as far as Śaṃtanu. In 1.90 Janamejaya asks Vaiśaṃ-

[1] I would like to acknowledge: Julia Leslie, who encouraged me to plot the patriline with women in capitals (although the implications of this will not be fully brought out here); all those who have commented on earlier versions of this paper or its ideas, especially Sîan Hawthorne; and the Arts and Humanities Research Council for its funding of my research and my trip to the Edinburgh conference. References are to the Poona *Mahābhārata* (1933–66); translations are those of van Buitenen (1973; 1975) unless otherwise stated or modified.

pāyana to tell the patriline again, and he does so. The first version was in *śloka* verse; the second is in prose, and continues past Śaṃtanu to include the principal characters in the Pāṇḍava story which Vaiśampāyana has already introduced (1.58–61) and will now tell at great length after giving the *vaṃśas*. The second version also comes right down to the present at the time of the telling, to include King Janamejaya himself, the listening great-grandson of the Pāṇḍavas, and his sons and grandson.

In what follows I chart the two versions of Janamejaya's ancestry side by side, and make some preliminary and inconclusive patrilineal observations based on their synopsis and the details and stories of the line's characters given at other points in Vaiśampāyana's address. In the discussion I mention various operational features of the patriline by way of introduction, and then I progress chronologically, focusing on selected points in the *vaṃśa*, and ending with the Pāṇḍavas. I do not cover the events occurring beyond Vaiśampāyana's Pāṇḍava narrative, i.e. after the retirement of the Pāṇḍavas and their wife Draupadī (leaving Arjuna, Pāṇḍava's grandson King Parikṣit and Yuyutsu under Kṛpa's supervision in Hāstinapura, with the Vṛṣṇi King Vajra in Indraprastha, 17.1). Those latter events are, I believe, the key to the understanding of the *Mahābhārata* as a patrilineal Bhārata story;[2] but before he begins to report Vaiśampāyana's narration, they are told so scantily by Ugraśravas – first to the *ṛṣis* while Śaunaka the *kulapati* is still busy tending his fires (in 1.1–3; the *ṛṣis* interrupt and ask Ugraśravas to wait and tell it properly when Śaunaka comes), and then to Śaunaka and the *ṛṣis* (ob-

---

[2] I call all the people picked out through time by the *Mahābhārata*'s central patriline the Bhāratas, following the text's title (and in honour of Bharata's *vaṃśa*-prefatorial prominence in the Śakuntalā story), even though the man from whom they would get this name appears only about half-way through it. For Vedic details of Bharata and other characters in the *vaṃśa*, see Witzel 2005 ('the MBh lineage of the Bharata-Kuru up to the eponymous Kuru and Pratīpa coincides more or less with the Vedic evidence', p. 40).

liquely in 1.4–54) – that they might be approached back-to-front, as it were, through Vaiśaṃpāyana's narration. To this eventual end, and notwithstanding the obvious rider that Vaiśaṃpāyana's narration as reported by Ugraśravas cannot properly stand alone (for it must already be conditioned most recently by intervening events), I hope that that narration – the vast majority of the *Mahābhārata*, continuous from 1.55.1 to 18.5.25 except for brief speeches from Janamejaya (and occasionally others, see Hiltebeitel 2006), and focused here around the idea of a *vaṃśa*'s lineal continuity as set out in 1.70–90 – may provide some tools for future use.

ŚLOKA VAṂŚA (1.70, cont'd 1.89) — **PROSE *VAṂŚA*** (1.90)

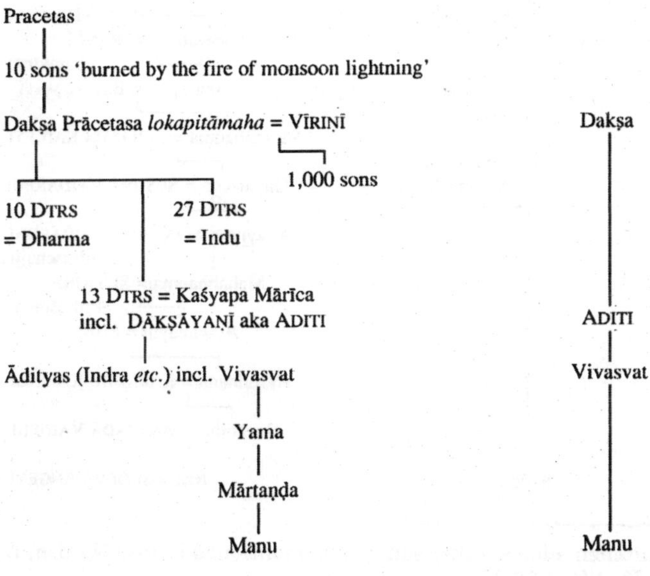

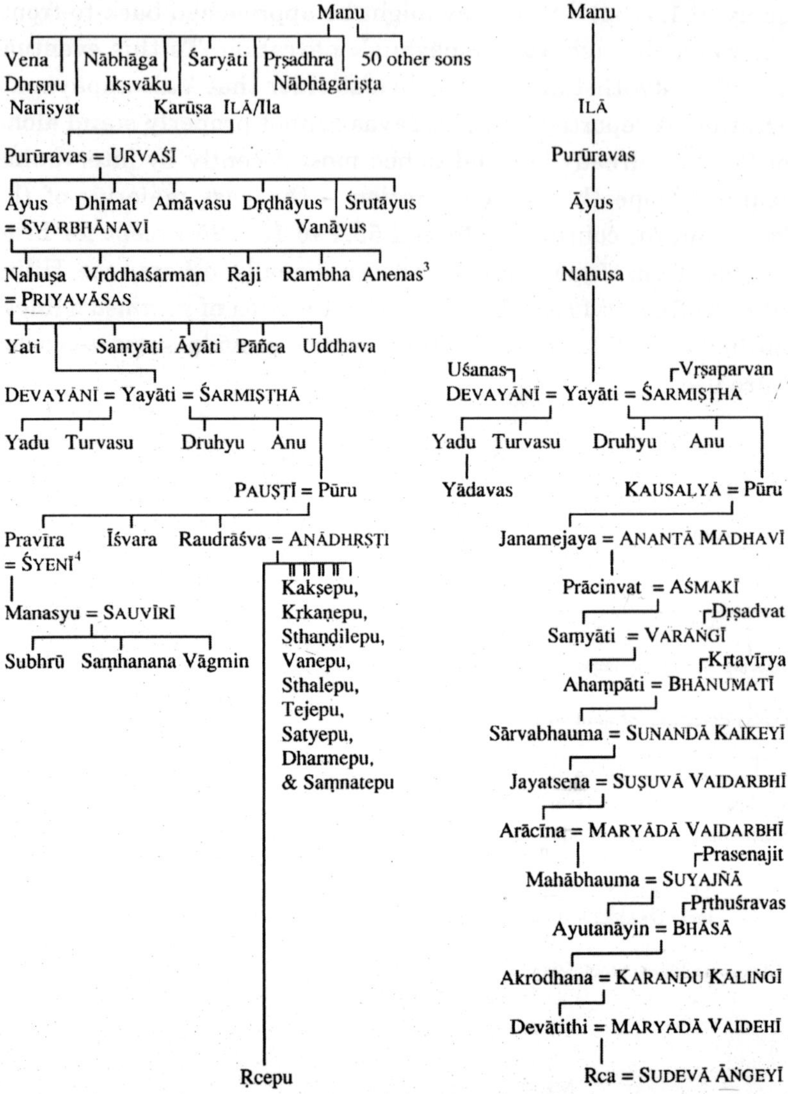

[3] Van Buitenen adds another son, Gaya, mentioned in mss Ñ1, Dn, D1, D2 and D4 (1973: 174; 1.70.23).

[4] Van Buitenen has 'Śūrasenī', as per mss K0, K4, Ñ, B, and D (except D5) (1973: 210; 1.89.6). This choice has consequences; see below at footnote 51.

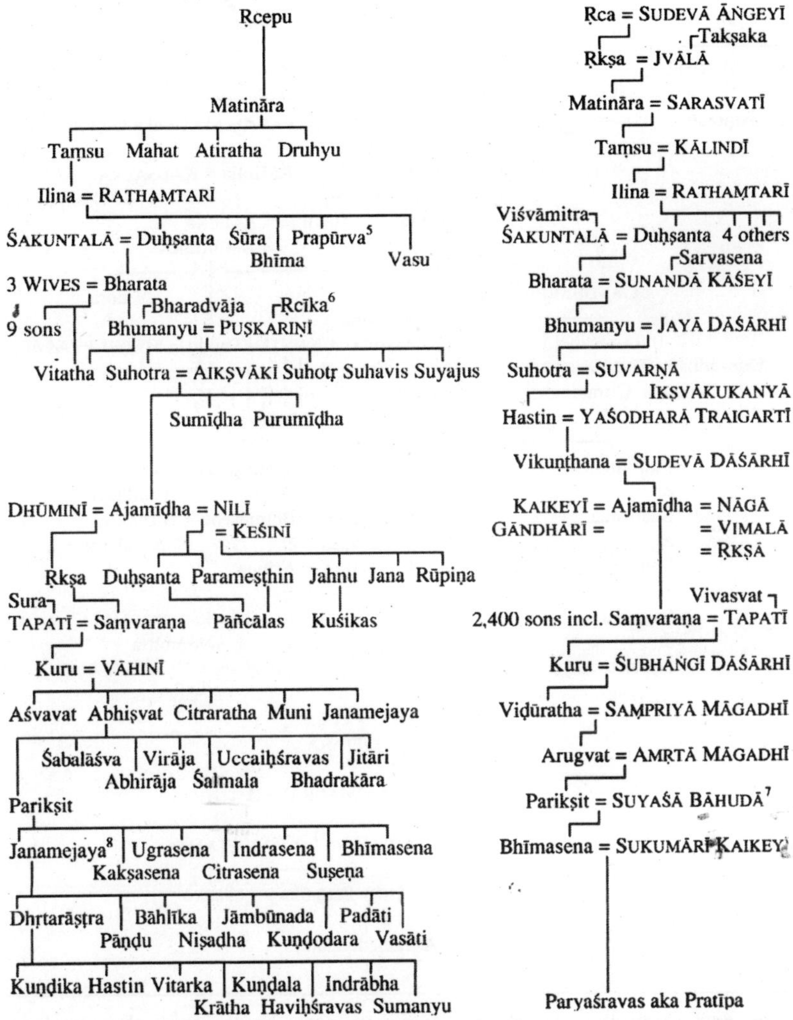

---

[5] Van Buitenen has 'Pravasu', as per mss Ñ2, B, Dn, D1, D2 and D4 (1973: 211; 1.89.15).

[6] Van Buitenen takes 'Ṛcīka' as another name of Bhumanyu (1973: 211; 1.89.21).

[7] Van Buitenen takes 'Bāhudā' as 'Bahuda's daughter' (1973: 214; 1.90.44).

[8] Gilles Schaufelberger's chart (at http://www.geocities.com/harindranath_a/maha/GENEALOGIES.01.pdf, accessed 12 October 2007) has Jana-

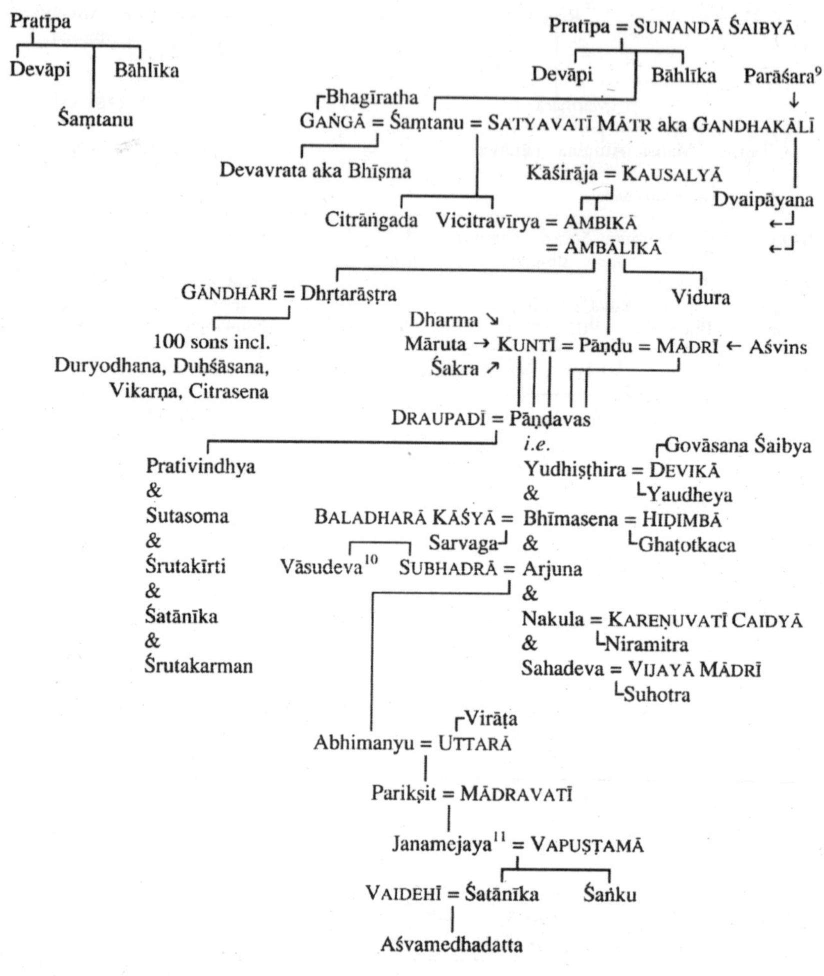

mejaya and unnamed others as Jitāri's younger brothers in the previous generation, and Kakṣasena as Parikṣit's eldest son. But 1.89.47 mentions that *janamejayādayaḥ sapta ... parikṣito 'bhavan putrāḥ*; in the next verse, Janamejaya is understood.

[9] Arrows indicate extramarital impregnation.

[10] For Vasudeva's line, see 7.119 (which Witzel [2005: 44–48] highlights in terms of the *somavaṃśa*).

[11] At 1.3.1 Ugraśravas says that this Janamejaya has three brothers: Śrutasena, Ugrasena, and Bhīmasena. The present and previous of this group is a matter for elsewhere.

ŚRĀDDHA AND PATRILINE

> As soon as his eldest son is born a man becomes a man with a son, and no longer owes a debt to his ancestors; that is why the (eldest) deserves to have the whole (estate). The son to whom he transfers his debt and by whom he wins eternity is the one born out of duty; people know that the others are born out of desire. (*Manusmṛti* 9.106–107, trans. Doniger & Smith)[12]

After death, a man lives in heaven on the *śrāddha* offerings of his male descendants, just as while alive he sustained his ancestors with *śrāddha* offerings. If the memorial line breaks, he and they are doomed. The basic ritual is of three *piṇḍa*s, offered by eldest son to father, grandfather, and great-grandfather (13.92.13); but the *vaṃśa* promises longer-term food.

In the case of kings, the memorial line is particularly potent, as it tracks not only ritual fatherhood, but also management of and responsibility for a certain chunk of Earth; that is to say, ideally it tracks a larger mimicry in which one does not merely offer *śrāddha* as one's *pitṛ*s did in their time, but also rules as they did, and in light of how they did. The patriline implies patrimony: management of a function. The material marker of the patrimony in the case of land inheritance tracks the biologically definitive nature of fatherhood and marks the *kṣatriya* case out from other *guru–śiṣya*-style transmissions.

Patrilineal heaven is ridiculed as temporary by some *Mahābhārata* passages (e.g. 1.85.4; 6.31.20–21 = *Bhagavadgītā* 9.20–21); also censured is the killing of animals for food (Wilhelm 1991; Chapple 1993; Bodewitz 1999). But according to the *vyādha* of Mithilā,

> With regard to meat-eating, the rule is told by the *muni*s: one eats always having given properly and faithfully for the gods and ancestors, and one is not corrupted by the food. Hence one becomes a non-

---

12 *jyeṣṭhena jātamātreṇa putrī bhavati mānavaḥ* I *pitṝṇām anṛṇaś caiva sa tasmāt sarvam arhati* II 106 II *yasminn ṛṇaṃ saṃnayati yena cānantyam aśnute* I *sa eva darmajaḥ putraḥ kāmajān itarān viduḥ* II 107 II

meat-eater; this is the *śruti*, it is heard. The *brahmacārin* visiting his wife in her season becomes a brahmin. (3.199.11–12, my translation)[13]

The patriline works by iterated ritual sacrifice, a feeding of the womb, a feeding of the departed, and a feeding of the non-departed patrilineal descendants and their guests with the leftovers. The whole of Vaiśaṃpāyana's address to Janamejaya is *in familia memoriam* at a meaty feast.[14]

In his lengthy post-war address to Yudhiṣṭhira, Bhīṣma discourses on the *śrāddha* ritual (13.23–24, 87–92; for Bhīṣma's first-hand *śrāddha* story, see 13.83.10ff.). Although the *śrāddha* ritual concept is associated with the phasing of the moon[15] – the basic *śrāddha* rites satisfy the *pitṛ*s for a period of one month – in 13.88 Bhīṣma gives an elaborate taxonomy of different meats which satisfy them for incrementally larger numbers of months, jumping after twelve months to twelve years for a meal of *vadhrīṇasa* (a tricky one according to Ganguli 1970: 145; Monier-Williams 1899: 917 gives for *vadhri* 'one whose testicles are cut out'), with rhinoceros (*khadga*) as the logical limit, providing inexhaustible *śrāddha* (as also may various substitutes in traditional style).[16]

In terms of the guests to be invited and fed, and the preparation of the food, Bhīṣma stresses an especial need to be dis-

---

[13] *atrāpi vidhir uktaś ca munibhir māṃsabhakṣaṇe* | *devatānāṃ pitṝṇāṃ ca bhuṅkte dattvā tu yaḥ sadā* | *yathāvidhi yathāśraddhaṃ na sa duṣyati bhakṣaṇāt* || 11 || *amāṃsāśī bhavaty evam ity api śrūyate śrutiḥ* | *bhāryāṃ gacchan brahmacārī ṛtau bhavati brāhmaṇaḥ* || 12 || This is more than an analogy, since √*bhuj* covers also objects of non-alimentary appetite: 'eat' and 'food' are overspecific translations.

[14] On patriliny and blood sacrifice, but with unfortunate neglect of Sanskrit materials, see Jay 1992.

[15] And the human fertility cycle; see again footnote 13, which should broaden our framing of 'blood sacrifice', particularly in light of Georg von Simson's work on lunar myth.

[16] At 1.1.203, 1.56.29, and 18.5.36, having even one *pāda* of the *Mahābhārata* recited at a *śrāddha* feeds the *pitṛ*s forever. See also 13.116 on the ethics of *śrāddha* meat.

criminating. One should invite the purest brahmins (see also 12.221.73); one should not invite persons who lack virility (13.24.14), or who are the sons of *putrikā*s (female genealogical links, 13.24.21), or 'who have been subjugated by their wives' (*strīpūrvāḥ*, 13.24.23, trans. Ganguli 1970: 121), or who have quarrels with their fathers (13.90.8). The host of the *śrāddha* should give gifts (which must not depend upon wealth earned by women, 13.24.33) to those who have good patrilineal wives (13.24.49). Menstruating women and women who are lacking ears (*vyaṅgitā karṇayoś*) should not attend; women brought for the occasion should be ones born in one's own *vaṃśa* (13.92.15). *Śrāddha* performance is not required for women; *pativratā* is its equivalent (13.46.12; 3.196.20). *Śrāddha* performance is productive of offspring (12.111.19; 13.57.12); Yudhiṣṭhira's *aśvamedha* included *śrāddha* (14.92.13).

The patrilineal system as witnessed by the *śrāddha* ordinance and by Vaiśaṃpāyana's *vaṃśa*s requires minimally one name per generation. Other names are by way of incidental wives, in-laws, and/or extra sons (who are not in the mainline here, but who might be so in the *vaṃśa*s of others[17]). It is precisely the generational economy of the notated arrangement, and its resistance to bifurcation in each generation either upwards (by involving maternal ancestors as liberally as paternal ones) or downwards (by splitting the inheritance between two sons), which makes it a distinctive form. It implies a self-selection effect – an ideology of self-replication – with extraordinary implications.[18]

[17] Such *vaṃśa*s would not be unilocational unless a kingdom be divided. If different *vaṃśa*s (e.g. Yādava and Paurava) be combined and collated, one might construct an inverted two-dimensional tree form (or a river form, depending on how one imagines the directionality of the form's composition).

[18] For patriliny and the economy of the self-same, see Hawthorne 2006: 283–300.

THE PATRILINEAL WIFE

Śakuntalā tells Duḥṣanta in his court, offering herself as his wife and her son as his:

> A husband enters his wife and is reborn from her – thus the old poets know this as a wife's wifehood ... She is a wife who is handy in the house ... who bears children...whose life is her husband...who is true to her lord ... A son, the wise say, is the man himself born from himself; therefore a man will look upon the mother of his son as his own mother ... A son stumbles and covered with dirt embraces his father – is there joy beyond that? ... From the *Veda*s themselves, as you know, the twiceborn recite these verses at the birth ceremony of their sons: 'From each limb thou hast come forth, thou are born from my heart, thou art born with the name of son, live thou a hundred autumns! For my nourishment lies with thee, and my eternal lineage – therefore live thou, my son, in all happiness for a hundred autumns!' (1.68.36, 39, 47, 52, 61–63)[19]

In respect of the wife he may take, from the patrilineal *kṣatriya*'s point of view there are two dangerous possibilities: that she would make it hard for him to have his son by her be his heir (if she is his first wife and this his first son); or that she would try to have him make his son by her his heir (if she is a subsidiary wife and he already has a dharmically earmarked heir). The patrilineal system, as extrapolated from the stories, *vaṃśa*s etc., thrives on total control over the fruits of her womb; and the offering of this on principle is the mark of a good family – her natal family – and a good woman (or girl; at 13.44.13 a bride of ten is recommended for a man of thirty, a bride of seven

---

[19] *bhāryāṃ patiḥ saṃpraviśya sa yasmāj jāyate punaḥ* I *jāyāyā iti jāyātvaṃ purāṇāḥ kavayo viduḥ* II 36 II ... *sā bhāryā yā gṛhe dakṣā sā bhāryā yā prajāvatī* I *sā bhāryā yā patiprāṇā sā bhāryā yā pativratā* II 39 II ... *ātmātmanaiva janitaḥ putra ity ucyate budhaiḥ* I *tasmād bhāryāṃ naraḥ paśyen mātṛvat putramātaram* II 47 II ... *paripatya yadā sūnur dharaṇīreṇuguṇṭhitaḥ* I *pitur āśliṣyate 'ṅgāni kim ivāsty adhikaṃ tataḥ* II 52 II ... *vedeṣv api vadantīmaṃ mantravādaṃ dvijātayaḥ* I *jātakarmaṇi putrāṇāṃ tavāpi viditaṃ tathā* II 61 II *aṅgād aṅgāt saṃbhavasi hṛdayād abhijāyase* I *ātmā vai putranāmāsi sa jīva śaradaḥ śatam* II 62 II *poṣo hi tvadadhīno me saṃtānam api cākṣayam* I *tasmāt tvaṃ jīva me vatsa susukhī śaradāṃ śatam* II 63 II See also Nārada's verses at *Aitareyabrāhmaṇa* 7.13.

for a man of twenty-one; at 3.290.3 young Kuntī is ashamed to menstruate while she is yet unmarried). It simplifies matters[20] if all rights to the girl are given away in marriage, by her first family and/or herself, to the groom's patriline.[21] Hence perhaps *Manusmṛti*:

> A man should not marry a girl who ... is named after a constellation, a tree, or a river, or who has a low caste name, or is named after a mountain, a bird, a snake, or has a menial or frightening name. (*Manusmṛti* 3.8–9, trans. Doniger & Smith)[22]

At 13.45.16 Bhīṣma (!) favours the purchase of brides in order definitively to mark the transfer of reproductive rights (Schmidt 1987: 43); or it may be marked by resolved violence, as in the *rākṣasa* marriage (Hara 1974; Jamison 1996: 218–235) or one of the more combative *svayaṃvara*s (1.96; 12.4). In the *Mahābhārata* compacts are often made between the bride's party and the groom's (sometimes one or both may be parties of one) stipulating the basis on which the marriage may proceed. I call these 'pre-nups'.

The *Mahābhārata* stories which begin with a man's (usually a king's) difficulties in obtaining a son (and Vedic texts leave us in no doubt that this was an ancient and basic ritual preoccupation) must be read in terms of obtaining, not just any son, but one who can be a good patrilineal heir. Particularly perilous is the spectre of the *putrikādharmiṇī*, a woman who is nominated by her father to produce a grandson as *his* heir, and so cannot be fully given away in marriage (13.44.14). The brotherless woman

---

20 In some ways. Consider the abuse of wives, daughters-in-law, and sisters-in-law.

21 Essential is Trautmann 1981: 277–293 on *kanyādāna*. The self-giving of a woman may occur after the father has failed to marry her off (Jamison 1996: 236–250; Schmidt 1987: 76–109).

22 *nodvahet kapilāṃ kanyāṃ nādhikāṅgīṃ na rogiṇīm | nālomikāṃ nātilomāṃ na vācāṭāṃ na piṅgalām || 8 || narkṣavṛkṣanadīnāmnīṃ nāntyaparvatanāmikām | na pakṣyahipreṣyanāmnīṃ na ca bhīṣaṇanāmikām || 9 ||* See also *Kāmasūtra* 3.1.11–12.

was suspected of being a potential *putrikā*, i.e. a bad patrilineal wife; as Uma Chakravarty says, 'Situations led her to become daring and also disgraceful' [23], and she is associated with Uṣas (*Ṛgveda* 1.124.7), dishonesty (*Ṛgveda* 4.5.5), red clothing (*Atharvaveda* 1.17.1), and even incest (*Ṛgveda* 3.31.1). Incest is an inevitable implication if a daughter has a son for whom the public role of 'father' is played by nobody but her own father. Prajāpati's being shot by Śiva's arrow whilst sexually engaged with his daughter in the form of a deer is a primordial motif in the Veda (*Ṛgveda* 1.71.5; *Aitareyabrāhmaṇa* 3.33; *Śatapathabrāhmaṇa* 1.7.4.1–17;[24] Kramrisch 1981: 3–50) and might charter a patrilineal custom of having a king in every generation.

DAKṢA, MANU

The *putrikā* mechanism is visible twice early in the Bhārata patriline, with Dakṣa's daughter Aditi and Manu's daughter (or partial daughter) Ilā.

Dakṣa's sons are not heirs apparently because, at Nārada's instigation, they renounced (1.70.6; see O'Flaherty 1975: 46–48 for the *Vāyupurāṇa* version; see also O'Flaherty 1973: 70–76). Dakṣa's *putrikā* daughters step in (see 1.59), the line towards Manu coming through Aditi. Though this might ordinarily imply patrilineal problems for Dakṣa's son-in-law, here the son-in-law is Kaśyapa, Dakṣa's patrilineal nephew (1.60.4–9; see also 12.200.19); the groom's patrilineal ancestors are effectively the same as the bride's, so their heavenly tenure will not be endangered. In the case of Manu, Ilā is said to be a bi-gendered single parent (Purūravas's mother and father, 1.70.16);[25] there are no in-laws.

---

[23] Chakravarty 2000: 181; see also Schmidt 1987: 30–44.

[24] In these Brāhmaṇa versions the daughter is (possibly) Uṣas.

[25] *purūravās tato vidvān ilāyāṃ samapadyata* I *sā vai tasyābhavan mātā pitā ceti hi naḥ śrutam* II 16 II

Dakṣa is famous for the disruption of his sacrifice by Śiva. The story is told at 7.173.42–51; 10.17–18; 12.274; 12, App. 28; 12.330.42–71; 13.76 and 13.145.11–23. Many of these versions evoke Śiva's punishment of Prajāpati's incest. None mention Dakṣa's daughter except the 13.76 version, in which she is not apparently Śiva's wife; this is a feature of the Purāṇic versions, which may have a slightly different thrust (see Pathak 1978; Kramrisch 1981: 322–340; Klostermaier 1991; Mertens 2002). In the *Mahābhārata* it is clear that Dakṣa commandeers his daughter and is forced to come to terms with Śiva, but the precise link is obscure. Nonetheless, the eventual *entente* between Dakṣa and Śiva fits the cross-cousin style of Aditi's marriage (and the *phalaśruti* at 12, App. 28.424–425).

> The lineage of Manu became famous as the lineage of Men, for from this Manu were born the Men – the brahmins, barons, and the others (1.70.11c–f).[26]

The *śloka vaṃśa* names ten of Manu's children. They include kings Śaryāti (3.121–125), Ikṣvāku (a prolific *kṣatriya* ancestor),[27] and Vena (who was assassinated in favour of good King Pṛthu, 12.59; Fitzgerald 2004: 130–135).[28] These kings are famous as patriarchs in their own right, as it were, rather than as sons of Manu. The *śloka vaṃśa* mentions fifty other Mānavas who killed each other. These might remind us of the mutually cannibalistic creatures Vaiśaṃpāyana mentioned earlier:

---

[26] *manor vaṃśo mānavānāṃ tato 'yaṃ prathito 'bhavat* I *brahmakṣatrādayas tasmān manor jātās tu mānavāḥ* II 11 II On Manu see Gonda 1985: 16–17.

[27] As if to compensate for Ikṣvāku's not carrying the line, at 12.336.44–50 Vaiśaṃpāyana says that Nārāyaṇa's *dharma* was passed down as follows: Nārāyaṇa → Brahmā → Dakṣa → Āditya → Vivasvat → Manu → Ikṣvāku → all and sundry.

[28] At 12.59.93–99 Vena's ancestry is given without mention of Manu: Viṣṇu → Virajas → Kīrtiman → Kardama → Anaṅga → Atibala (married Sunīthā, *mṛtyos ... duhitā ... mānasī*) → Vena.

> From the creatures who, hungry for food, began devouring each other Adharma was born, destroyer of all beings. His wife ... had three loathsome sons, forever bent on evil deeds, Fear, Panic, and Death, destroyer of the creatures. (1.60.52–53)[29]

Similarly perhaps the need for the *avatāra* appearances and the cosmic Kurukṣetra clean-up (see Brodbeck 2006b; 2009a), which may also be associated with the defective, pre-Śiva creation of 10.17.12–17. In view of the patrilineal defectiveness also of Dakṣa's sons and Vena's first posthumous son Niṣāda (12.59.101–103), we see a pattern – manifestation / problem (of excessive identification or difference) / stabilising, regulative sacrificial remedy – familiar from Vedic Brāhmaṇa texts (Smith 1989: 50–69), which conflicts with the norm of primo-geniture (Bhattacharya 1989: 156, 380–382).

Later, Mārkaṇḍeya tells the Pāṇḍavas Manu's story. A small fish in the River Vīriṇī asks Manu for help, for big fish eat little ones (3.185.8; Vīriṇī links Manu and Dakṣa, being also Dakṣa's wife). Manu nurtures the fish, which tells him to build an ark and collect *bījāni sarvāṇi* (3.185.30, 34). When the flood comes,[30] the fish tows the ark to a Himālayan peak, reveals itself to be Brahmā, and tells Manu to repopulate the world.

> [Manu] was unsure about the manner of creating and performed many austerities; then, aided by his great asceticism, Manu himself proceeded to create all the creatures rightly, bull of the Bhāratas. (3.185.51e–52)[31]

---

[29] *prajānām annakāmānām anyonyaparibhakṣaṇāt* I *adharmas tatra saṃjātaḥ sarvabhūtavināśanaḥ* II 52 II *tasyāpi nirṛtir bhāryā nairṛtā yena rākṣasāḥ* I *ghorās tasyās trayaḥ putrāḥ pāpakarmaratāḥ sadā* I *bhayo mahābhayaś caiva mṛtyur bhūtāntakas tathā* II 53 II

[30] Manu's plight can be compared and contrasted with that of the Vṛṣṇis in the Mausalaparvan: the men die off (von Simson 2007) and Dvārakā is flooded (Couture 2003); but Abhimanyu is a Bhārata, not a Vṛṣṇi.

[31] *pramūḍho 'bhūt prajāsarge tapas tepe mahat tataḥ* II 51 II *tapasā mahatā yuktaḥ so 'tha sraṣṭuṃ pracakrame* I *sarvāḥ prajā manuḥ sākṣād yathāvad bharatarṣabha* II 52 II

In the *Śatapathabrāhmaṇa* version (1.8.1.1–11) Manu ritually produces a daughter from the water – Iḍā – and breeds through her (*iḍā* appears in the *Ṛgveda* as *iḷā*, Monier-Williams 1899: 164). In the *Mahābhārata* the fish matches Ilā in the *vaṃśas*, despite its self-identification as Brahmā.

YAYĀTI

In the *śloka vaṃśa* Yayāti is not Nahuṣa's eldest; the first son's name may suggest that he was a renouncer. The prose *vaṃśa* adds regarding Yayāti a *śloka* (1.90.9) quoted from the *anuvaṃśa* ('chronicle'), detailing which of Yayāti's sons were born to which mother.[32]

In his first story in the hiatus of the *śloka vaṃśa*, Yayāti is hunting, looking for deer (*mṛgalipsuḥ*, 1.73.14); tired and thirsty, he finds Devayānī in a dry well. Another time, *mṛgalipsur* (1.76.4), tired and thirsty, he meets her again, and she proposes. Her father Śukra approves, absolving Yayāti of the sin of *varṇasaṃkara* (they are brahmins), but makes the pre-nuptial condition that he not sleep with Śarmiṣṭhā, an *āsurī*, Devayānī's lifelong slave. In time Yayāti breaks the pre-nup and is cursed by Śukra (1.78.30) to old age – unless he can find a supremely dutiful son.[33] By the time he has done so (in the shape of Pūru, a son by Śarmiṣṭhā) a thousand years have passed. The brahmins object because Pūru is the youngest, but are persuaded of his obedience and allow his anointing (1.80); so Yayāti, after breaking his pre-nup, eventually gets his heir.

We may superimpose this story of Yayāti's curse and remedy with the immediately following story of Yayāti being thrown out of heaven and then returning there (1.81–88). While falling

---

[32] *yaduṃ ca turvasuṃ caiva devayānī vyajāyata* | *druhyuṃ cānuṃ ca pūruṃ ca śarmiṣṭhā vārṣaparvaṇī* ||

[33] In the *śloka vaṃśa* at 1.70.78 Śukra curses Yayāti, but it is not mentioned that he is Devayānī's father.

earthwards, Yayāti discourses to the four sons of his (unmentioned-in-the-*vaṃśas*) daughter, explaining his fall as an inevitable consequence of the finitude of merit. They offer him theirs; he refuses; and they all go to heaven! Another version of this story told at 5.104–121 explicitly states that Yayāti's restoration was effected through his daughter's sons (5.119.27, 120.16–17, 121.1; see also 3.205.17; 13.6.30). It tells how Yayāti loaned Gālava, a young brahmin, his daughter Mādhavī so that Gālava could pimp her to various kings as an heir-producer (Dasgupta 2000: 52–54). After this she chose the forest as a husband (at her *svayaṃvara* at the holiest *tīrtha*, the vulva of the earth where the Gaṅgā and Yamunā meet, 3.83.65–80) and became a doe. When Yayāti falls from heaven and tries to refuse her sons' offers of merit, she intervenes and insists that he accept. In this version Mādhavī the doe might seem to be a bad patrilineal wife, and we might fear for the *pitṛ*s of those who paid for heirs from her. Or perhaps those kings would have their heirs, but Yayāti would be named in their *vaṃśa*s as a father-in-law. This latter possibility might keep Yayāti himself in heaven *pro tempore*, but his ancestors would lose out.[34] Far more patrilineally satisfactory is the resolution at 1.80: Pūru becomes Yayāti's heir, so Yayāti is rescued anyway, as a full link, with his *pitṛ*s. His disobedient elder sons produce the Yādavas, Yavanas, Bhojas and *mleccha*s (1.79, 80.26, 90.10; compare *Aitareyabrāhmaṇa* 7.18).

MATINĀRA, SAṂVARAṆA: HUNTING AND *SATRAS*

In the *śloka vaṃśa* after Pūru, through two wives, Śyenī (apparently a bird; see 1.60.54–67) and Anādhṛṣṭi (an *apsaras*), the line splits, the senior branch being overtaken by the junior in the subsequent generations. This also occurs later with Dhṛta-

[34] This might be connected to Nahuṣa's fall from heaven (1.70.24–27; 3.175–178; 5.17; 12.254; 12.260), and/or to Purūravas's being cursed to death (1.70.17–21), both of which are remedied (see 13.6.31).

rāṣṭra's senior line and Pāṇḍu's junior, and here as there we see this marked by *rājasūya* and *aśvamedha* performances from the junior line in the takeover (1.89.10). The prose *vaṃśa* by contrast has Pūru with a single son, Janamejaya, who nonetheless performs three *aśvamedhas* (1.90.11); the line then proceeds with many more generations and no obvious problems until it catches up with the *śloka* version in the person of Matinära (of the junior branch). At that point the prose *vaṃśa* tells that Matināra, son of snake Takṣaka's daughter, performed a twelve-year Sarasvatī *satra*, after which Sarasvatī gave him a son.[35] That this *satra* could be a spell away from the kingdom is suggested by the movement implied in the names of Matināra's wife Sarasvatī and Taṃsu's wife Kālindī (i.e. Yamunā). This could be a counterpart to the patrilineal hiccup in the *śloka vaṃśa*.[36]

Nine or ten generations later, just before Kuru who ploughed the field (9.52), we find Saṃvaraṇa. According to the *śloka vaṃśa* (his patrilineal precariousness is perhaps suggested in the prose by his being one of Ajamīḍha's 24,000 sons) Saṃvaraṇa was driven out by ten armies led by his cousin the king of the Pāñcālas[37] and settled in hilly woodlands by the Sindhu; but in time, with the help of his *purohita* Vasiṣṭha, he became *saṃrāj*, re-took the ancestral seat, and offered *mahāyajñas* (1.89.29–41; compare *Ṛgveda* 7.18, 33, 83).

The *gandharva* Citraratha tells Arjuna *et al.* an elaborate version of Saṃvaraṇa's story (1.160–163). Out hunting in highlands, his hungry, thirsty and exhausted horse dies, and Saṃvaraṇa meets Tapatī, daughter of the sun, and is incapacitated by desire. She refers him to her father, who approves the match

---

[35] This Matināra's son Taṃsu and his son Ilina are underlined by the second *anuvaṃśa* quotation (1.90.28): *taṃsuṃ sarasvatī putraṃ matinārād ajījanat* | *ilinaṃ janayām āsa kālindyāṃ taṃsur ātmajam* ||

[36] Note also Ayutanāyin's performance of *puruṣamedhas* (1.90.19).

[37] This Pāñcāla king is Duḥṣanta's son, so this may connect with the Bharata business (on which see below).

(at Vasiṣṭha's suggestion; but it has been Sūrya's plan from the start); but while the couple then make love on the mountain for twelve years, Saṃvaraṇa's kingdom suffers a drought which ends only when Vasiṣṭha comes to fetch them back to the capital, where King Saṃvaraṇa now performs sacrifices for twelve years.

In Citraratha's version there is no sign of rival cousins except in the form of the exiled listening Pāṇḍavas. Saṃvaraṇa's royal negligence is here strongly associated with *kāma*, and he may remind us of Ṛśyaśṛṅga (3.110–113), whose lust leads him to service another patriline before his own. In Saṃvaraṇa's case, as in Yayāti's (Yayāti considers his marriage with Devayānī to have been a *dīrghasatra*, 1.70.38; see also 3.129.3 for Yayāti's *sadasyas*), there is the motif of hunting, which may be associated with time away from one's erstwhile *paśus*. Ṛśyaśṛṅga, a brahmin and the son of a doe, is no hunter, but the categorisation of hunting – along with gambling/dicing, boozing, and womanising – under the rubric of *kāma* is clear.[38] When Yayāti finally installs Pūru after his thousand-year spree, it is like Saṃvaraṇa's return, and Matināra's. Saṃvaraṇa and Matināra are also nominally linked through the recurrence of Ṛkṣa, Matināra's father in the prose *vaṃśa*, in connection with Saṃvaraṇa (in the *śloka vaṃśa* as his father who intervenes as an extra king after Ajamīḍha, and in the prose *vaṃśa* as his possible mother).

On *Mahābhārata satras*, we refer to Minkowski (1989; 1991) and Hiltebeitel (2001: 118–176) and their sources. The text's *satras* include also those of Agastya (1.109.14; 14.95; Reich 2001: 158–165); Parāśara (*rākṣasa satra*, aborted, 1.172); Drupada (after Droṇa has halved his holdings, 3.119.20; Brodbeck 2006a; 2009b); Kṛṣṇa (twelve-year Sarasvatī *satra*, 3.13.13); Marutta (3.129.16; see also 14.7.4); Bṛhaddyumna (3.139); Baka Dālbhya

38 2.61.20–21; 3.14.7; 5.33.73–74; 12.28.31, 59.59–61, 138.26, 277.26; see also 5.43.9, 88.55; see 3.125.8 and 13.141.28–29 for the aetiology; see also *Manusmṛti* 7.47–50; *Arthaśāstra* 8.3; Kane 1973: 53–55; Wilhelm 1991; Allsen 2006; van Buitenen 1973: 447; *Saṃyuttanikāya* 56.11.

(9.40.12–13, against Dhṛtarāṣṭra's kingdom; see also 3.27; Koskikallio 1999: 305–307, 324–331); etc.; and also those of the *devas* on another level.[39] *Brahmasatra* is another name for the *kapotavrata* (12.235.4); *satreṇa* means simply 'in disguise', and is frequently used in the Virāṭaparvan; and of course there is the Kurukṣetra *raṇasatra*[40] and the *sarpasatra*. The *satra* is a collective:

> The man who makes a gift of earth unto one that is emaciated and cheerless and destitute of the means of life and languishing with weakness, and who thereby supplies one with the means of subsistence, becomes [a *satrin*]. (13.61.26, trans. Ganguli)[41]

13.61.77–79 mentions *bhūmisatra* to reinstate an exiled king (*bhūmipālaṃ cyutaṃ rāṣṭrād*).

DUḤṢANTA, ŚAKUNTALĀ, BHARATA, SUHOTRA

As mentioned earlier, the Duḥṣanta-Śakuntalā story precedes the *vaṃśas* at 1.62–69 and includes a speech by Śakuntalā in Duḥṣanta's court. The *śloka vaṃśa* gives details of Bharata's sons:

> Bharata begot nine sons on his three wives, but the king did not approve any one of them, for they were not of his stature.[42] Thereupon Bharata offered up grand sacrifices and received a son from

---

[39] Sometimes thousand-year *satras*, 1.189; 2.3.13–14; 3.80.116–117, 129.1; 6.7.46; 9.36–37, 40.3; 12.327.49; 13.18.16 with deer; 13.85.10, 25 with dispute over whose the offspring are; 13.140.16; etc.

[40] 3.242.14; 12.99.16–17; Feller 2004: 253–293; Minkowski 2001: 175; which Bhīṣma says Kṛṣṇa saw as the *mahīsatra*, 13.143.16; Brodbeck 2009a.

[41] *kṛśāya mriyamāṇāya vṛttimlānāya sīdate* I *bhūmiṃ vṛttikarīṃ dattvā satrī bhavati mānavaḥ* II 26 II

[42] Perhaps some of these sons were of use to their maternal kin. Chatterjee mentions a tradition established by Bharata that 'the throne would be inherited only by the most qualified person in the kingdom, and not solely on the basis of primogeniture' (Chatterjee 2005: 145). This is an exaggeration but is perhaps suggested by this verse, and is in line with the impression given by Chopra & Chopra 1988–90 (here I am grateful for e-mail communication from Pradip Bhattacharya, November 2005).

Bharadvāja by the name of Bhumanyu, O Bhārata. The scion of the Pauravas [i.e. Bharata] deemed himself Bhumanyu's father and consecrated him Young King, O best of the Bhāratas. Then the king himself had a little son, Vitatha, and this Vitatha became a son of Bhumanyu. By [Ṛcīka's Puṣkariṇī, Bhumanyu] had Suhotra, Suhotar, Suhavis, and Suyajus. The eldest of the princes, Suhotra, inherited the kingdom; he offered up many Soma Sacrifices such as the Royal Consecration and the Horse Sacrifice. (1.89.17–22)[43]

The prose *vaṃśa* has no problem with Bharata's sons: only Bhumanyu is mentioned, as if he were Bharata's natural son; and Bhumanyu's only son here is Suhotra, whose son Hastin goes on to build Hāstinapura (1.90.34–36). But the prose *vaṃśa* includes in connection with Duḥṣanta two *śloka*s (the *anuvaṃśa* is not mentioned here) almost identical to ones already presented in the long version of the Śakuntalā story (1.90.31–32, 69.29–30),[44] where they are spoken by the heavenly voice intervening at the end of the court scene to verify that Bharata is Duḥṣanta's child and insist that he accept Śakuntalā and keep his pre-nup.

Earlier, Duḥṣanta (one of five sons) goes hunting. His spree is very bloody, rendering an entire forest *hatamahāmṛgam*, with only buffalo (*mahiṣa*s) remaining (1.63).[45] In another forest he

[43] *bharatas tisṛṣu strīṣu nava putrān ajījanat* | *nābhyanandanta tān rājā nānurūpā mamety uta* || 17 || *tato mahadbhiḥ kratubhir ījāno bharatas tadā* | *lebhe putraṃ bharadvājād bhumanyuṃ nāma bhārata* || 18 || *tataḥ putriṇam ātmānaṃ jñātvā pauravanandanaḥ* | *bhumanyuṃ bharataśreṣṭha yauvarājye 'bhyaṣecayat* || 19 || *tatas tasya mahīndrasya vitathaḥ putrako 'bhavat* | *tataḥ sa vitatho nāma bhumanyor abhavat sutaḥ* || 20 || *suhotraś ca suhotā ca suhaviḥ suyajus tathā* | *puṣkariṇyām ṛcīkasya bhumanyor abhavan sutāḥ* || 21 || *teṣāṃ jyeṣṭhaḥ suhotras tu rājyam āpa mahīkṣitām* | *rājasūyāśvamedhādyaiḥ so 'yajad bahubhiḥ savaiḥ* || 22 ||

[44] *bhastrā mātā* [1.69.29] OR *mātā bhastrā* [1.90.31] *pituḥ putro yena jātaḥ sa eva saḥ* | *bharasva putraṃ duḥṣanta māvamaṃsthāḥ śakuntalām* || 29/31 || *retodhāḥ putra unnayati naradeva yamakṣayāt* | *tvaṃ cāsya dhātā garbhasya satyam āha śakuntalā* || 30/32 || 'The mother is the father's water sack. He is the father by whom the son is born. Support your son, Duḥṣanta, do not reject Śakuntalā. A son who has seed, O king of men, saves from Yama's realm. You are the planter of this child, Śakuntalā has spoken the truth.' A water-bottle is something a thirsty hunter might want.

[45] Kuru (as ploughman; compare 12.59.119, 126; Bailey 1981), Duḥṣanta, and the Pāṇḍavas (variously burning and uprooting trees) are asso-

finds Śakuntalā alone at Kaṇva Kāśyapa's *āśrama*; she tells him of her origin (father Viśvāmitra, mother *apsaras* Menakā), abandonment, protection by birds, and adoption by Kaṇva. Duḥṣanta marries her there and then, agreeing to her pre-nup whereby their son will inherit his kingdom; they consummate the marriage; then he goes home alone (1.64–67).

One senses in Śakuntalā's pre-nup – particularly in light of the later stories of Śaṃtanu (1.94) and Daśaratha (3.261)[46] – the suspicion of competition for Duḥṣanta's kingdom in the next generation. At this stage Duḥṣanta worries about Kaṇva (who was away at the time of the marriage), and what he might wreak with his *tapas*. Note that if, for example, Śakuntalā were to be Kaṇva's *putrikā*, then by agreeing to the pre-nup Duḥṣanta would just have made a mess of his own patriline. Śakuntalā's assurances in court in the next scene (1.68–69) that she is and will be *pativratā* could be fake, and Bharata a Trojan horse.[47] Duḥṣanta repudiates Śakuntalā because (he tells his ministers afterwards)

> if I had taken [Bharata] as my son on her word alone, suspicion would have been rife among the people and he would never have been cleared of it. (1.69.36)[48]

But perhaps all along (or by now) Duḥṣanta would have other sons, and not need an heir.

At 3.209.1–9 Mārkaṇḍeya relates, in a genealogy of fires descended from Bṛhaspati, an interesting circumstance: both Bha-

---

ciated with economic development of land (see also 12.133.14–15, 23–24); Duḥṣanta is associated by implication with a high standard of wife.

[46] In the Daśaratha story, Bharata is the son whose mother engineers that he be made heir instead of Rāma; but the Bharata in that story refuses the throne, ruling only as proxy and until Rāma returns.

[47] Kaṇva approves the match, without patrilineal designs upon Śakuntalā; but Duḥṣanta has not yet met him.

[48] *yady ahaṃ vacanād eva gṛhṇīyām imam ātmajam | bhaved dhi śaṅkā lokasya naivaṃ śuddho bhaved ayam* || 36 || See also Rāma's repudiation of Sītā (3.275).

radvāja and Bharata are Śamyu's sons (see also Brodbeck 2009b). This might be linked to the *śloka vaṃśa* where Bharata has problems getting a good heir, and Bharadvāja gives him Bhumanyu. Bharata is Duḥṣanta's only child in both *vaṃśa*s, but in the generations after Bharata the *śloka* version presents a situation whereby Suhotra, the natural son of Bharata's adopted-from-Bharadvāja son Bhumanyu, might compete for the throne with Vitatha, Bhumanyu's adopted son and also Bharata's latter-day natural one. That Suhotra and his brothers compete and prevail is perhaps suggested by their performance of ritual *savas* including *rājasūya* and *aśvamedha* (*satra*s and *raṇayajña*s could also be called *savas*; for Suhotra see also 1.1.166; 12.29.22, 220.50; in connection with Yudhiṣṭhira, Baka Dālbhya and company, 3.27.24). In the prose *vaṃśa*, the building of Hāstinapura may perhaps be viewed as consolidating the result of a dispute. In the end, Duḥṣanta's hunting success matches his patrilineal success; but elsewhere Ajamīḍha's son Duḥṣanta's junior branch (i.e. the Pāñcālas) fails to take the line.

ŚAṂTANU, BHĪṢMA (VASU)

In the *śloka vaṃśa*, the five brother-packed generations after Kuru feature a Parikṣit, and his son Janamejaya (and brothers), and Dhṛtarāṣṭra and Pāṇḍu as brothers; and the line is cut short, ending with Dhṛtarāṣṭra's sons. An unlinked fragment containing Pratīpa and his sons is as if tacked on at the end. Vaiśampāyana blithely mentions that there were many others in Bhārata's line and in Manu's *ailavaṃśa* (1.89.54–55); and Janamejaya must cry *encore*. In the prose *vaṃśa* Kuru has a different wife, but a few generations later we find Parikṣit, who marries the River Bāhudā and is followed by Bhīmasena (Janamejaya's youngest brother in the *śloka* version; see also footnote 11) – whose son is Pratīpa, thus linking to Śaṃtanu and continuing thereafter down to the Pāṇḍavas, the next Parikṣit, and beyond.

Śaṃtanu is not Pratīpa's eldest. Both *vaṃśas* mention that Devāpi renounced (1.89.53, 90.47; see also 9.38.32–39.10; 5.147.14–28, where Dhṛtarāṣṭra says Devāpi had a skin disease and Bāhlīka went to live with his mother's natal family; see also 12.315.20). The prose *vaṃśa* adds in connection with Śaṃtanu an etymologising *anuvaṃśa śloka*:

> Every old man he touched with his hands felt joy and became young again. Therefore they know him as Śaṃtanu. (1.90.48)[49]

Śaṃtanu's story, as Vaiśaṃpāyana tells it after concluding the prose *vaṃśa*, begins in heaven (1.91). Mahābhiṣa Aikṣvākava posthumously lusts for Gaṅgā and is cursed by Brahmā to rebirth; he will be Pratīpa's son (Śaṃtanu). Gaṅgā is independently enjoined to go to earth by the Vasus, who have also been cursed to birth amongst mortals and wish to minimise the term: she is to produce then drown them, saving only one (1.91.18–20). She appears and proposes to Pratīpa, who refuses her, saying he eschews dalliances with *asavarṇā* women (1.92.6). But he blesses in advance her marriage to his unborn son, and her pre-nuptial condition that her husband not interfere in her doings or know her true identity (1.92.14–15). Before he anoints Śaṃtanu and retires, Pratīpa gives him patrilineally interesting instructions to look out for, love, and never question a celestial lady (1.92.20–22).

King Śaṃtanu is a keen hunter. Three hunting trips are narrated, whose intended chronology is uncertain. The first-presented of these, like Duḥṣanta's hunting trip, is bloody and results in a delayed heir who lives first with his mother. Out killing deer and buffalo (unlike Duḥṣanta; *mṛgān mahiṣāṃś ... vinighnan*, 1.92.25), Śaṃtanu meets and marries Gaṅgā, agreeing to her pre-nup (her father is mentioned in the *vaṃśa*, but not in the story). She produces and drowns seven sons; Śaṃtanu objects at the eighth, breaking the pre-nup; she explains herself,

---

[49] *yaṃ yaṃ karābhyāṃ spṛśati jīrṇaṃ sa sukham aśnute* I *punar yuvā ca bhavati tasmāt taṃ śaṃtanuṃ viduḥ* II 48 II

and leaves with the living son, promising to return him (and mentioning that Vasiṣṭha Āpava stipulated a long and childless life for him). Years later, tracking a shot deer along the Gaṅgā (*mṛgaṃ viddhvā*, 1.94.21),[50] Śaṃtanu finds his son, whom Gaṅgā returns to him. He is installed as *yuvarāja*.

On Śaṃtanu's other hunting trip (1.120) one of his companions finds twin babes in the wood, and Śaṃtanu adopts them as Kṛpa and Kṛpī. Śaradvat Gautama, their birth-father, comes to find them at Hāstinapura, tells Śaṃtanu of their *gotra*, and trains Kṛpa. Śaṃtanu's hunting concerns the question of his heir (see also 12.125–126, 297); Kṛpa is said by Śaṃtanu to be his son (1.120.16; see Allen in this volume), but Śaradvat claims paternity (1.120.19; see also 1.121.12; Karve 1965: 74–76). Kṛpa is a loyal royal servant and remains to supervise Parikṣit and Yuyutsu when the Pāṇḍavas retire (17.1); in the end he is (like Bharadvāja) a brahmin, and the hunt on which Śaṃtanu found the twins is probably appropriately bloodless. Our interpretation may depend on whether we think the incidents in the Kṛpa business – the finding of the twins by Śaṃtanu, and then by Gautama – take place before Śaṃtanu meets Gaṅgā, or while he is married but with perishing sons, or while Gaṅgā is away with the living son, or after Gaṅgāputra has been anointed *yuvarāja*.

Bhīṣma tells his own hunting story to Yudhiṣṭhira after a question about Kṛṣṇa-Nārāyaṇa's animal forms (12.202). He got waylaid at Mārkaṇḍeya's *āśrama*, listening to Kaśyapa's tale of Viṣṇu-as-a-boar. It is as if Śaṃtanu's son opts out of hunting, as he opts out of the patriline: he abdicates as heir and swears celibacy as the pre-nuptial condition of Śaṃtanu's second marriage (to Satyavatī, 1.94), and, unlike Yayāti, Duḥṣanta and Śaṃtanu, he respects his pre-nup absolutely (1.97.13–18, 24; Chatterjee

---

[50] The deer may have used a riverine passage to throw off its pursuer (see Cummins 1988: 39).

2005; Harzer 2005; Fitzgerald 2007). But if Bhīṣma be charged with (abetting his father's *kāma*-induced) patrilineal irresponsibility resulting in fratricidal mayhem and near destruction of the patriline, we note that his patrilineal fecklessness was foretold (by the Vasus, 1.91.21, and by Āpava as Gaṅgā relates, 1.93.35–38) and that the Nārāyaṇa whose tales kept him from the hunt is the same Nārāyaṇa by whose oblique power the war takes place (the Earth-rescue frame is detailed already at 1.58–61) and by whose *satyakriyā* the next Parikṣit is revived.

Bhīṣma is not a link in the *vaṃśa*; he is a Vasu fallen from above. So is King Vasu Uparicara, the first instalment of whose story is told by Vaiśaṃpāyana much earlier, at 1.57, where Satyavatī is described as Vasu's daughter (*vāsavīṃ kanyāṃ*, 1.57.57).

King Vasu of Cedi, a Paurava and keen hunter (1.57.1), established the annual ritual of Indra's *maha* pole (1.57.23–24; Gonda 1967; Kuiper 1979: 157–171; Lidova 2002–03). A *saṃrāj*, he had five sons, but they became kings in other kingdoms, not his (1.57.30). Vasu takes a wife Girikā, the daughter of a mountain and a river; as he is set to conceive a son in her, his ancestors intervene and order him to go hunting. On Vasu's hunt, as on Bhīṣma's, the hunter gets distracted: thinking of Girikā, Vasu ejaculates, catches his semen (*retas*, *śukra*) on a leaf, and sends it home for her via a kite (*śyena*); it falls into the Yamunā[51] and is swallowed by *apsaras* Adrikā in the form of a fish; the fish, caught, produces human twins; one becomes King Matsya; the other, Satyavatī, is given to the fisherfolk (*dāśas*).

Vasu apparently does not know these children as his, nor do they know him as their father. But Satyavatī's sons are Vasu's daughter's sons. If we identify Satyavatī's brother King Matsya with Virāṭa of the Virāṭaparvan, then 1.57 allows us to locate

[51] After a struggle; another kite wants it. This is a riff on Pūru's sons in the *śloka vaṃśa* (and/or *vice versa*).

Vasu as Parikṣit's hidden ancestor through Uttarā, Parikṣit's mother, who is Virāṭa's daughter, as well as through Satyavatī.

Vasu fell out of heaven (12.322–324; 13.6.34, 116.54–55; 14.94.19–22).[52] He lived in a hole, but later returned to an undiminishable location due to his Nārāyaṇa-*bhakti* (12.323.56–57; see Ganguli's footnote), as also did Bhīṣma (12.51.13; 13.153–154). Vasu is named in the *śloka vaṃśa* as Duḥṣanta's youngest brother (perhaps Bharadvāja's ghost!); so Vasu's ancestors, like Bhīṣma's, are safe in this *vaṃśa* even though he is not a direct patrilineal link.

PĀṆḌU

There are several accounts of Pāṇḍu's hunting exploit. In the *Mahābhārata*'s first *adhyāya* Ugraśravas, beginning the first summary of Vaiśaṃpāyana's story, mentions Pāṇḍu to the *ṛṣis* (1.1.67–75). Pāṇḍu left the kingdom, settled with his wives in a forest, and hunted. 'He came to dire grief by killing a deer when it mated' (*mṛgavyavāyanidhane kṛcchrāṃ prāpa sa āpadam*, 1.1.68). The Pāṇḍavas are conceived by gods, and later brought to Dhṛtarāṣṭra's court at Hāstinapura. Whereas Śakuntalā announces Bharata alone (so must force a heavenly voice), the Pāṇḍavas are announced as such by *ṛṣis* who accompany them; rumours of their illegitimacy soon fade.

The prose *vaṃśa* (the *śloka* version has finished, or paused until 1.91) tells of the production of Dhṛtarāṣṭra and Pāṇḍu by *niyoga* (Dhand 2004; Doniger 1995; G. Sutherland 1990; Schmidt 1987: 44–75), mentions Dhṛtarāṣṭra's sons, and continues:

> Pāṇḍu had two wives, Kuntī and Mādrī, both gems among women. Once, Pāṇḍu went hunting and saw a seer covering a doe, coupling with her. As he was jumping the doe, before he had reached the sap

[52] The question of vegetarianism/*ahiṃsā* in Vasu's case may be compared with the question of celibacy in Bhīṣma's. For these questions combined, see again 3.199.12 (footnote 13 above).

> of his lust and was sated, Pāṇḍu shot him with an arrow. Wounded by the arrow, he said to Pāṇḍu, 'Because you, who follow the Law and know the sap of lust, have shot me before I reached the sap of lust, therefore you too will at once return to the five elements when you are in the same state, before reaching the sap of lust.' Pāṇḍu paled and, avoiding the curse, did not lie with his wives. (1.90.63–66)[53]

Pāṇḍu knows that childless men have no *lokas*, so the Pāṇḍavas are born by *niyoga*; Pāṇḍu dies with Mādrī; and *ṛṣis* take Kuntī and the sons to Hāstinapura.

After the *vaṃśas* (1.102ff.), Pāṇḍu leaves the kingdom to his blind brother and goes with his wives to the southern slopes of the Himālaya. Dhṛtarāṣṭra acquires 100 sons, and wonders about the succession. Janamejaya asks Vaiśaṃpāyana about the Pāṇḍavas, and Vaiśaṃpāyana tells again of Pāṇḍu's hunt. Here the mating hart is said to be a *mṛgayūthapa* ('leader of its herd', 1.109.5), and is shot by five arrows;[54] but it is actually a *ṛṣiputra* living as a deer (1.109.26–27; see Thite 1972: 202–206), who

---

[53] *pāṇḍos tu dve bhārye babhūvatuḥ kuntī mādrī cety ubhe strīratne* II 63 II *atha pāṇḍur mṛgayāṃ caran maithunagatam ṛṣim apaśyan mṛgyāṃ vartamānam* I *tathaivāplutam anāsāditakāmarasam atṛptaṃ bāṇenābhijaghāna* II 64 II *sa bāṇaviddha uvāca pāṇḍum* I *caratā dharmam imaṃ yena tvayābhijñena kāmarasasyāham anavāptakāmaraso 'bhihatas tasmāt tvam apy etām avasthām āsādyānavāptakāmarasaḥ pañcatvam āpsyasi kṣipram eveti* II 65 II *sa vivarṇarūpaḥ pāṇḍuḥ śāpaṃ pariharamāṇo nopāsarpata bhārye* II 66 II See also 1.173; 3.182, 205–206; Piatigorsky 1993: 109–150; Leslie 1998; Goldman 1978: 358–359.

[54] Van Buitenen's translation of 1.109.6 is, I suspect, potentially confusing, as was pointed out to me by Paul de Villiers: 'Pāṇḍu shot both the doe and the buck with five ... arrows' (*tāṃ ca mṛgīṃ taṃ ca ... nirbibheda śarais ... pāṇḍuḥ pañcabhir*). The reason for both creatures to be mentioned here in the accusative can be given by the verb, *nir-√bhid*, which means 'to split apart' as well as 'to pierce, wound'; the implication may be just that Pāṇḍu's salvo succeeded in separating the pair (by hitting the male). This interpretation would encourage us to revise also van Buitenen's awkward reading of 1.109.8 – 'Still coupled with his doe, he fell instantly to the ground', *saṃsaktas tu tayā mṛgyā ... kṣaṇena patito bhūmau* – by suggesting that *saṃsakta* operates only to identify the male, not to assert his ongoing coition (the male deer is actually a *ṛṣiputra*, and this dual or indefinite identity makes it convenient to signify him with a nominative verbal participle rather than a more substantive noun).

identifies himself[55] and argues with and curses Pāṇḍu before dying. Pāṇḍu tries to excuse his deed in terms of *rājadharma* and says that deer has been sacrificially offerable fair game ever since Agastya's *satra* (1.109.12–15; see also 13.116.56–57, 117.16–19); but this is no ordinary deer.

Pāṇḍu blames himself for falling prey to *kāma* and *vyasanam*, and resolves for *mokṣa*: 'I shall seek the Law of the wind, beholden to no one' (*na vaśe kasya cit tiṣṭhan sadharmā mātariśvanaḥ*, 1.110.18). His wives suggest he take them too and thus reach heaven (1.110.26); indeed, if he were to go solo, he would mayhap die childless (see Brodbeck 2007: 151–152). The three travel to the Hundred-Peak Mountains; but rather than proceeding northwards alone, Pāṇḍu thinks of sons, and suggests *niyoga* to Kuntī.

Pāṇḍu is a patrilineal success. He obtains a five-fold heir, which correlates with his hunt: one son per hit. As Śakuntalā's being *pativratā* was crucial in the case of Duḥṣanta, so Kuntī's and Mādrī's being *pativratā* is crucial in the case of Pāṇḍu. The hart killed while mating evokes the aforementioned Vedic motif of Śiva's punishing Prajāpati's incest, and may indicate the conquest or non-existence of the *putrikā*-minded father-in-law.

THE PĀṆḌAVAS

The Pāṇḍavas are forced into exile twice by Duryodhana and his advisors; for and after the Vāraṇāvata firehouse (1.131–198), and through the dicing after Yudhiṣṭhira's *rājasūya* (i.e. principally for the Āraṇyakaparvan and the Virāṭaparvan, but effectively 2.52–12.38). In the first case, exiled with their mother, they return with a *purohita* and a wife and the kingdom is divided; in the second case, exiled with their wife, they return

[55] Kiṃdama seems to indicate that the doe was not also a person transformed, which is unclear in Vaiśampāyana's version at 1.109.7 (*ṛṣiputras ... bhāryayā saha ... mṛgarūpeṇa saṃgataḥ*).

with weapons and fight. During both exiles they hunt, and they hear the discourses of *ṛṣis* etc. (including stories of their ancestors and other kings).

During the first exile, Hiḍimbā (1.139–143), sent by her brother Hiḍimba to hunt the Pāṇḍavas, is overcome by *kāma* at the sight of Bhīma; he is self-controlled. She says to Kuntī, 'I tried to carry him off, but I could not master him' (*apanetuṃ ca yatito na caiva śakito mayā*, 1.142.9). Bhīma kills her brother and they marry (the first Pāṇḍava marriage) after she agrees to Yudhiṣṭhira's pre-nup that she return Bhīma to his brothers every night. The marriage is temporary, but Ghaṭotkaca, its product, though ordinarily resident with his mother, is loyal and mortally invaluable to the Pāṇḍavas (3.144–145; 7.148–154). Perhaps Hiḍimbā is a somewhat reformed cannibal.

During the first exile, the Pāṇḍavas hunt as a group (1.135.19, 144.1); and when they win Draupadī it is by Arjuna's five-arrow trick. The Pāṇḍavas' pre-nuptial condition this time is that Draupadī marry all five; Drupada, after deliberation with Vyāsa, agrees. After the Pāṇḍavas have returned and established Indraprastha, Arjuna to defend the integrity of the marriage goes on a twelve-month tour (see Allen 1996), marrying along the way Ulūpī at Gaṅgādvāra (1.206), Citrāṅgadā in Kaliṅga (1.207), and Subhadrā in Dvārakā (1.213.12). Of these three, in the *vaṃśa* only Subhadrā is mentioned.[56] The marriage to Subhadrā is instigated by her brother Kṛṣṇa, who suggests an abduction. Yudhiṣṭhira gives his consent; Arjuna does the deed; her kin give their approval in retrospect.

This alliance between the Vṛṣṇis and the Pāṇḍavas encourages almost immediately the Khāṇḍava massacre (1.214–

---

[56] For this reason I do not set out the stories of Ulūpī and Citrāṅgadā here, notwithstanding their later ramifications (see below; the prose *vaṃśa* draws a line by saying *ity ete ekādaśa pāṇḍavānāṃ putrāḥ*, 1.90.89, but several of those mentioned are left hanging).

225; Takṣaka's wife is killed and his son cursed[57]) and the *rājasūya* which leads to the dicing, the second exile, and the takeover of Hāstinapura by the Pāṇḍavas (rather than their coexistence as junior-line neighbours). The *rājasūya* is encouraged also by Pāṇḍu's posthumous message to Yudhiṣṭhira via Nārada:[58]

> 'You can conquer the earth. Your brothers obey you,' he said, Bhārata, 'offer up the great sacrifice of the Consecration!' (2.11.66)[59]

The Pāṇḍava coup makes Pāṇḍu a full link in the Hāstinapura *vaṃśa*. As for the Vṛṣṇis, when they perish *en masse* in the Mausalaparvan they might regret having given Subhadrā away. She is *pativratā*, no *putrikā*, so after the death of the Draupadeyas and her son Abhimanyu, his son Parikṣit is a Pāṇḍava–Kuru heir. But the Pāṇḍavas' Hāstinapura takeover also allows the Vṛṣṇi remnant to settle in Indraprastha under Vajra.

First there is the twelve-plus-one-year exile. The Pāṇḍavas hunt together, with purified (*śuddha*) arrows (3.37.41, 79.8, 157.7, 175.11; see also 3.37.31–33; 3.47; 3.244; 3.251.11–13, 257.9), to feed themselves and Draupadī and to feed the brahmins according to the ordinances (*vidhivad*) 'as if they were beloved sons, kinsmen or co-uterine brothers' (*putrān iva priyāñ jñātīn bhrātṝn iva sahodarān*, 3.47.9, my translation). Draupadī's sons are safe in Kāmpilya (3.23.46), and the only further nubile is Uttarā. Her father gives her; and the self-controlled Arjuna, to his eventual patrilineal advantage, accepts her for Abhimanyu.[60]

---

[57] Takṣaka, who will go on out with Vaiśaṃpāyana's narration to bite Parikṣit and narrowly escape death at the *sarpasatra* (see Minkowski 1991: 397), is Ṛkṣa's old father-in-law in the prose *vaṃśa*.

[58] Nārada says of the *rājasūya* that 'war follows upon it' (*yuddhaṃ ... pṛṣṭhagamanam*, 2.11.69).

[59] *samartho 'si mahīṃ jetuṃ bhrātaras te vaśe sthitāḥ* I *rājasūyaṃ kratuśreṣṭham āharasveti bhārata* II 66 II On the *rājasūya*, see also 3.241.26–27.

[60] Compare Pratīpa's acceptance of Gaṅgā for his son. Things go wrong in both cases; but Parikṣit is revived to breed at least.

In the Āraṇyakaparvan the Pāṇḍavas are imperilled through hunting four times.

- At 3.154 the others are briefly abducted by the *rākṣasa* Jaṭāsura when Bhīma is out hunting alone (Arjuna is away also, visiting Indra).
- At 3.175–178 Bhīma, hunting alone, is captured by his ancestor Nahuṣa, who is living as a snake and only returns to heaven when Yudhiṣṭhira answers his questions. Nahuṣa's fall-and-return could be allied with Yayāti's (as mentioned above, footnote 34), and/or with the Pāṇḍavas'; the snake connection is also interesting (and may link to the substance of the questions), but is out of range at present.
- At 3.248–256, while her husbands are hunting without her, Jayadratha abducts Draupadī; they rescue her efficiently, and after Bhīma has humiliated him Jayadratha is dismissed. The episode is significant not least because Jayadratha's wife is Dhṛtarāṣṭra's daughter Duḥśalā, who represents the possibility of an alternative senior-line heir to thwart the junior even if Duryodhana and his brothers be killed. But Duḥśalā is *pativratā*, even after Jayadratha's death (14.77); and although Jayadratha occasions the death of Abhimanyu as a direct result of his humiliation here (3.256.24–29; 7.41.11–18), by then Abhimanyu has already passed on his seed.
- At 3.295–298 the Pāṇḍavas chase a deer and, peeling off from the group, die one by one; Yudhiṣṭhira wins them back to life, again by answering questions (see Shulman 1996; West 2005–06). The questioner here is Dharma, Yudhiṣṭhira's seminal father.

The exile's term elapses, Duryodhana refuses to return the Indraprastha kingdom, and war results, with the death of almost everyone. Dhṛtarāṣṭra's sons and their sons are killed, as are Draupadī's sons,[61] and the newlywed Abhimanyu, and Ghaṭot-

[61] On Draupadī and *pativratā* see S. Sutherland 1989; Black 2007: 66–73; Malinar 2007; Patton 2007; Hiltebeitel 2007.

kaca, and Arjuna's son by Ulūpī (i.e. Irāvat; see 6.86); and all Pāṇḍava women are cursed barren (10.13.18; 10.15.16–18, 31–33). Pāṇḍava-sons Yaudheya, Sarvaga, Niramitra and (this) Suhotra are unmentioned except in the prose *vaṃśa* (but see for 'the Yaudheyas' paying homage to Yudhiṣṭhira, 2.48.13; killed by Arjuna, 7.18.16, 136.5; 8.4.46; killed by Yudhiṣṭhira, 7.132.25; 'Sarvaga' names the wind and more usually Nārāyaṇa; for a Niramitra Traigarta killed by Sahadeva see 7.82; see also 12.237.16, *niramitrasya kiṃ bhayam*). No Pāṇḍava daughters are apparent. Arjuna's son Babhruvāhana might be heir to the Pāṇḍavas were he not also son of the *putrikā* Citrāṅgadā, promised to her patriline by Arjuna's pre-nup agreement with her father (1.207).

With Śikhaṇḍin's help Arjuna has killed Bhīṣma, the non-seminal but invincible superintendent of the Kuru patriline; and, in retrospectively explicit connection with that fact, during Arjuna's travels with Yudhiṣṭhira's *aśvamedha* horse Arjuna is himself killed by Babhruvāhana (14.78–82; see Fitzgerald 2007). Potential patrilineal disaster is represented by the massacre at Kurukṣetra, by the felling of the almost immortal Bhīṣma, and by the phantom death of Arjuna at the hands of his remaining but unavailable son; but even before Arjuna's revival (by a snake!) Kṛṣṇa has revived Parikṣit (14.62–69; see 10.16.8–17),[62] thus rescuing the ancestors (Bhīṣma's patriline; see Allen 2007) with Arjuna as a full link, and with Yudhiṣṭhira an eldest son overtaken. Yudhiṣṭhira and Bhīṣma and Dhṛtarāṣṭra are explicitly in this position in different ways; their lineal non-centrality has quite some didactic lining for Janamejaya to ponder.

---

[62] Kṛṣṇa's aunt Kuntī reports at 14.65.23–24 that the as yet unborn and unnamed Parikṣit, though it is clear that Pāṇḍava blood runs in his veins, was intended by Abhimanyu to grow up amongst the Vṛṣṇis; this invokes the bilocal upbringing of e.g. Bharata, Bhīṣma, and the Pāṇḍavas themselves, but is a question of Parikṣit's *father*'s matrilocale and patrilocale, not his own. Note that Yudhiṣṭhira suggests Kṛṣṇa, not he, be *yajamāna* of the *aśvamedha*, but Kṛṣṇa refuses (14.70.20–25).

REFERENCES

ALLEN, Nick J. 1996. The hero's five relationships: a proto-Indo-European story. In: Julia Leslie (ed.), *Myth and Mythmaking. Continuous Evolution in Indian tradition*: 1–20. Richmond: Curzon.

——— 2007. Bhīṣma as matchmaker. In: Brodbeck & Black 2007: 176–188.

——— (in this volume). Bhārata genealogy: the core parental-generation males.

ALLSEN, Thomas T. 2006. *The Royal Hunt in Eurasian History*. Philadelphia: University of Pennsylvania Press.

BAILEY, Greg M. 1981. Brahmā, Pṛthu and the theme of the earth-milker in Hindu mythology. *Indo-Iranian Journal* 23: 105–116.

BHATTACHARYA, Pradip 1989. *Themes and Structure in the Mahābhārata. A Study of the Ādi Parva*. Calcutta: Dasgupta and co.

BLACK, Brian 2007. Eavesdropping on the epic: female listeners in the *Mahābhārata*. In: Brodbeck & Black 2007: 53–78.

BODEWITZ, Henk W. 1999. Hindu *ahiṃsā* and its roots. In: Jan E. M. Houben & Karel R. van Kooij (eds.), *Violence Denied. Violence, Non-violence and the Rationalization of Violence in South Asian Cultural History:* 17–44. Leiden: Brill.

BRODBECK, Simon 2006a. Ekalavya and *Mahābhārata* 1.121–128. *International Journal of Hindu Studies* 10.1: 1–34.

——— 2006b. Myth and ideology of the imperial *kṣatriya*: viewing the *Mahābhārata* from here and now. *Journal of Vaishnava Studies* 14.2: 93–102.

——— 2007. Gendered soteriology: marriage and the *karmayoga*. In: Brodbeck & Black 2007: 144–175.

——— 2009a. Husbands of Earth: *kṣatriya*s, females, and female *kṣatriya*s in the *Strīparvan* of the *Mahābhārata*. In: Robert P. Goldman & Muneo Tokunaga (eds.), *Epic Undertakings*. (Papers of the 12th World Sanskrit Conference, 2): 33–63. Delhi: Motilal Banarsidass.

——— 2009b. The Bhāradvāja pattern in the *Mahābhārata*. In: Petteri Koskikallio (ed.), *Parallels and Comparisons. Proceedings of the Fourth Dubrovnik International Conference on the Sanskrit Epics and Purāṇas, September 2005*: 137–179. Zagreb: Croatian Academy of Sciences and Arts.

BRODBECK, Simon & Brian BLACK (eds.) 2007. *Gender and Narrative in the Mahābhārata*. London: Routledge.

BUITENEN, J. A. B. van (trans.) 1973. *The Mahābhārata*, vol. 1 (book 1). Chicago: University of Chicago Press.

——— (trans.) 1975. *The Mahābhārata*, vol. 2 (books 2 and 3). Chicago: University of Chicago Press.

CHAKRAVARTY, Uma 2000. Vedic daughter. *Annals of the Bhandarkar Oriental Research Institute* 81: 179–189.

CHAPPLE, Christopher Key 1993. *Nonviolence to Animals, Earth, and Self in Asian Traditions*. Albany: State University of New York Press.

CHATTERJEE, Gautam 2005. The ethical foundations of Bhīṣma's promises and dilemmas. In: T. S. Rukmani (ed.), *The Mahābhārata. What is Not Here is Nowhere Else (yannehāsti na tadkvacit)*: 145–162. Delhi: Munshiram Manoharlal.

CHOPRA, Ravi & Baldev R. CHOPRA (dir., prod.; Rahi Masoom Reza, scriptwriter) 1988–90. *Mahabharat*, television version, 94 episodes.

COUTURE, André 2003. Dvārakā: the making of a sacred place. In: Phyllis Granoff & Koichi Shinohara (eds.), *Pilgrims, Patrons, and Place. Localizing Sanctity in Asian Religions*: 224–248. Vancouver: University of British Columbia Press.

CUMMINS, John 1988. *The Art of Medieval Hunting. The Hound and the Hawk*. London: Weidenfeld and Nicolson, 1988 (reprint: Edison, N.J.: Castle Books, 2003).

DASGUPTA, Madhusraba 2000. Usable women: the tales of Ambā and Mādhavī. In: Mandakranta Bose (ed.), *Faces of the Feminine in Ancient, Medieval, and Modern India*: 48–55. New York: Oxford University Press.

DHAND, Arti 2004. The subversive nature of virtue in the *Mahābhārata*: a tale about women, smelly ascetics, and God. *Journal of the American Academy of Religion* 72.1: 33–58.

DONIGER, Wendy 1995. Begetting on margin: adultery and surrogate pseudomarriage in Hinduism. In: Lindsey Harlan & Paul B. Courtright (eds.), *From the Margins of Hindu Marriage. Essays on Gender, Religion, and Culture*: 160–183. New York: Oxford University Press.

DONIGER, Wendy & Brian K. SMITH (trans.) 1991. *The Laws of Manu*. London: Penguin.

FELLER, Danielle 2004. *The Sanskrit Epics' Representation of Vedic Myths*. Delhi: Motilal Banarsidass.

FITZGERALD, James L. (trans. & ed.) 2004. *The Mahābhārata*, vol. 7 (Books 11 and 12a). Chicago: University of Chicago Press.

——— 2007. Bhīṣma beyond Freud: Bhīṣma in the *Mahābhārata*. In: Brodbeck & Black 2007: 189–207.

GANGULI, Kisari Mohan (trans.) 1970. *The Mahabharata of Krishna-Dwaipayana Vyasa translated into English prose from the original Sanskrit text*, 4 vols. Delhi: Munshiram Manoharlal (First published by Pratap Chandra Roy and Sundari Bala Roy; Calcutta: Bharata Press, 1883–96; reprint: Delhi: Munshiram Manoharlal, 1993).

GOLDMAN, Robert P. 1978. Fathers, sons and gurus: oedipal conflict in the Sanskrit epics. *Journal of Indian Philosophy* 6: 325–392.

GONDA, Jan 1967. The Indra festival according to the Atharvavedins. *Journal of the American Oriental Society* 87: 413–429 (reprinted in *Selected Studies*, vol. 4; Leiden: Brill, 1975).

——— 1985. *Fatherhood in the Veda*. Turin: Indologica Taurinensia.

HARA, Minoru 1974. A note on the *rākṣasa* form of marriage. *Journal of the American Oriental Society* 94: 296–306.

HARZER, Edeltraud 2005. Bhīṣma and the *vrātya* question. In: T. S. Rukmani (ed.), *The Mahābhārata. What is Not Here is Nowhere Else (yannehāsti na tadkvacit)*: 163–178. New Delhi: Munshiram Manoharlal.

HAWTHORNE, Sîan Melvill 2006. Origins, genealogies, and the politics of identity: towards a feminist philosophy of myth. PhD thesis, University of London.

HILTEBEITEL, Alf 2001. *Rethinking the Mahābhārata. A Reader's Guide to the Education of the Dharma King*. Chicago: University of Chicago Press.

——— 2006. The *Nārāyaṇīya* and the early reading communities of the *Mahābhārata*. In: Patrick Olivelle (ed.), *Between the Empires. Society in India 300 BCE to 400 CE*: 227–255. New York: Oxford University Press.

——— 2007. Among friends: marriage, women, and some little birds. In: Brodbeck & Black 2007: 110–143.

JAMISON, Stephanie W. 1996. *Sacrificed Wife / Sacrificer's Wife. Women, Ritual, and Hospitality in Ancient India*. New York: Oxford University Press.

JAY, Nancy 1992. *Throughout Your Generations Forever. Sacrifice, Religion and Paternity*. Chicago: University of Chicago Press.

KANE, P. V. 1973. *History of Dharmaśāstra. Ancient and Mediæval Religious and Civil Law*, vol 3. (Government Oriental Series, B.6.) 2nd edition. Poona: Bhandarkar Oriental Research Institute.

KARVE, Irawati 1965. *Kinship Organization in India*. 2nd edition. Bombay: Asia Publishing House.

KLOSTERMAIER, Klaus 1991. The original Dakṣa saga. In: Arvind Sharma (ed.), *Essays on the Mahābhārata*: 110–129. Leiden: Brill.

KOSKIKALLIO, Petteri 1999. Baka Dālbhya: a complex character in Vedic ritual texts, epics and Purāṇas. *Studia Orientalia* 85: 301–387.

KRAMRISCH, Stella 1981. *The Presence of Śiva*. Princeton: Princeton University Press.

KUIPER, F. B. J. 1979. *Varuṇa and Vidūṣaka. On the Origin of the Sanskrit Drama*. Amsterdam: North Holland Publishing Co.

LESLIE, Julia 1998. A bird bereaved: the identity and significance of Vālmīki's *krauñca*. *Journal of Indian Philosophy* 26.5: 455–487.

LIDOVA, Natalia R. 2002–03. Indramahotsava in late Vedic and early epic traditions. *Journal of the Asiatic Society of Mumbai* 77–78: 85–108.

Mahābhārata, Critical Edition. Ed. by V. S. Sukthankar, S. K. Belvalkar, P. L. Vaidya *et al.* 19 vols. Poona. Bhandarkar Oriental Research Institute, 1933–66.

MALINAR, Angelika 2007. Arguments of a queen: Draupadī's views on kingship. In: Brodbeck & Black 2007: 79–96.

MERTENS, Annemarie 2002. What defines a stage? The Purāṇic myth of Satī's death in transition. In: Mary Brockington (ed.), *Stages and Transitions. Temporal and Historical Frameworks in Epic and Purāṇic Literature. Proceedings of the Second Dubrovnik International Conference on the Sanskrit Epics and Purāṇas, August 1999*: 359–376. Zagreb: Croatian Academy of Sciences and Arts.

MINKOWSKI, Christopher Z. 1989. Janamejaya's *sattra* and ritual structure. *Journal of the American Oriental Society* 109: 401–420.

——— 1991. Snakes, *sattras* and the *Mahābhārata*. In: Arvind Sharma (ed.), *Essays on the Mahābhārata*: 384–400. Leiden: Brill.

——— 2001. The interrupted sacrifice and the Sanskrit epics. *Journal of Indian Philosophy* 29: 169–186.

MONIER-WILLIAMS, Monier 1899. *A Sanskrit-English Dictionary*. Oxford: Clarendon Press (reprint: 1976).

O'FLAHERTY, Wendy Doniger 1973. *Asceticism and Eroticism in the Mythology of Śiva*. London: Oxford University Press.

O'FLAHERTY, Wendy Doniger (trans.) 1975. *Hindu Myths. A Sourcebook Translated from the Sanskrit*. London: Penguin.

PATHAK, Madhusudan M. 1978. Dakṣayajñavidhvaṃsa-episode in Purāṇas: a comparative study. *Purāṇa* 20.2: 204–223.

PATTON, Laurie L. 2007. How do you conduct yourself? Gender and the construction of a dialogical self in the *Mahābhārata*. In: Brodbeck & Black 2007: 97–109.

PIATIGORSKY, Alexander 1993. *Mythological Deliberations. Lectures on the Phenomenology of Myth*. Ed. by Audrey Cantlie. London: School of Oriental and African Studies, University of London.

REICH, Tamar C. 2001. Sacrificial violence and textual battles: inner textual interpretation in the Sanskrit *Mahābhārata*. *History of Religions* 41.2: 142–169.

SCHMIDT, Hanns-Peter 1987. *Some Women's Rites and Rights in the Veda*. Poona: Bhandarkar Oriental Research Institute.

SHULMAN, David Dean 1996. The Yakṣa's questions. In: G. Hasan-Rokem & D. Shulman (eds.), *Untying the Knot. On Riddles and Other Enigmatic Modes*: 151–167. New York: Oxford University Press.

SIMSON, Georg von 2007. Kṛṣṇa's son Sāmba: faked gender and other ambiguities on the background of lunar and solar myth. In: Brodbeck & Black 2007: 230–257.

SMITH, Brian K. 1989. *Reflections on Resemblance, Ritual, and Religion*. (reprint: Delhi: Motilal Banarsidass,1998).

SUTHERLAND, Gail Hinich 1990. *Bīja* (seed) and *kṣetra* (field): male surrogacy or *niyoga* in the *Mahābhārata*. *Contributions to Indian Sociology* (N.S.) 24.1: 77–103.

SUTHERLAND, Sally J. M. 1989. Sītā and Draupadī: aggressive behavior and female role-models in the Sanskrit epics. *Journal of the American Oriental Society* 109.1: 63–79.

THITE, Ganesh Umakant 1972. Animalism in ancient India. *Journal of the Oriental Institute, Baroda* 21.3: 191–209.

TRAUTMANN, Thomas R. 1981. *Dravidian Kinship*. Cambridge: Cambridge University Press.

WEST, Emily Blanchard 2005–06. An Indic reflex of the Homeric *Cyclopeia*. *The Classical Journal* 101.2: 125–160.

WILHELM, Friedrich 1991. Hunting and the concept of *dharma*. In: Julia Leslie (ed.), *Rules and Remedies in Classical Indian Law* (Panels of the Seventh World Sanskrit Conference, 9): 7–16. Leiden: Brill.

WITZEL, Michael 2005. The Vedas and the epics: some comparative notes on persons, lineages, geography, and grammar. In: Petteri Koskikallio (ed.), *Epics, Khilas, and Purāṇas. Continuities and Ruptures. Proceedings of the Third Dubrovnik International Conference on the Sanskrit Epics and Purāṇas, September 2002*: 21–80. Zagreb: Croatian Academy of Sciences and Arts.

Bhārata Genealogy: The Close Parental-Generation Males

N. J. ALLEN

Since the main protagonists of the *Mahābhārata* are Duryodhana and the Pāṇḍavas, the parental generation in question is that of their parents, Dhṛtarāṣṭra and Pāṇḍu. But these are the eldest and second of the *three* half-brothers fathered by Vyāsa when the Bhārata line is at risk of extinction: the youngest half-brother is of course Vidura. Although Vidura marries, his children are insignificant, so genealogically he seems unnecessary; and although he is often mentioned in the main story (except in the War Books – he is no fighter), nothing he does has any lasting effect on the course of events. So why are there three half-brothers, rather than two?

Anyone interested in Indo-European cultural comparativism might at once ask: do the three brothers represent Dumézil's three functions? Dumézil himself had a different interpretation, which I mention briefly at the end of the paper; but that need not prejudice us.

Definitions of the three functions can be worded in various ways, but here is a brief formulation. The first function relates to sacred and jural wisdom, the second to physical force and war, the third to wealth, fecundity and abundance. The functions are often manifested by characters or events appearing in texts in this standard order, which expresses their descending hierarchy, but the order may also be reversed. Since the birth order, Dhṛtarāṣṭra–Pāṇḍu–Vidura, mirrors the ranking of their mothers (respectively Ambikā the elder princess, Ambālikā the younger princess and Ambikā's unnamed slave-girl), one wonders first whether the functions may here be represented in the standard descending order. But this is out of the question: the eldest brother does not exhibit sacred and jural wisdom, while the youngest does. So let us consider the reverse order.

Not only does Vidura's name mean 'wise, intelligent', but he incarnates the god Dharma, as we are told repeatedly. Moreover, it is interesting to look at how Vidura is introduced. The main story of the epic starts with the ancestry and birth of the epic's purported composer Vyāsa (1.57); but then a single *śloka* on the birth of Bhīṣma (76) is followed at once by an account of the birth of Vidura (77–81). The remaining 25 *śloka*s of this *adhyāya* cover the birth of 27 other individuals or groups, among whom Dhṛtarāṣṭra and Pāṇḍu (both in *śloka* 95) occupy positions 14 and 15. Here then, Vidura's birth is presented before the birth of his older brothers and at greater length – as if it were somehow more important.[1]

The birth story itself, told again more fully in 1.101, concerns a famous *ṛṣi* called Aṇī-Māṇḍavya. Accused of helping some robbers, the sage is impaled on the orders of a king, but is later freed. The torture he has suffered turns out to be in a deeper sense a punishment inflicted by the god Dharma for a sin that Māṇḍavya committed as a child – he had impaled insects or a bird. Since the punishment was so disproportionate to the offence, Māṇḍavya cursed the god to take birth in the womb of a slave, as 'a wise, dharmic and sinless body having the form of Vidura' (1.57.81).

---

[1] Similarly, the passage starting in 1.2.78 passes directly from the birth and life of Bhīṣma to the birth of Vidura (81) and only thereafter mentions the older brothers (82).

So, not only is Vidura the god Dharma incarnate, but his whole birth story is pervaded by the category *dharma*. A *ṛṣi* is by his very nature devoted to *dharma*. The unnamed king who orders the impalement is exercising his normal function of administering *dharma* in the sense of law, and the god himself, the ultimate cause of the punishment, is doing just the same, albeit for a different offence. The hasty judgment of the king and the excessive severity of the god expose them both to the charge of attempting brahmanicide – a truly heinous offence against *dharma*. Finally, in punishing the god, the brahman too is practising *dharma* in the sense of justice.

Throughout his life Vidura consistently supports justice and good sense. Before the war he helps the Pāṇḍavas when Duryodhana tries to incinerate them in the lacquer house, and later he repeatedly tries to avert the conflict by urging compromise. Immediately after the war he delivers to Dhṛtarāṣṭra a substantial philosophico-religious discourse on suffering and rebirth (11.2–7). His unique death accords with his life. Having retired to the forest to practise austerities, he is approached by Yudhiṣṭhira. Using his cognitive/religious yogic skills, Vidura transfuses his limbs, breaths and senses into his nephew (15.33.25), thereby strengthening him. But the pious Yudhiṣṭhira is also an incarnation of Dharma, so the event merely confirms the pre-existing spiritual identity of uncle and nephew. Thus the strong case for Vidura as first-functional is reinforced by the long-established arguments that Yudhiṣṭhira represents the same function.

Following the reverse birth order we come next to Pāṇḍu. When selected by Kuntī at her *svayaṃvara* he is described as powerful, lion-toothed, elephant-shouldered and bull-eyed (1.105.2), and the happy pair are compared to Indra and Paulomī. Pāṇḍu then acquires his second wife Mādrī and, 'endowed with strength and energy', decides to conquer the world (1.105.7). The 'lion-like' king takes his large army and wins a string of victories. Comparing him to Indra, the conquered kings bring him copious and varied tribute. He returns in triumph to his capital, where his mother embraces him as happily as Indra's wife embraces Indra (1.106.4).

Now begins his second career: he departs to the jungle to devote himself to hunting. When his great-grandson Parikṣit is likened to him (1.36.9, cf. 1.45.20), it is as a strong-armed bowman, the greatest on earth, and addicted to hunting. Accompanied by his two wives, Pāṇḍu looks like Indra's elephant with two she-elephants. The heroic king, an expert on weapons, ranges the forest equipped with sword, bow and arrows and breastplate (1.106.9–10). It is while hunting that he shoots a mating deer – in fact a hermit in disguise – and thus incurs the curse which will lead to the gods impregnating his wives and to his own premature death. His death, while making love to Mādrī, is interesting. Overcome by the force of his lust (*kāmabalāt*), he takes her by force (*balād*), in spite of her writhing as forcefully as she can (*yathābalam*); in fact the word *bala* occurs five times in four *ślokas* (1.116.7d–10).[2] Compare the final moments of Vidura, absorbed in his yoga. In short, Pāṇḍu is a king who, having achieved his ambitions as a warrior, devotes himself to the peacetime activity that comes closest to war. He is compared several times to Indra and apparently to no other god. Though not listed as an incarnation in book 1, he is later linked by Vyāsa with Indra's companions, the Maruts, as also is second-functional Bhīma (15.39.9, 11);[3] and after his death he lives with Indra (15.27.8; 18.5.12). All of this makes Pāṇḍu a rather straightforward candidate for interpretation as second-functional.

As for Dhṛtarāṣṭra, a prominent feature of his life is that he and his wife Gāndhārī have one hundred sons and one daughter.

---

[2] Note also Pāṇḍu's epithet *mahābalaḥ* in 1.57.95.

[3] It is among the Maruts that Bhīma sits post mortem, next to his divine father Vāyu, if we follow the Vulgate (18.4.5, with 17*).

Although they all incarnate demons, this impressive number exhibits the abundance and fecundity typical of the third function. Moreover, he seems also to have been wealthy. During the exile of the Pāṇḍavas he distributed manifold largesse (15.8.14–15). He himself claims that his son was brought up in the lap of luxury (*'tyantasukhasaṃvṛddho lakṣmyā*, 7.100.25) and that his army is well paid (7.89.21–23). When Bhīṣma, Droṇa and Kṛpa explain why they are fighting for the Kauravas, each in turn says that he is held captive by the Kauravas because of their wealth (*baddho 'smy arthena kauravaiḥ*, 6.41.36).[4]

Consider also the hundred-and-first son. Yuyutsu is born from an unnamed Vaiśya servant who looks after the king while Gāndhārī is undergoing her prolonged and difficult pregnancy (1.107.35–36). Unlike the one hundred, Yuyutsu fights for the Pāṇḍavas, joining them at the last moment. This means, as Yudhiṣṭhira presciently remarks, that Yuyutsu will survive to perform Dhṛtarāṣṭra's *śrāddha* offerings and to continue his line (*piṇḍaś ca tantuś ca*, 6.41.93). Yuyutsu is indeed the first to offer water at his father's death ritual (15.47.12), and it is he, the Vaiśyāputra, who will manage the Bhārata kingdom while Parikṣit is a minor (17.1.6–7). Thus Dhṛtarāṣṭra, who already appears third-functional for other reasons, has as continuator of his line a son whose mother is from the third-function *varṇa*.[5] In contrast, Arjuna's son Parikṣit, who continues the line of the Pāṇḍavas, has as mother the princess Uttarā.

---

4 Before retiring to the forest Dhṛtarāṣṭra deluges the Earth with extraordinary quantities of largesse (15.20.13); but on this occasion the wealth is given him by Yudhiṣṭhira.

5 Consider also the following. When Duryodhana is recovering from the humiliation of the Cattle Inspection, he wants to celebrate a *rājasūya*, but is told by his family priest to substitute a *vaiṣṇava* ritual, for which he is to construct and use a golden plough (3.241.30). This recalls the Scythian origin myth, where a golden plough and yoke represent the third function (Dumézil 1968: 446–448). Although the *purohita* sometimes seems to be addressing Duryodhana, the ritual is performed for Dhṛtarāṣṭra; so the episode perhaps adds a little further weight to our interpretation of the king.

Finally, at their death (as is foreseen by Nārada), Dhṛtarāṣṭra and Gāndhārī go to the domain of Kubera or Dhaneśa, God of Wealth (15.27.11; 18.5.11–12; cf. Biardeau 2002, vol. 2: 841–842). Kubera represents the third function (for instance, among the Lokapālas; cf. Allen 1999b: 247–248). His realm is described as *durāsada* 'hard to approach', and apparently no other *Mahābhārata* humans go there at death. Thus the post-mortem associations of the triad – respectively with Dharma, Indra and Kubera – accord neatly with their function.

The three half-brothers do not *invariably* show so neat a fit with the functions. Thus, when they were students, Pāṇḍu excelled in archery, Dhṛtarāṣṭra in bodily strength, and Vidura in respect of *dharma* (1.102.19–20); and the eldest brother's strength is confirmed when, after the war, he crushes an iron image of Bhīma, exhibiting the force of ten thousand elephants (11.11.18). Such strength is of course typical of second – rather than third-functional characters – an anomaly that may or may not be dismissible simply as 'noise' in the manifestation of the functions. Moreover, hints can be found of the reverse functional ranking – the one implied by the birth order and the mother's status. Gāndhārī marries by the first-functional mode (Allen 2007), and she incarnates the goddess Mati ('Intelligence', 1.61.98); and Vidura, 'the Steward', distributes food and gifts at the *vaiṣṇava* ritual (3.242.20–22). Even so, *overall*, Vidura is characterised by his wisdom, as is Pāṇḍu by his warrior qualities, while Dhṛtarāṣṭra is linked with fecundity, wealth, Vaiśyas and Kubera.[6] The contradictions and reversals in the

[6] Since the first two functions quite often stand together in opposition to the third, note that both Vidura (when he was Dharma) and Pāṇḍu are cursed by sages, while Dhṛtarāṣṭra is not. However, in some respects the first and third brothers stand together and apart from the second. Thus Vidura lives with Dhṛtarāṣṭra in Hāstinapura, while Pāṇḍu soon leaves it; and the pair are the sons of Ambikā and her slave substitute, while Pāṇḍu is the son of Ambālikā.

application of the functional schema may reflect the general cosmic disorder in the lead-up to the Great War.

On finding a trifunctional pattern the next question should be whether it is a substructure within a pentadic pattern, the additional elements – one ranking above the triad, one below – both being in some sense heterogeneous to it (Allen 1999a). In this case a higher-ranking element is easily found. Vyāsa fathers the triad as requested by his mother, but he also has one further son, namely Śuka. Śuka was the first pupil to learn the epic from his father, its composer (1.1.63), and the main version of Vyāsa's birth story ignores the triad and only mentions *svaṃ ātmajam* Śuka (1.57.74).

While the triad is fathered only on request by a reluctant Vyāsa, Śuka's case is quite different (12.310–311). After a century of austerities directed to Śiva, the sage is promised the son he wants, who will be as pure as the five elements and devoted to Brahmā. Accordingly, at the sight of the *apsaras* Ghṛtācī, Vyāsa ejaculates onto his firesticks, and the infant is born from them, to the acclaim of the cosmos. From childhood onwards the boy's only real interest is in achieving *mokṣa*, and after various adventures he undertakes a journey to this end, which involves identification with Brahmā (12.320). Brahmā, the Creator or the Absolute, sometimes represents the fourth function in its positive aspect (Allen 1999b), and so, I believe, does *mokṣa*, the desirable condition of being outside or beyond ordinary life, with its round of births and deaths. Moreover, Śuka shows exactly the mixture of homogeneity and heterogeneity required of fourth-function representatives.

No further half-brothers are available to represent the fourth function in its negative aspect, but a strong candidate is Dhṛtarāṣṭra's brother-in-law. Śakuni, Gāndhārī's brother and Duryodhana's maternal uncle, is an important character, who is born as a scourge of *dharma* (*dharmahantṛ*) by the anger of the gods (1.57.93). As Mani puts it (1975: 670), Śakuni 'pulled the strings for all the evil actions of Duryodhana'. In particular, he proposes the gambling matches in book 2 and thus becomes the 'root of the Kurus' wicked behaviour' (*anayasya mūlam*, 9.27.59). His name means 'bird', often one of ill omen (Scheuer 1982: 42–44). Together with Karṇa and Duḥśāsana he belongs to the inner circle around Duryodhana, who himself must represent the fourth function negative (Allen 1999b: 251). Moreover, if the latter is born from a portion of Kali, Śakuni incarnates Dvāpara (1.61.72, 15.39.10) – Dvāpara and Kali being respectively the third and fourth (i.e. last) in the descending set of *yugas* and throws of dice.[7] These two sinners are also linked in another way. Śakuni, beheaded by his nephew Sahadeva, the youngest Pāṇḍava, is the last victim of the ordinary fighting in the Great War (9.27.57), and the only important Kaurava warrior still alive is Duryodhana, whose death forms a separate episode of the same book.

* * *

From a methodological point of view the pentad we have tried to analyse is far from an ideal case. The triad fathered by Vyāsa at his mother's request is presented by the text as an explicit, well bounded set or totality, but this cannot be said of the pentad, or of either 3+1 quartet. Instead, the pentad was constructed analytically: we put together four-half-brothers and the brother-in-law of one of them, and we omitted two other brothers-in-law – Kuntī's brother Vasudeva (father of Kṛṣṇa) and Mādrī's brother Śalya (nothing is said of Vidura's wife having a brother).

On the other hand, a number of considerations support the analysis.

[7] Cf. González-Reimann 2002. If both these warriors are linked to time, it is because time is the ultimate destroyer (often personified as (Mahā-)Kāla or Śiva). After his death Śakuni is absorbed back into Dvāpara (18.5.18).

(1) The other brothers-in-law are not serious contenders for a place in the pentad: Vasudeva has too little importance in the plot of the epic, and Śalya, like his sister, represents the third function – for several reasons, including his place in the sequence of Kaurava marshals (Allen 2005: 27–28).

(2) Let us, just for a moment, dismiss the idea of hierarchically organised functions and simply try to evaluate the members of our pentad, as if we were listeners at an early recitation of the epic. In terms of relative moral status they rather clearly form an ascending hierarchy, from Śakuni, then Dhṛtarāṣṭra, at the bottom, to Vidura, then Śuka, at the top.

(3) Returning to the functional perspective, one notes that the figure proposed for the bottom slot in the hierarchy (F4-) relates to the triad genealogically through the lowest ranking member of the triad (F3), rather than through one of the others.

(4) Though the two aspects of the fourth function stand at opposite ends of the hierarchy, their representatives quite often show similarities. In this case the similarity lies in the bird-names, which set the pair apart from the triad. *Śakuni* can mean a bird in general, and at the end of the war, Śakuni's mouldering corpse is surrounded by birds (*śakuntāḥ*, 11.24.25 – cf. the note by Fitzgerald 2004: 68). *Śuka* means specifically a parrot: Ghṛtācī took this form after being seen by Vyāsa, and at his birth Śuka is circled clockwise by various species of bird, including parrots (12.311.4, 20).

(5) Both adjacent generations express the pentadic ideology, and although neither case is absolutely straightforward, their analytical problems differ from those of the parental generation. The presence of the same pattern in the other generations does not prove that the pentadic analysis of the intervening one is correct, but renders it more likely. In the following tabulation, based on Allen (1999a for the children's generation, and 2005), some entries may seem surprising, but each is supported by arguments that I take to be both untendentious and adequate.

F4-	F3	F2	F1	F4+
Vyāsa	Vicitravīrya	Citrāṅgada	Kṛpa	Bhīṣma
Śakuni	Dhṛtarāṣṭra	Pāṇḍu	Vidura	Śuka
Duryodhana[8]	twins	Bhīma	Yudhiṣṭhira	Arjuna

(6) Confidence in the tabulation is strengthened by one further argument. Regarded globally, the relation between the generations is one of succession: a parental generation gives way to its successor. But in two of the columns this global relationship is realised in a more specific form. Vidura transfuses himself into his successor;[9] and, more abstractly, Śakuni, the embodiment of the *dvāparayuga* or throw, gives way to the embodiment of the succeeding *yuga* or throw. Śakuni fits well vertically, as well as horizontally.

* * *

In conclusion, what I have attempted is an interpretation of the *Mahābhārata* parental generation in terms of the Indo-European pentadic ideology – an idea for which there is now considerable evidence, both from the *Mahābhārata* and elsewhere. Though deeply indebted to Dumézil, the interpretation differs radically from the one he offers (Dumézil 1968: 145–175), under which Vidura relates to Aryaman, Pāṇḍu to Varuṇa, and Dhṛtarāṣṭra

---

[8] Karṇa too belongs in this slot, but can be omitted here. I also omit Kṛṣṇa, who belongs in the same slot as his brother-in-law Arjuna.

[9] Two observations. (1) The transfusion recalls the 'rite of transfer' (*sampradāna*), in which a moribund father hands on various components of his being to his son – see e.g. *Kauṣītaki-Upaniṣad* 2.14 (Olivelle) and Sprockhoff 1979: 386–396. (2) Vidura's post mortem destination is the '*santānaka* worlds' (15.33.32) – the name derives from *santāna* 'continued succession, continuity'.

to Bhaga. Space is lacking to explore and criticise his reasoning, and to consider how the criticisms bear on his theories about myth/epic transposition, eschatology, and 'Minor Sovereigns' (i.e. the substructure of the first function). On the other hand, the linking of Dhṛtarāṣṭra with the third function accords with another Dumézilian argument. Dumézil often argued that, in terms of myth, a complete IE society only arises after a conflict that pits third-function representatives against their superiors. No doubt a full interpretation of the Great War will need to take this into account.

For the moment I have tried to provide further evidence not only of the pervasiveness of the pentadic ideology but also of the thoughtful and sophisticated planning that went into the Great Epic. We can now see that, far from being 'unnecessary', a Vidura figure is essential to the full pentadic pattern. But there is much more to be said about this generation. I have discussed elsewhere the marriages of the three half-brothers, all arranged by Bhīṣma (Allen 2007); and I hope one day to explore the parallels between them and the sons of Iapetos in early Greek tradition.

REFERENCES

ALLEN, N. J. 1999a. Arjuna and the second function: a Dumézilian crux. *Journal of the Royal Asiatic Society* (Ser. 3) 9: 403–418.

—— 1999b. Hinduism, structuralism and Dumézil. In: E. C. Polomé (ed.), *Miscellanea Indo-Europea*: 241–260. Washington: Institute for the Study of Man.

—— 2005. Romulus et Bhîshma: structures entrecroisées. *Anthropologie et Sociétés* 29.2: 21–44.

—— 2007. Bhīṣma as matchmaker. In: Simon Brodbeck & Brian Black (eds.), *Gender and Narrative in the Mahābhārata*: 176–188. London: Routledge.

BIARDEAU, Madeleine (trans.) 2002. *Le Mahābhārata*. 2 vols. Paris: Seuil.

DUMÉZIL, G. 1968. *Mythe et épopée*, vol. 1. Paris: Gallimard.

FITZGERALD, James L. (trans. & ed.) 2004. *The Mahābhārata*, vol. 7 (Books 11 and 12a). Chicago: University of Chicago Press.

GONZÁLEZ-REIMANN, L. 2002. *The Mahābhārata and the Yugas. India's Great Epic Poem and the Hindu System of World Ages*. New York: Peter Lang.

MANI, Vettam 1975. *Puranic Encyclopaedia*. Delhi: Motilal Banarsidass.

SCHEUER, J. 1982. *Śiva dans le Mahābhārata*. Paris: PUF.

SPROCKHOFF, J. F. 1979. Die alten im alten Indien: Ein Versuch nach brahmanischen Quellen. *Saeculum* 20: 374–433.

Duryodhana's Truths: Kingship and Divinity in *Mahābhārata* 5.60

ANGELIKA MALINAR

The fifth Book of the *Mahābhārata*, the Udyogaparvan (UdP), deals with the failed efforts to negotiate peace and the arguments for and against war. It narrates diplomatic missions, in particular that of Kṛṣṇa, that address what many epic characters regard as a no-win situation, since they find themselves torn between contradictory claims of duty and different loyalties towards each other. Duryodhana figures prominently in these debates, since he is shown as a warrior and king who no longer defines his role in terms of the family-clan and kinship laws (*kuladharma*) alone. When faced with the choice between his responsibility towards his clan and his duties as a king responsible for his kingdom and all people, he chooses the latter. By stepping out of the framework of familial obligations he forces the other members of the family to do the same, even if they would prefer to keep peace. This attitude provokes opposition from his surrounding, especially from his father Dhṛtarāṣṭra, although he cannot oppose his son. Since Dhṛtarāṣṭra's concerns circle around the disastrous consequences of war and the strength of the Pāṇḍavas, he time and again warns Duryodhana that he will probably lose the war. Criticism is also voiced by his mother Gāndhārī who points out that her son lacks one essential quality of a sovereign, that is, the capacity to control himself. Apart from Vidura, she appears in the UdP, as a *dharmadarśinī*, a teacher of a royal *dharma* that includes certain ascetic qualities of self-restraint that result from 'conquering oneself' (5.127.28). Gāndhārī (5.127.21) declares that a king must first of all be able to control himself before he can claim to control others. He must conquer himself like a country:

> A kingdom, you intelligent man, cannot be obtained, protected, nor enjoyed as the whim takes you, bull of the Bharatas. For the one who is not in control of his senses does not keep his kingdom for long. However, the wise man who controls his senses, he guards his kingdom. Desire and anger drag a man away from his interests; by defeating these two enemies a king conquers the earth. ... Who masters just himself at first as if he were a country, does not seek to master his councilors and enemies in vain. (5.127.21–23, 28)[1]

These statements correspond to similar views voiced in the *Bhagavadgītā* (BhG) and in other parts of the epic; sometimes in connection with Yoga and Sāṃkhya teachings. In the passage quoted before, self-control is intrinsically connected to a king's capability to permanently rule his kingdom. This connection is corroborated in a passage in the *Arthaśāstra*, where the 'conquest of the senses' (*indriyajaya*, 1.6) is regarded as one of the qualifications a king has to acquire and cultivate through continuous practice.

While Duryodhana's close relatives are critical of his intransigence and recklessness, his opponents and enemies pass more shattering judgments. He is accused being the killer of the family (*kulaghna*, 5.123.8, 21) out of selfishness and greed. For some, Duryodhana is even guilty of theft since he has taken property that does not really belong to him (5.29.28). Especially those who

---

[1] *na hi rājyaṃ mahāprājña svena kāmena śakyate | avāptu rakṣituṃ vāpi bhoktuṃ vā bharatarṣabha* || 21 || *na hy avaśyendriyo rājyam aśnīyād dīrgham antaram | vijitātmā tu medhāvī sa rājyam abhipālayet* || 22 || *kāmakrodhau hi puruṣam arthebhyo vyapakarṣataḥ | tau tu śatrū vinirjitya rājā vijayate mahīm* || 23 || ... *ātmānam eva prathamaṃ deśarūpeṇa yo jayet | tato 'mātyān amitrāṃś ca na moghaṃ vijigīṣate* || 28 || Similar views are voiced by Vidura (5.34.55–56). Both, *kāma* and *krodha* are regarded as enemies of self-restraint (*paripanthin*) in BhG 3.37 too.

side with the Pāṇḍavas are sceptical about the chances for a peaceful solution or even openly demand to wage war right away. Draupadī and Kṛṣṇa, for instance, emphasize Duryodhana's evil-mindedness. Especially Kṛṣṇa is turned into Duryodhana's direct opponent during his diplomatic mission, which results in an assertion of claims of power, supremacy and righteousness on both sides that culminates in Duryodhana's order to arrest Kṛṣṇa and the latter's yogic display of his multiple allies in his body (cf. 5.129). Although ambiguities and doubts remain also with regard to the role of Kṛṣṇa,[2] it comes as no surprise, that, for most scholars and the pro-Pāṇḍava tradition of bardic reciters, Duryodhana appears as the perfect villain, the dark, demonic force that causes destruction. This indeed echoes a view voiced in some, probably late, passages of the epic, in which Duryodhana is regarded as the incarnation of the demon Kali. Also in the *Arthaśāstra* Duryodhana is cited as an example for a bad king (cf. 1.6.8).

However, the epic authors and redactors do not just play off the 'good guys' against the bad ones; rather, on each side one finds dark stains and ambiguities, although, in the end, the Pāṇḍavas shine more brightly.[3] This ambiguity contributes to the narrative depth of the epic, which does not emerge as a by-product of the contradictions that result from the many additions in the course of the epic's textual history just by coincidence. Rather, different perspectives, voices and even epic traditions seem to have co-existed. The many different versions of the epic,

---

[2] See Hiltebeitel (1976: 42) who points out that Kṛṣṇa's 'alleged peace mission' brought the conflicts into the open (cf. also Deshpande 1991: 336 and Dev 1989, however, without discussing earlier studies). Kalyanov (1979) reads the UdP as an illustration of diplomatic techniques and regards Kṛṣṇa (too) simply as a peace supporter.

[3] This holds especially true for the trickery and disregard for the codex of fighting Kṛṣṇa and the Pāṇḍavas display in the course of the epic. For other views on Duryodhana and the Kauravas see, for instance, Holtzmann (1892–95) and Gitomer (1992) on *Veṇisaṃhāra*.

including those in which the Kauravas, especially Duryodhana and Karṇa, are made heroes, as well as its dramatic adaptations, testify to the complexity of both the plot and its characters. Also in the extant UdP Duryodhana is made by the epic narrators a figure not only to condemn or to make fun of, but also to reflect upon.

This can be seen in a speech put into his mouth in chapter sixty of the UdP which I shall analyse with regard to the concept of sovereignty proposed by Duryodhana and his self-perception as a 'speaker of truth'. The speech is part of a longer conversation between Duryodhana and his father Dhṛtarāṣṭra, in which the blind king once again tries to prevent his son from waging war against his cousins (5.56–59). Dhṛtarāṣṭra argues that the Pāṇḍavas are invincible because the gods are on their side and therefore Duryodhana's position is weak (cf. 5.59.8ff). In his reply, a furious Duryodhana refutes this theory of 'divine support' by pointing to its logical flaws, also giving us his ideas about kingship. He rejects the alleged superiority of the gods by considering the sources of divine power and status (*devatvam*). The following two interpretations of divinity are refuted: (1) gods are gods because they represent the values of renunciation and are detached from human emotions; (2) divinity is connected to sacrificial reciprocity. This reasoning is followed by Duryodhana's self-proclamation as the overlord of all beings, who is capable of dealing with matters human as well as divine. Special emphasis is put on Duryodhana's mastery of water and the power of his 'truth-speaking'. The speech concludes with a praise of Duryodhana's mightiness (*māhātmya*, 5.60.23).

ARGUING DIVINITY

Duryodhana's argument against divine interference starts by reflecting on ascetic achievements as the means of proving the divinity of gods (*devaprāmāṇya*). He tells his father:

> Lose your fear, best of kings, when you think that the Pāṇḍavas (sons of Pṛthā) were invincible because the gods are their friends! 'The gods obtain their status as gods (*devatvam*) because they are not subject to love and hatred, malice and greed and due to their indifference to living creatures' – this is what Dvaipāyana Vyāsa, Nārada and Rāma Jāmadagni once told us. Gods never act like men, out of love, greed, compassion or hatred. So if Agni, Vāyu, Dharma, Indra or the two Aśvins were to act according to their sympathies, the sons of Pṛthā should not have come to grief. Therefore never at any time must you harbour this worry, since they always have an eye to divine affairs only. (5.60.2–7)[4]

Duryodhana considers here a concept of divinity that is based on ascetic values. According to this notion, gods are gods because they are perfect ascetics, since they are not subject to love, hatred, and all the other emotions characteristic of human beings. Gods are gods because they are not involved in worldly affairs. Like ascetics they are indifferent towards living beings and must not take sides. The very suffering of the Pāṇḍavas proves that the gods do not interfere and act according to their liking. Rather, interference would contradict their status as ascetics since detachment and partiality in a violent conflict do not go together. In the very moment these ascetic gods take sides they would lose their divine power and become useless for the Pāṇḍavas. With this sophisticated reasoning Duryodhana has ruled out that ascetic indifferences could serve as a model for judging social actions. He points to the paradox of the power produced by ascetic practice. The successful ascetic cannot return to the world without losing his achievements, his ascetic power

---

[4] *aśakyā devasacivāḥ pārthāḥ syur iti yad bhavān* । *manyate tad bhayaṃ vyetu bhavato rājasattama* ॥ 2 ॥ *akāmadveṣasaṃyogād drohāl lobhāc ca bhārata* । *upekṣayā ca bhāvānāṃ devā devatvam āpnuvan* ॥ 3 ॥ *iti dvaipāyano vyāso nāradaś ca mahātapāḥ* । *jāmadagnyaś ca rāmo naḥ kathām akathayat purā* ॥ 4 ॥ *naiva mānuṣavad devāḥ pravartante kadācana* । *kāmāl lobhād anukrośād dveṣāc ca bhāratarṣabha* ॥ 5 ॥ *yadi hy agniś ca vāyuś ca dharma indro 'śvināv api* । *kāmayogāt pravarteran na pārthā duḥkham āpnuyuḥ* ॥ 6 ॥ *tasmān na bhavatā cintā kāryaiṣā syāt kadācana* । *daiveṣv apekṣakā hy ete śaśvad bhāveṣu bhārata* ॥ 7 ॥

cannot be used for a social cause because it exists only as long as it is not used.

Duryodhana then considers whether gods' involvement in human affairs could prove their divinity. This involvement is brought about by their participation in ritual transaction and in consequence is confined by the structure of ritual reciprocity (5.60.8–19):

> Even if the proof of being a god (*devaprāmāṇya*) was in the case of gods to be established on the basis of their sympathies or by criteria of hatred or greed, love, this will not help. Even if Agni Jātavedas would have enclosed the worlds from all sides, wishing to burn them down, he will have to sing hymns of praise when he has been invoked by me (through ritual formulas). (5.60.8–10)[5]

Duryodhana uses here the well-known characteristic of Vedic gods that they are related to human beings through the ritual transactions of the Vedic fire sacrifice. It is a relationship of mutual dependency (*paraspara*); gods and men live from each other (see Bailey 1983: 26–32; cf. also BhG 3.10–13). Thus, when men call gods to their sacrificial fire through ritual invocations, the gods are not just accepting an invitation, they are also attending for their own sakes. In addition, they are bound to the correctness and truth of the ritual language: when invoked correctly, they respond. According to Duryodhana, this dependence makes it highly improbable that the gods can ever become the permanent allies of the Pāṇḍavas. Since the gods cannot reject Duryodhana's sacrifice for moral reasons, they have to respond to his ritual invocations too. Agni in particular, the god of the sacrificial fire, cannot help transporting the sacrificial offering to its addressee. It is the correctness of the invocation that matters, not the respectability of the invoking priest or sacrificer. In con-

---

[5] *atha cet kāmasaṃyogād dveṣāl lobhāc ca lakṣyate* | *deveṣu devaprāmāṇyaṃ naiva tad vikramiṣyatu* || 8 || *mayābhimantritaḥ śaśvaj jātavedāḥ praśaṃsati* | *didhakṣuḥ sakalāṃl lokān parikṣipya samantataḥ* || 9 || *yad vā paramakaṃ tejo yena yuktā divaukasaḥ* | *mamāpy anupamaṃ bhūyo devebhyo viddhi bhārata* || 10 ||

sequence, Duryodhana declares that, even if one of these gods would like to take sides with the Pāṇḍavas, he would bind them to his sacrifice by ritual invocations and make them his servant. It seems that Duryodhana's reasoning mirrors a cultural milieu in which participation in ritual exchange implies the subjection of the gods to the patron's goals and priestly invocations. Ritual is primarily considered to be an instrument for achieving the goals of patrons and priests.

Duryodhana's claim of sacrificial skill is not confined to this speech, but is also the topic of another speech in 5.57. The view of sacrifice presented here is rather different from the one discussed before (and perhaps older) and helps characterizing Duryodhana's arguments in 5.60. The speech in 5.57 is not about the dependency of the gods and the power of the patron-priest, but rather explores the 'sacrifice of battle' (*raṇayajña*). Duryodhana together with Karṇa are the consecrated patrons (*dīkṣitau*) while Yudhiṣṭhira is accorded the position of the animal to be offered (*paśuṃ kṛtvā*) and the weapons serve as ritual utensils (5.57.12–18):

> The chariot is the altar, the sword the sacrificial spoon, the club the ladle, the armour the sacrificial assembly, the four horses are the four priests, the arrows the *darbha* grass, fame the oblation. Having offered up ourselves to Manu Vaivasvata in this battle, O king, we will win and return, covered with glory, the enemies slain. (5.57.13–14)[6]

His intention is to perform a sacrifice in which he functions as the patron who reins in the horses who take the role of the 'four priests'. However, at the same time he is aware that this is a type of sacrifice in which he himself may turn out to be the victim (*ātmayajñena*, 5.57.14). While he intends to make his opponent the victim, he has also to relinquish something. This

---

[6] *ratho vedī sruvaḥ khaḍgo gadā sruk kavacaṃ sadaḥ* | *cāturhotraṃ ca dhuryā me śarā darbhā havir yaśaḥ* || 13 || *ātmayajñena nṛpate iṣṭvā vaivasvataṃ raṇe* | *vijitya svayam eṣyāvo hatāmitrau śriyā vṛtau* || 14 ||

tyāga, his act of relinquishment, is that he is willing to give up his life too (5.57.16). The sacrifice is addressed to Manu Vaivasvata, the protector of creation and *dharma*. However, the actual sacrificial activity of offering is a contest, a battle which not only aims at killing the victim, but to find out who that victim would be. The survivor is the winner in that he reaps the fruits of sacrifice by making his opponent the sacrificial victim and thus the substitute of himself as pointed out by Biardeau:

> Each warrior hopes to sacrifice the enemy killed by him as his substitute, or he will be sacrificed as the substitute of the enemy killed by him (Biardeau 1997: 83).

She further argues that the 'sacrifice of battle' is feasible because the sacrificer and the victim are closely connected to each other, not only because each sacrificial is a substitute of the patron, but usually also property to be given up:

> the victim somehow belongs to the sacrificer, is something of his properties or someone of his own kin. ... That is why the epic war must be ... a war between brothers. (Biardeau 1997: 84 and note 13)

The possibility that he himself may become the victim is acknowledged by Duryodhana:

> Either I will be the ruler of the earth, having killed the Pāṇḍavas, or they, the sons of Paṇḍu will enjoy the earth when they kill me. I have given up my life, O king. (5.57.16–17a)[7]

This willingness for 'self-sacrifice' is not heard of in 5.60 which not only points to the different function of the speech as an attempt to reassure Dhṛtarāṣṭra, but also to a different interpretation of ritual. While 5.57 interprets the battle by drawing on the

---

[7] *ahaṃ hi pāṇḍavān hatvā praśāstā pṛthivīm imām* I *māṃ vā hatvā pāṇḍuputrā bhoktāraḥ pṛthivīm imām* II 16 II *tyaktaṃ me jīvitaṃ rājan* ... II In her account of 5.57 Jatavallabhula (1999: 73) neglects the fact that Duryodhana declares to have given up his life while bracing himself for two possible outcomes of the 'sacrifice of battle'.

paradigm of sacrifice,[8] 5.60 emphasizes its being controlled by the patron. Duryodhana's superiority is based on his ritual supremacy; he is not shown as participating in a complex and risky ritual performance, but as its master. 5.60 demonstrate Duryodhana's ritual power that allows him controlling the gods invited to and consecrated for the sacrifice. The key concept here is the correct invocation, the use of the 'true language' of *mantra*s in his ritual invocations (*abhimantraṇa*) that he intends to use as a language of command. This matches the interpretation of Duryodhana's position given at the end of 5.60: his words will come true and be efficient, because he is true as the 'speaker of truth' (*satyavāk*).

Taking Duryodhana's arguments together, the conclusion is that those who are usually deemed powerful 'gods', be they the gods of ritual or ascetic-like gods, are in fact rather powerless. What does he propose instead? Duryodhana declares that, as the king, he is the superior godhead because he is neither dependent on sacrificial food nor indifferent towards his kingdom and the peoples. In this way, Duryodhana is made to present a particular blend of ascetic and ritual aspects of divinity; he is turned into a god-king whose duty is to defend the territory and protect his subjects. For this purpose he is ready to use all his ritualistic skills and *tejas*, his 'fiery might' (van Buitenen) as a warrior. He declares himself to be the overlord of the gods and demons as the master of ritual. Apart from his *tejas*, his command of 'ritual invocation' (*abhimantraṇa*) is regarded as the instrument and source of his power:

---

8 Heesterman (1985) assumes an intimate connection between war and the original Vedic sacrifice, and the depiction of the 'sacrifice of battle' resembles in some regards Heesterman's views on the original structure of Vedic sacrifice. However, according to Biardeau & Malamoud (1976) this parallelism between battle and sacrifice serves to legitimate the violent duty of a warrior in a way that resembles the legitimacy of violence in a sacrificial context – that is, ritual violence is to be regarded as non-violence.

> Supreme indeed is the fiery might (*tejas*) the celestials possess, but my own surpasses that of the gods, know it, Bhārata. I will steady the earth, when it cleaves asunder and the peaks of [its] mountains, with my ritual invocations, while the world is watching. Again and again I shall stop the terrifying, thundering gale and avalanche of rocks that spell the destruction of the sentient and insentient beings, the standing and moving, out of compassion with the living creatures, while the world is watching. (5.60.10–13)[9]

In this passage Duryodhana describes his power over the earth, especially in times of distress and natural calamities. Since he embodies supreme sovereignty as a king, his power must be demonstrated with regard to the actual kingdom, the earth. This is signified not primarily by pointing to the prosperity of his kingdom, but to his capability to protect is from instability of all kinds, such as earthquake or rock fall. However, he does not so much present himself as the consecrated 'husband of the earth' (see Hara 1973) human kings are supposed to be, but as a cosmic actor who displays his power in times of distress and danger 'while the world is watching'. This description rather echoes the visible display of divine cosmic power which is a prominent feature of theophanies, such as, for instance, in BhG ten and eleven. His exalted position is also indicated in the motive mentioned for his activities, 'compassion for the creatures' (*bhūtānām anukampayā*, 5.60.13). While Kṛṣṇa declares to destroy 'out of compassion' (10.11) ignorance for those who approach him, compassion for the creatures or 'all beings' is mentioned as a quality of a righteous king in several instances also elsewhere in the epic (cf., for instance, 5.88.18, 93.6; 12.59.84, 64.26, 65.2, 88.22) and is praised as a virtue in Buddhist texts.

---

[9] *yad vā paramakaṃ tejo yena yuktā divaukasaḥ* | *mamāpy anupamaṃ bhūyo devebhyo viddhi bhārata* || 10 || *pradīryamāṇāṃ vasudhāṃ girīṇāṃ śikharāṇi ca* | *lokasya paśyato rājan sthāpayāmy abhimantraṇāt* || 11 || *cetanācetanasyāsya jaṅgamasthāvarasya ca* | *vināśāya samutpannaṃ mahāghoraṃ mahāsvanam* || 12 || *aśmavarṣa ca vāyuṃ ca śamayāmīha nityaśaḥ* | *jagataḥ paśyato 'bhīkṣṇaṃ bhūtānām anukampayā* || 13 ||

DURYODHANA AND THE WATERS

Duryodhana's supreme royal power not only provides stability of the earth, but also extends to the earthly and heavenly waters. This is of special importance for the legitimacy of his rule and their conveyance in the unction ceremony (*abhiṣeka*) during the royal consecration is one of its pivotal moments (see Heesterman 1957; Proferes 2007). The 'waters' referred to in the ritual as well as in Duryodhana's speech are the rivers, the rain and the ocean. 'Winning' and 'ruling' the rivers is essential for the prosperity of the kingdom and it is a sign of the king's power. However, what is emphasised in the following passage is not the water's fertility, but Duryodhana's mastery of the waters when he sets out to expand his kingdom through military expeditions:

> Chariots and foot soldiers will march over water that I made solid. I am the one and only mover of gods, demons and living creatures (5.60.14).[10]

Royal, magic and yogic powers are blended when Duryodhana claims that he can 'solidify' water (*stambhitāsu apsu*, 5.60.14). What may sound like a vague allusion to one of the miracles Jesus is said to have performed when he was walking on the water and invited apostle Peter to join him (Matthew 14.22–33) is in fact a specific skill accorded also in other texts to gods, kings, *yogins* as well as Buddhist ascetics and to those who perform a 'truth act' (*satyakriyā*). The mastery over water has different forms and aspects, only one of them is 'solidification' (see also Brown 1928). The *Yogasūtrabhāṣya* on *sūtra* 3.42, for instance, describes the capacity to cross water as a consequence of acquiring the supernatural power of 'lightness' (*laghutva*) which is brought about by controlling the usual connection

---

[10] *stambhitāsv apsu gacchanti mayā rathapadātayaḥ* | *devāsurāṇāṃ bhāvānām aham ekaḥ pravartitā* || 14 || van Buitenen suggests that the water is 'frozen', which seems too much to restrict the process described here to the imagery of an 'icy' surface, if understood literally. See below.

between the *yogin*'s body and space (*ākāśa*): 'due to this lightness, he (the *yogin*) stays with both feet upon water' (*laghutvāc ca jale pādābhyāṃ viharati*). In Buddhist texts walking on the water is one of the six *siddhis*, supernatural powers, acquired by a *buddha* or by monks progressed to advanced stages of meditative practice. However, this power does not signify the highest achievements and is usually surpassed by crossing water through levitation or flying.[11] Conversely, the theme appears often in stories about a miraculous crossing of water by kings and also queens on horses or chariots, sometimes accompanied by their troupes. In the epic this is described as resulting from a process of stabilisation or solidification expressed by using forms of the verbal root √*stambh* (√*staḅh*), 'to make firm'. It is a process that affects the element water directly, while in the case of yogic 'lightness' the transformation happens on the side of the *yogin* as the ability of levitation or flying. Such solidification can be brought about by different means, including spells, *mantras*, a *satyakriyā* act or *māyā*, 'magic' power. This may explain why crossing water due to solidification may not necessarily be a sign of yogic achievements. However, it is rather often ascribed to successful, virtuous and truthful kings who may overcome any obstacles especially during military expeditions. In the present context, attention may be drawn to two epic passages which are now to be found in the appendices to the Droṇaparvan provided in the critical edition. The first passage deals with the ideal king and cultural hero Pṛthu Vainya who turned the earth into a habitable place that allows not only survival, but a pleasant life. His subjects dwelled safely wherever they wanted and there were no divisions in this old, primordial kingdom (*pravibhāgo na*

---

[11] Brown (1928: 18) notes with regard to the gradation of 'levitational accomplishments' in Buddhist texts: 'it is only lesser beings, lay disciples, that walk on water'. Higher beings prefer flying. In his otherwise seminal study Brown does not discuss incidents of 'sealing' the water nor Duryodhana.

rāṣṭrāṇaṃ purāṇam cābhavat tadā, Mbh 7, App. I, No. 8, line 777). One of his fundamental, life-procuring deeds is the 'stabilisation' of the waters, which here means that he lays down their course towards the ocean (*saṃstambhayann āpaḥ samudram abhiyāsyataḥ*, Mbh 7, App. I, No.8, line 779). Duryodhana near the end of his speech will announce that he will destroy his enemies, whom he compares (5.60.24d–25) to rivers that meet their end in the ocean. Still closer to Duryodhana's self-portrayal is the other passage about king Dilīpa who is depicted as a 'speaker of truth' (*satyavādin*) and a formidable sacrificer. It is regarded as miraculous (*adbhuta*) that the wheels (of the chariot) of the fighting king would not sink into the waters (*apsu yudhyamānasya cakre na paripetatuḥ*, Mbh 7, App. I, No. 8, line 520). It seems that it is not by chance that Duryodhana proclaims after he has ascribed this skill to himself that he is the 'procurer' of all conditions of gods and demons and his mastery of the waters is not only a sign of successful expansion of his kingdom, but also one of the reasons why his subjects have nothing to fear.

However, Duryodhana's eagerness to proclaim his mastery over the waters can be connected to the larger epic context on still another level: the (inter)personal level of the epic narrative. Two incidents can be considered here: while the first relates a humiliation, the second confirms Duryodhana's power over the element. The first incident can be regarded as one of the 'Urszenen', primary scenes, of the epic conflict and is one of the causes of Duryodhana's increased hatred of the Pāṇḍavas and the following invitation to the dice game. This incident and its aftermath are related in chapters 43–46 of the Sabhāparvan (Mbh 2.43–46). During his visit to the newly built *sabhā* of Yudhiṣṭhira in Indraprastha, Duryodhana is laughed at by the Pāṇḍavas because he becomes a victim of 'optical illusions'. In particular he has a problem distinguishing solid ground from liquid water, or to be more precise distinguishing crystal from water:

> One time Dhṛtarāṣṭra's son came, in the middle of the hall, upon a crystalline floor, and assuming it was water, the flustered king gathered up his robe (2.43.3–4b).[12]

Despite the embarrassment Duryodhana tries to save his face and continues walking around even though in a rather bad mood (*durmanas*, *vimukha*, 2.43.4). Next, he mistakes a pond full of crystal-clear water for solid ground:

> Then he took a pond with crystal-clear water adorned with crystalline lotuses, for solid ground and fell into the water with his clothes on' (2.43.5).[13]

Let alone having made a fool of himself and being jeered at by Yudhiṣṭhira's servants, Duryodhana has still to swallow another humiliation. He is given new clothes that he obviously has to put on in open view which causes the following reaction:

> When mighty Bhīmasena saw him that way, as did Arjuna and the twins, they all burst out laughing (2.43.7).[14]

Having now lost face, Duryodhana 'did not look at them'. Still, the devastating situation is not over:

> Once more he pulled up his robe and trod on the ground as through about to cross over (water), and again all those people laughed at him (2.43.9).[15]

---

[12] *sa kadācit sabhāmadhye dhārtarāṣṭro mahīpatiḥ* I *sphāṭikaṃ talam āsādya jalam ity abhiśaṅkayā* II 3 II *svavastrotkarṣaṇaṃ rājā kṛtavān buddhimohitaḥ* I

[13] *tataḥ sphāṭikatoyāṃ vai sphāṭikāmbhujaśobhitām* I *vāpīṃ matvā sthalam iti savāsāḥ prāpataj jale* II 5 II

[14] *tathāgatam tu taṃ dṛṣṭvā bhīmaseno mahābalaḥ* I *arjunaś ca yamau cobhau sarve te prāhasaṃs tadā* II 7 II Mentioning Bhīma here may allude to another water related incident in the cousins' youth: Bhīma used to bully the sons of Dhṛtarāṣṭra and was especially fond of 'water games' including dragging the boys under the water and holding them down until they came close to drowning. One night a revengeful Duryodhana tried to drown a sleeping Bhīma whom he had tied up. However, the fetters were no match for Bhīma and he safely emerged from the water (cf. Mbh 1.177.15–35).

[15] *punar vasanam utkṣipya pratariṣyann iva sthalam* I *āruroha tataḥ sarve jahasus te punar janāḥ* II 9 II

When he later bitterly recalls this humiliation that causes him severe pain and anger (what in the colophons is called *saṃtāpa*) he is set on revenge and, in consequence, the dice game is schemed. I think it is not by chance that Duryodhana is shown praising himself for having control over water to the extent of being able to solidify (*stambhita*) it. This reference takes up one of the important threads of the epic plot and corresponds to Duryodhana's main concern that he will no longer be humiliated and laughed at by the Pāṇḍavas.

The other incident demonstrating Duryodhana's connection to the waters and his ability to 'seal' it off is told in Mbh 9 when he hides in a pond (*hrada*) after the battle and Aśvatthāman's nocturnal raid on the camp of the Pāṇḍavas. Being under water, he is, at first, not spotted by the Pāṇḍavas who seek to hunt him down, driven by their 'desire to kill Duryodhana' (*duryodhana-vadhepsayā*, 9.29.5). Interestingly with regard to his claims in 5.60, his sojourn in the water is described as a result of the 'solidification' of the water brought about by his own *māyā*, his magic power:

> Searching with all their means, they (the Pāṇḍavas) could not see the lord of the earth. For he had, with the club in his hand, escaped in immense speed, then entered the pond and solidified the water with his own magic power. (9.29.6c–7)[16]

Apart from the temporary invisibility this hiding-place procures for Duryodhana, the other effect of his 'sealing off' the surface of the pond is obviously that others cannot enter the water. This aspect of the process otherwise goes without saying because the waters are solidified so that chariots and horses do not sink. Now, the water is sealed off against intruders or other visitors

---

[16] *yatnato 'nveṣamāṇās tu naivāpaśyañ janādhipam* ॥ 6 ॥ *sa hi tīvreṇa vegena gadāpāṇir apākramat* । *taṃ hradaṃ prāviśac cāpi viṣṭabhyāpaḥ svamāyayā* ॥ 7 ॥ The 'solidification' of the water is also confirmed when Bhīma reports that Duryodhana has been discovered (9.29.43: *asau duryodhano rājan vijñāto mama lubdhakaiḥ* । *saṃstabhya salilaṃ śete ...*).

and is confirmed when Duryodhana leaves the pond and 'splits the sealed water apart' (*bhittvā stambhitaṃ toyaṃ*, 9.31.36). This situation is also suggested when Duryodhana is joined by some of his remaining allies (Kṛpa etc.): 'when they sat down at the pond, in which the king was lying, they addressed the inaccessible (dangerous) king who slept in water' (9.29.10).[17] This description emphasizes that Duryodhana 'in the water' is unassailable not only by using the adjective *durdharṣa*, which connotes both his inviolability and dangerousness, but also by depicting him resting (lying) and sleeping.

In the case of Viṣṇu his sleeping on the still floating or at least moving water (one of the reasons he is supported by the snake-god Ananta-Śeṣa) signifies, *inter alia*, his being in charge of a future creation, since he as the cosmic creator contains the living beings in himself. He is unassailable as the reigning cosmic king and may float on the water, while Duryodhana sleeps safely in the water because he has sealed the water above him. He is the king in hiding and still a potential, sleeping threat, since he is the last mighty survivor of the battle. This depiction may also link to the myths and imagery of 'Agni in the water' (see Oldenberg 1917: 112–119) and of Apāṃ Napāt (see ṚV 2.35), the 'water born' who strengthened his power by staying in the water and 'burns without fuel'. Such a connection would also make sense with regard to the 'sacrifice of battle' that Duryodhana in 5.57 claims to have 'spread out' as the consecrated patron. According to the traditional notion, Agni, the sacrificial fire, is kindled with the fire of the sacrificer. Hiding in the water, the patron takes also his fire with him and this could indicate on a symbolical level of the sacrificial imagery that the sacrifice is not over, that the patron's fire is not extinguished and indeed, the victim is not yet killed (see above). Conversely, his hiding also

---

[17] *te taṃ hradaṃ samāsādya yatra śete janādhipaḥ* | *abhyabhāṣanta durdharṣaṃ rājānaṃ suptam ambhasi* || 10 ||

shows that Duryodhana's sacrificial and royal potentials, his fire and 'fiery might' (*tejas*, 5.60.27), are only 'sleeping'. This is exactly what Kṛpa and the others imply when they ask Duryodhana to 'get up' and fight against Yudhiṣṭhira, in order to 'either enjoy the earth or reach heaven' (9.29.11).[18] The fiery character of Duryodhana 'in the water' is also emphasised when he eventually emerges out of the water, as Saṃjaya reports to Dhṛtarāṣṭra: 'your son emerged shining like the sun' (*udatiṣṭhata putras te pratapan raśmimān iva*, 9.31.36).[19]

Duryodhana's sojourn in the water is emphasized throughout the passage[20] as something astonishing and this is also pointed out in another comment by Saṃjaya. Comparing the coolness and purity of the water of the pond to the ocean he notes:

> the water there had been sealed by your son by means of his magic power (*māyā*). In a highly astonishing way using divine power he entered the water and lay there hardly visible for anyone, O king. (9.29.54–55)[21]

A similar view is voiced by Kṛṣṇa when the Pāṇḍavas eventually reach the pond. He stresses that Duryodhana has not to fear anybody:

---

[18] *rājann uttiṣṭha yudhyasva sahāsmābhir yudhiṣṭhiram* I *jitvā vā pṛthivīṃ bhuṅkṣva hato vā svargam āpnuhi* II 11 II

[19] Other comparisons are that his body 'shines like a mountain streaming water' (9.31.46). When he stirs the water in preparation for splitting the surface he is compared with a mighty snake (*mahānāga*, 9.31.33) and a *nāga* king (9.31.35). On the connection between Duryodhana, Dhṛtarāṣṭra and the *nāgas* see von Simson (1984: 220–222) who also refers to a passage in the *Rājataraṅginī* on a son of Aśoka deemed to be able to 'solidify water' too.

[20] Cf. 5.29.26c, 27b, 28d, 30d, 31c, 32c, 43 (sealed the water), 45, 54 (sealed), 55, 56, 61 (sealed), 64 (sealed); 5.30.2 (sealed). This shows the importance of the situation and, also the fascination of theme.

[21] *śītāmalajalaṃ hṛdyaṃ dvitīyam iva sāgaram* I *māyayā salilaṃ stabhya yatrābhūt te sutaḥ sthitaḥ* II 54 II *atyadbhutena vidhinā daivayogena bhārata* I *salilāntargataḥ śete durdarśaḥ kasyacit prabho* II 55 II Although the compound *daivayogena* can also simply mean 'accidentally' this seems not to suit the context, which stresses that Duryodhana uses *māyā*, see also Kṛṣṇa's comment in the following.

> Watch the magic power Dhṛtarāṣṭra's son has used on the waters, having sealed off the water, there is no danger for him from a human being. For, having created this divine wonder-work, he has entered the water! (9.30.3c–4)[22]

Conversely, when Duryodhana finally gives up his shelter, he stirs the water and then 'splits the water apart' (*bhittvā*, 9.31.36).

These intertextual aspects of Duryodhana's ability to solidify the waters in 5.60.14 show that literary contexts are closely connected to the broader theoretical issues and arguments in the epic. In the following stanzas Duryodhana's claims to supremacy are expanded upon when he declares that the rivers will carry him into any country he wishes to conquer:

> Whichever countries I invade for whatever purpose with my grand-armies, the waters will take me there, wherever I want (5.60.15).[23]

His power over earth and water emphasises his protective and expansive capability, which is one fundament of the prosperity of his kingdom and extends to the rain that never falls unwelcome (5.60.17). Therefore his subjects prosper and obey the laws: '... All my subjects are most law-abiding and there is nothing that plagues them' (5.60.16).[24]

'SPEAKER OF TRUTH'

In declaring himself as the supreme sovereign it is obvious that Duryodhana's loyalty does not belong to the family but to his territory. Authority and responsibility are only based on himself and the truth of his speech. Proof of the righteousness of his rule

---

[22] *paśyemāṃ dhārtarāṣṭreṇa māyām apsu prayojitām* II 3 II *viṣṭabhya salilaṃ śete nāsya mānuṣato bhayam* I *daivīṃ māyām imāṃ kṛtvā salilāntargato hy ayam* II 4 II

[23] *akṣauhiṇībhir yān deśān yāmi kāryeṇa kenacit* I *tatrāpo me pravartante yatra yatrābhikāmaye* II 15 II

[24] *bhayāni viṣaye rājan vyālādīni na santi me* I *mattaḥ suptāni bhūtāni na hiṃsanti bhayaṃkarāḥ* II 16 II

is success, that is, the prosperity of his kingdom. He views himself in the position of a god-king who does not reckon with anyone above or beyond himself, rather he regards himself to be the 'mover' (*pravartitā*) of all beings. On this basis he declares that he cannot be defeated and he again points out that the Pāṇḍavas would not have suffered, if the gods were able to prevent it (5.60.18–19). Drawing on his reputation as a 'speaker of truth' (*satyavāk*) Duryodhana promises his father that there is no rescue for those he hates:

> Neither gods nor Gāndharvas, neither demons nor Rākṣasas are able to save the ones I hate – this is the truth I tell you. Whatever good or evil I have ever wished upon friends and enemies alike has never failed to come true. So, when I say 'This will happen', it has never before turned out to be otherwise. Therefore I am known as the 'speaker of truth'. This mightiness of mine, witnessed by the world, heard of in every quarter, I declare to you, for cheering you up, not in order to boast, king; since I have never praised myself before, for to praise oneself is the act of the mean. (5.60.20–24)[25]

Pledging to defeat the enemies, the imagery of the rivers is once again employed:

> Those defeated by me will perish with their allies upon reaching me, like rivers perish upon reaching the ocean' (5.60.25d–26).[26]

This time the rivers are compared with the enemies and Duryodhana with the ocean in which they will vanish. This echoes not only the famous passage in *Chāndogya-Upaniṣad* 6.10.1 (cf. also *Muṇḍaka-Upaniṣad* 3.2.8; BhG 2.70) but also BhG 11.28 when Arjuna compares the destruction of the warriors in the

---

[25] *naiva devā na gandharvā nāsurā na ca rākṣasāḥ* | *śaktās trātuṃ mayā dviṣṭaṃ satyam etad bravīmi te* || 20 || *yad abhidhyāmy ahaṃ śaśvac chubhaṃ vā yadi vāśubham* | *naitad vipannapūrvaṃ me mitreṣv ariṣu cobhayoḥ* || 21 || *bhaviṣyatīdam iti vā yad bravīmi paraṃtapa* | *nānyathā bhūtapūrvaṃ tat satyavāg iti māṃ viduḥ* || 22 || *lokasākṣikam etan me māhātmyaṃ dikṣu viśrutam* | *āśvāsanārthaṃ bhavataḥ proktaṃ na ślāghayā nṛpa* || 23 || *na hy ahaṃ ślāghano rājan bhūtapūrvaḥ kadācana* | *asad ācaritaṃ hy etad yad ātmānaṃ praśaṃsati* || 24 ||

[26] ... *śrotāṃsi vijitān mayā* || 25 || *saritaḥ sāgaraṃ prāpya yathā naśyanti sarvaśaḥ* | *tathaiva te vinaṅkṣyanti mām āsādya sahānvayāḥ* || 26 ||

'blazing mouth' of Kṛṣṇa in his form as deadly time with the 'dashing waters of the many rivers rushing directly into the ocean' (*yathā nadīnāṃ bahavo 'mbuvegāḥ samudram evābhimukhā dravanti*). This imagery complements the mastery over the waters that help the king on his military expeditions by turning him into the mighty confluence, the ocean, of all hostile rivers, that is of all those peoples and other kings who would not recognize his sovereignty. The speech ends with a final self-appraisal:

> In me are superior insight, greatest splendour and the greatest prowess, superior knowledge and superior Yoga. I surpass them (the Pāṇḍavas and their allies) all. (5.60.27)[27]

This proclamation of Duryodhana's superior power is called his *māhātmya*. The word *māhātmya* is used not only by Duryodhana, but also elsewhere in the epic, as the designation for the mightiness of powerful beings as well as for the praise of this mightiness in composed speech. More often than not both meanings converge in the epic. However, someone's mightiness as well as its praise *māhātmya* are usually referred to or uttered by somebody else.[28] Very rarely an individual speaks about his own *māhātmya* (apart from 5.60.23, see 5.15.19) although others may understand a self-appraisal as a *māhātmya* as is the case in BhG 10, when Kṛṣṇa is praising himself as a mighty *īśvara*, which is then referred to by Arjuna as *māhātmya* (BhG 11.2, see Malinar 2007). In contrast to this, 5.60 is a case of a self-glorification,

---

[27] *parā buddhiḥ paraṃ tejo vīryaṃ ca paramaṃ mayi* | *parā vidyā paro yogo mama tebhyo viśiṣyate* || 27 ||

[28] In most of the sixty-six passages in which the word *māhātmya* is used in the epic (most frequently in books one, three, five, twelve and thirteen), the protagonists speak (about) someone's *māhātmya* (both in the sense of mightiness as a quality and a speech about it) or listen to it; as is the case with *māhātmya* of Vāsudeva-Kṛṣṇa (1.1.6; 6.33.2, 62.35, 64.12; 9.18.18; 14.16.5), Viṣṇu (5.103.35; 12.203.10, 271.2ff.), Rudra (7.57.72), Hariścandra (2.11.52), Sāgara (3.104.5) or of women and *pativratās* (3.196.2, 6 and 3.206.32). Occasionally they are also asked to 'see' it (cf. 5.15.19, here Nahuṣa speaks about his own *māhātmya*; 12.326.25).

that is, a type of speech otherwise known as *ātmastuti* when used to describe Vedic hymns in which the gods speak about themselves in the first person (see Thompson 1997). This indicates that the use of the first-person in praise is accepted in the case of gods, a structure also corroborated by BhG 10. While human beings may praise not only gods, but also each other (see for instance, Duryodhana's praise of Bhīṣma in 5.153.11–14), self-glorification is not endorsed and often rejected as 'boasting'. These different types of speech are at play in 5.60 when Duryodhana is shown to be concerned about a possible misunderstanding of his speech: he finds it necessary to distinguish his speech from 'bragging' (*ślāghā*).[29] This seems appropriate since the text displays certain characteristics of 'self-praise' (*ātmastuti*) and boasting that may serve to read Duryodhana's claim as signs of his vanity. Indeed, his losing the war demonstrates the limits of his power and proves his 'words of truth' wrong. However, Duryodhana seems to be somehow aware of the risk he takes in speaking his *māhātmya*. Since he does not perceive himself to be a god he needs to distinguish himself from the bragging and boasting of other warriors.

Yet, in contrast to the 'hollow words' of boasting, Duryodhana puts himself at stake when he offers his 'truth speaking' in the past as proof for the truth and efficacy of his speech in the future. His *pramāṇa*, his means of proof is that he declares himself a *satyavāk*, a speaker of truth. Thereby the utterance of his *māhātmya* is turned into an act of 'truth speaking' and can be compared with a so-called *satyakriyā* act. The content of his 'truth' is that nobody can be saved from his hatred and he declares that this will also be true in the future. His 'truth' is about

[29] The colophons seem to agree and call this speech Duryodhana's *māhātmya*. Boasting is normally prohibited for a warrior as improper for an *ārya* (cf. 7.166.34: *āryeṇa tu na vaktavyā kadācit stutir ātmanaḥ*; cf. also 12.156.15). Duryodhana is accused of bragging by Kṛṣṇa in 5.71.14, 17 as is king Dambodhava who is eager to fight against Nara and Nārāyaṇa in 5.94.

the future itself, it consists of the utterance 'it will be' (*bhaviṣyati*, 5.60.22). What will be is what he wants to happen. For this claim Duryodhana does not offer any conditional evidence, as is sometimes the case in 'truth acts' when the truth of an assertion is turned into the precondition of the speech's efficacy. Rather, his projection of the future is based on the factuality of the past, in that he asserts that everything he ever wanted to happen to both, friends or enemies, did in fact happen. This is in accordance with the Vedic concept of 'truth' which includes 'also the congruence of a spoken word and a corresponding fact in the future reality' (Söhnen-Thieme 1995: 240; emphasis in the original). The anticipation of future events or actions opens 'truth speeches' to other forms of utterances such as promises, oaths or boasting. Thompson points out that 'truth speaking' as a 'truth act' implies that the effect is produced by the speaker himself and is not dependent on other agents or activities but the utterance itself (Thompson 1998: 138). This stands in contrast to oaths, in which other agents are invoked in order to solemnize the declaration, but not to promising or boasting. With regard to its extension into the future Duryodhana's truth-declaration is structurally most closely connected to a promise (cf. Hopkins 1932: 317–318). In support of his promise that those he hates will perish Duryodhana refers to his command over future events in the past which demonstrates his power of turning his intentions into reality. This resemblance of the 'truth-declaration' to a promise may also explain why it is important to state that it is not mere boasting. A promise, being a prediction about an individual's future accomplishments, is always threatened by the uncertainty with regard to the details of that very future and this is one reason why too unrealistic a promise or one left unfulfilled may be considered mere boasting. In Duryodhana's case the situation is further complicated in that the intention of speech is not only to project the destruction of the enemies, but to 'cheer up' his worried father. All these features confirm

Thompson's general observation that a declaration of truth serves to 'affirm one's authority'; it is an

> instrument of magic, which is as much as to say an instrument of rhetoric, by means of which to diminish the influence of one's rivals, and to increase one's own influence, as an agent whose words have power over both gods and men (Thompson 1998: 145–146).

Given the similarities with other types of 'true speech' that need to be explored further before drawing more precise distinctions (see Thompson 1998: 141), Duryodhana's speech cannot be easily classified as either 'truth act', promise or boasting, and it cannot be ruled out that this may even be the author's intention, because in spite of Duryodhana's assurance that he is not boasting, his speech remains ambiguous as it is blending a *māhātmya* with a promise and the claim of truth-speaking and -acting (*satyakriyā*). Seen against the background of the epic narrative, which proves his claims wrong, the ambiguity may even be a poetic necessity as Duryodhana will end up without any power over gods who are 'mighty lords of *yoga*' or Pāṇḍavas playing tricks and speaking half-truths. Conversely, doubts may be raised with regard to the 'truth' of Duryodhana's view on his past accomplishments. However, this line of questioning cannot be pursued further here and pronouncing judgement on what is truly and really true would probably amount to remaking Akira Kurusawa's *Rashomon*.

All these features also distinguish Duryodhana's position and the scope of his power from that of his main opponent in the UdP, from Kṛṣṇa, and his depiction as a god in the BhG can also be read as a rejection of interpretations of absolute royal power such as are voiced in 5.60. The difference between the two concerns not only the structure of speaking about their *māhātyma*, but also its fundament and efficacy: While Krishna turns to visible manifestation of his divine, yogic sovereignty (*rūpa aiśvara*), Duryodhana bases his claim on his mastery of ritual language, his truthfulness as a king and a human being, his 'fiery might', his knowledge and rather unspecified accomplishments in *yoga*. In contrast to the efficacy of ascetic powers (*tapas*) he has criticised before, he represents an understanding of *satya*, which is based on the utterance of his 'truths', but also on the knowledge of ritual invocations (*mantra*). Thompson points out that *satya* is a

> magical notion, one which is roughly equivalent ... with the term mantra, a term which is clearly understood in both Buddhist and Vedic communities alike, to be a speech of great power (Thompson 1998: 131).

This connection is also explicitly made by Duryodhana when he declares that his '*abhimantraṇa*', the ritual invocation of correct *mantra*s, will work as a spell cast on the gods who will not support his enemies, but will be under his command like weapons.[30]

FINAL REMARKS

In this speech Duryodhana's reluctance to give up parts of his kingdom in the name of family bonds has been given a reason. He has left the discourse of family law and has instead depicted himself as the overlord of all beings, expanding the power of king by appropriating ritual and ascetic power. He justifies his intransigence by drawing on his specific understanding of a territory-based kingship and focussed on the persona of the king. For Duryodhana, defending the territorial integrity of his kingdom and keeping his riches at all costs are part of his duties as a god-king. There is nothing to negotiate about. The epic will narrate his defeat and, in this regard, the BhG functions as a text that sets out to prove his logic wrong. A sign of the presence of a 'higher power' whose provenance is fully revealed in the

[30] This aspect of mantric consecration is prominent in the other instances in which forms of the verbal compound *abhi-mantr* are used (the substantive *abhimantraṇa* is only used in 5.60.11) in the epic. It is almost exclusively used for describing the consecration of weapons or referring to already consecrated weapons (cf., for instance, 3.17.25; 7.121.29; 9.16.46).

BhG, is already inserted in the final part of the UdP, when Kṛṣṇa shows himself as a powerful, divine being. Duryodhana's claim of overlordship is in the UdP not rejected on a conceptual level; instead he is either morally discredited or the superiority, both physical (*bala*) and moral (*dharma*), of the Pāṇḍavas and Kṛṣṇa is pointed out. Two reasons are given for this superiority: on the one hand, it is assumed that they have a god-like status and are supported by the gods; on the other hand, Kṛṣṇa is depicted as the highest god and the 'mighty lord of *yoga*' (see Malinar 2007, *passim*). The few texts that present Kṛṣṇa in this way most probably mirror the monotheistic teachings of the BhG. Similarly to the extant BhG, the extant UdP depicts Kṛṣṇa as both the epic hero and the highest god. In this way, Duryodhana and Kṛṣṇa are juxtaposed by pointing to the latter's hidden identity as the highest god (cf. 5.65–67; 5.83) and by even revealing it near the end of the UdP (5.129, see Malinar 1996: 87–93).

When asking for a possible cultural-historical context for these considerations and the presentation of Duryodhana's view of kingship, the following could be considered, although much depends on the date ascribed to the text. One possibility is the Buddhist concept of the *cakravartin* who rules the earth as the 'turner' (*pravartitā*) of the 'wheel of (Buddhist) *dharma*'. Holtzmann (1892–95) identified Duryodhaṇa with Aśoka and that view has been modified by von Simson (1984), who suggests identifying Aśoka and Dhṛtarāṣṭra. Another context could be the criticised concept of an absolute royal power focussed on the king with his appropriation of priestly functions and his power over nature. Such a view of kingship is in some later texts ascribed to the ruthless and cruel Nanda kings and even Aśoka criticises 'former kings' for their disregard for laws and the abuse of their powers. Both models suggest a post-Aśokan date for the text, which I think to be likely. Another aspect of the text is its relationship to the BhG, which does not only present a theology, but also a statement on undesirable concepts of absolute kingship and sovereignty such as presented by Duryodhana (cf. Malinar 2007). It seems that Duryodhana's *māhātmya* is connected to the depiction of Kṛṣṇa as the 'supreme god' in the BhG and thus helps in clarifying the discourse on kingship and power discernable in both texts. While the historical priority of the concept of kingship presented by Duryodhana seems highly probable, to infer from it the text-historical priority of 5.60 seems to stretch the evidence. Duryodhana's depiction can be a reaction to the BhG that serves to intensify the conflict between Duryodhana and Kṛṣṇa in the UdP, but it could also be the opposite, that the BhG is also a reply to 5.60. In the extant text, it certainly refutes Duryodhana's truths and it cannot be ruled out that both texts belong to the same redactional process.

When looking into the literary, that is epic context of this text, Duryodhana's praise of his power is related to the depiction of other *īśvaras* and their *aiśvarya*, especially Kṛṣṇa's in the BhG, that is, the power to rule not only over a kingdom, but also over the elements of nature. In this regard, 5.60 seems connected to BhG 10 in particular (see also the use of *anukampā* in both texts). The text is thus connected to the depiction of other powerful rulers, to gods, yogis and *siddhas*, which is indicated in the text, when Duryodhana claims his *yoga* to be the best. However, this claim remains in the case of Duryodhana rather unspecific and his mastery over the waters is, as has been shown, not connected to *yoga*, but to his 'truth-speaking' and, in Mbh 9, to his *māyā*. Yet, his mightiness is related to the display of yogic powers in other parts of the epic, which is an aspect of *yoga* that has been rather neglected in favour of the ascetic and liberating aspects of the practice (see however Hopkins 1901; Malinar 1996; 2007). Duryodhana's truths consist in blending magic and ritual power in the persona of the king, who rules as the king over all other creatures by 'speaking truth' and commands the gods through the use of *mantras*, both being powers he allegedly uses for procuring prosperity for himself and his kingdom. It is, however, another magic power that will stay with him to the very end: his mastery of the waters which is one of the themes characterising Duryodhana not only as an epic character, but also as interpreter and spellbinder of royal power.

REFERENCES

BAILEY, G. M. 1983. *The mythology of Brahmā*. Delhi: Motilal Banarsidass.

BIARDEAU, M. & Ch. MALAMOUD 1976. *Le sacrifice dans l'Inde ancienne*. Paris: Presses Universitaires de France.

BIARDEAU, M. 1997. Some remarks on the links between the Epics, the Purāṇas, and their Vedic sources. In: G. Oberhammer (ed.), *Studies in Hinduism, Vedism and Hinduism*: 69–177. Wien: Verlag der Österreichischen Akademie der Wissenschaften.

BROWN, W. N. 1928. *The Indian and Christian Miracles of Walking on the Water*. Chicago: The Open Court Publishing Company.

DEV, A. 1989. La guerre de Kurukṣetra n'aura pas lieu: Udyoga reconsidered. In: B. K. Matilal (ed.), *Moral Dilemmas in the Mahābhārata*: 77–96. Delhi: Motilal Banarsidass.

GITOMER, D. 1992. King Duryodhana: the Mahābhārata discourse of sinning and virtue in the Epic and drama. *Journal of the American Oriental Society* 112.2: 222–232.

HARA, M. 1973. The king as husband of the Earth. *Asiatische Studien / Études Asiatiques* 27: 97–114.

HEESTERMAN, J. C. 1957. *The Ancient Indian Royal Consecration. The Rājasūya Described According to the Yajus Texts and Annoted*. 's-Gravenhage: Mouton.

——— 1985. *The Inner Conflict of Tradition. Essays in Indian Ritual, Kingship, and Society*. Chicago: University of Chicago Press.

HILTEBEITEL, A. 1976. *The Ritual of Battle. Kṛṣṇa in the Mahābhārata*. Ithaca: Cornell University Press.

HOLTZMANN, A. 1892–95. *Das Mahābhārata und seine Theile*. 4 vols. Kiel: Haeseler.

HOPKINS, E. W. 1901. Yoga-technique in the Great Epic. *Journal of the American Oriental Society* 22: 333–379.

——— 1932. The oath in Hindu Epic literature. *Journal of the American Oriental Society* 52: 316–337.

JATAVALLABHULA, D. Feller. 1999. Raṇayajña: the Mahābhārata war as a sacrifice. In: J. E. M. Houben & K. R. van Kooij (eds.), *Violence Denied. Violence, Non-Violence and the Rationalization of Violence in South Asian Cultural History*: 69–103. Leiden: Brill.

KALYANOV, V. I. 1979. On Kṛṣṇa's diplomatics in the Mahābhārata. *Indologica Taurinensia* 7: 299–308.

MALINAR, A. 1996. *Rājāvidyā: Das königliche Wissen um Herrschaft und Verzicht. Studien zur Bhagavadgītā.* (Purāṇa Research Publications, Tübingen, 5.) Wiesbaden: Harrassowitz.

——— 2007. *The Bhagavadgītā. Doctrines and Contexts.* Cambridge: Cambridge University Press.

PROFERES, T. 2007. *Vedic Ideals of Sovereignty and the Symbolism of the Agnicayana Ritual.* Harvard: American Oriental Society.

OLDENBERG, H. 1917. *Die Religion des Veda.* Stuttgart: Cotta.

SIMSON, G. von 1984. The mythic background of the Mahābhārata. *Indologica Taurinensia* 12: 191–223.

SÖHNEN-THIEME, R. 1995. On the concept and function of *satya* ('truth') in ancient Indian literature. In: C. Galewicz (ed.), *International Conference on Sanskrit and Related Studies, September 23–26, 1993 (Proceedings)*: 235–244. Cracow: Enigma Press.

THOMPSON, G. 1997. Ahaṃkāra and ātmastuti: self-assertion and impersonation in the Ṛgveda. *History of Religions* 37: 141–171.

——— 1998. On truth-acts in Vedic. *Indo-Iranian Journal* 41: 125–153.

Bhīma's Quest for the Golden Lotuses (*Mahābhārata* 3.146–153 and 3.157–159)

DANIELLE FELLER

INTRODUCTION

This article is the sequel to another article, entitled 'Hanumān's jumps and their mythical models' (Feller 2009). In that article, I attempt to demonstrate that Hanumān's jumps, as they are described in the Epic, have at least two mythical models, namely: Garuḍa's flight to heaven to steal the *soma*, and the flight of the winged mountains, especially that of mount Maināka. In this light, it appears that Hanumān fits into a lineage of subversive heroes who go against Indra's authority. In the present article, I propose to read the story of the Saugandhikāharaṇaparvan in the same light, and see in Bhīma the true successor of his brother Hanumān and of the latter's predecessors.

THE STORY

The Saugandhikāharaṇaparvan (Mbh 3.146–153[1]) tells a rather well-known story which I will therefore only briefly summarise here. After many years of exile, in the course of their pilgrimage, the Pāṇḍavas and their wife Draupadī are sojourning in the holy hermitage of Badarī in the Himalayas. One day, Draupadī finds

[1] All references to the *Mahābhārata* and the *Rāmāyaṇa* are to the Critical Editions.

a fragrant golden lotus brought there by the wind. She requests Bhīma to get more of the same flowers, and Bhīma complies with her wish at once. He climbs the steep slopes of the charming Mount Gandhamādana,[2] which are forbidden to human beings, wreaking havoc as he proceeds with great speed. Like a rutting elephant (*mattavāraṇa*, four times in 3.146.31), he destroys everything on his way. Roaring like a lion and making a huge noise, he frightens all the denizens of the mountain. Like a hurricane, he uproots trees and plantains on his way, slays lions and tigers, and generally contrives to make as much noise as possible (3.146.6–58).

A huge monkey called Hanumān, who is sleeping peacefully in a plantain grove, wakes up due to the din made by Bhīma. He whips his tail around, producing a noise like Indra's thunderbolt (3.146.60), which is echoed by the mountain. Hearing this, Bhīma decides to investigate the origin of the sound. Seeing the monkey in the plantain grove, Bhīma rushes towards him, roaring loudly. Opening his eyes very slowly, the monkey rebukes him for his unseemly behaviour, and advises him not to proceed any further, for the mountain is forbidden to human beings. Instead, he offers him roots and fruit to eat (3.146.81).

Bhīma angrily tells him to get out of his way, boasting that he is the son of the wind, and the younger brother of Hanumān. Smiling, Hanumān tells him that he is too old and weak to get up, but that Bhīma can jump over him. This however, Bhīma refuses to do, since it would mean insulting the highest being, present in all creatures. Then Hanumān advises him to simply lift his tail and remove it out of his way. Bhīma, in a rather insulting manner, first tries to lift it with his left hand, then with both, then with all his might. But the tail is too heavy. Humbled, he begs Hanumān's forgiveness, and asks him who he really is.

---

[2] For remarks on the geographical location of this mountain, see Grünendahl 1993: 128.

Hanumān now reveals his true identity, and Bhīma is overjoyed to meet his older brother. He begs him to display the huge form he had taken to jump over the ocean in his search for Sītā. First Hanumān refuses, saying that in this age he can no longer assume that huge form, and proceeds to lecture Bhīma on time-spans, or *yugas*. But Bhīma insists, and in the end Hanumān assumes his gigantic, dazzling form, till Bhīma, unable to bear the sight of it, begs him to reduce it (3.149.12). After some more admonishing, Hanumān shows his younger brother the way to the golden lotuses and they part (3.147.1–150.15).

Bhīma goes on his way and finally comes upon a large river (*nadī*),[3] covered with a forest of pure golden lotuses, shining like the rising sun (3.150.25–26). In the distance, on Mount Kailāsa, he sees the wondrous abode of Kubera, the god of riches, who is the rightful owner of the lotuses. But when Bhīma tries to pluck them, hordes of *rākṣasas*, the guardians of the place, attack him. Bhīma easily defeats them, and then plunges into the water to pluck as many fragrant lotuses as he can (3.152.21–22). Some *rākṣasas*, who have managed to escape, go and tell Kubera what is happening. But Kubera only laughs, and says that Bhīma should be allowed to take the flowers, since they are for Draupadī (3.152.24).

As Bhīma is plucking the wonderful lotuses, ominous portents appear: a great wind rises, a meteor falls, the earth trembles, the sun is eclipsed, a shower of stones falls from the sky, the directions look bloody, and everything is wrapped in darkness. Seeing these omens, Yudhiṣṭhira is filled with apprehension. When he realises that Bhīma has disappeared, and when Draupadī tells him on what quest she has sent him, he decides to find him at once. They all climb up to the lotus pond. Seeing the numerous *rākṣasa*-corpses strewn around Bhīma, Yudhiṣṭhira gently scolds him for doing this deed and tells him not to do so

---

3 Here it is called a *nadī*, later a *nalinī* or a *puṣkariṇī* (lotus-pond).

again. He accepts the lotuses from Bhīma, however (3.153.28), and they all bathe and play in the pond, and spend some time there, with Kubera's consent.

This episode has a double, which is narrated very shortly afterwards (Mbh 3.157.14–3.159).[4] One day, Garuḍa carries away a big snake living in a lake near the Pāṇḍavas' place of residence. Then divine five-coloured[5] flowers (*puṣpāṇi divyāni ... pañcavarṇāni*, 3.157.17) fall from the mountain-top. Again Draupadī tells Bhīma that she would like to own those flowers, adding that it would be a pleasure to go and visit the mountain-top. Again Bhīma goes at once to get the flowers and has fearful fights with the *rākṣasas* who guard the pond. He kills Kubera's good friend, a *rakṣas* named Maṇimat. This time Kubera is enraged when he hears of it and rushes to the battlefield. But when he arrives, Yudhiṣṭhira and the others[6] have joined Bhīma, and Kubera's anger vanishes: moreover, he has remembered in the meanwhile that Maṇimat's death at Bhīma's hands had been preordained, and that it was Maṇimat's punishment for spitting on a *ṛṣi*. He forgives Bhīma, after admonishing him. The flowers are not mentioned any more. The second episode is quite a lot shorter than the first, and it entirely lacks Bhīma's meeting with his elder brother Hanumān. I do not propose here to discuss the possible reasons why this episode is duplicated, since this topic has been discussed by other scholars before.[7]

---

[4] Between the two episodes, the story is told how the Pāṇḍava brothers and Draupadī are abducted by a *rākṣasa* named Jaṭāsura, one day when Bhīma is away.

[5] What these five colours are, or what they symbolise, is not mentioned in the text. We may note, however, that the *soma*, especially the *soma*-juice, is described in the *Ṛgveda* as golden, brown, red, yellow, or white (when mixed with milk). Thus, without taking into account the combinations, we obtain five colours. See Geldner 1951: 1–2.

[6] But without Draupadī, whom they have entrusted to the sage Ārṣṭiṣeṇa.

[7] According to Grünendahl (1993: 110), this shorter second episode is the original one, and the more elaborate first one is a duplication. According

THE MYTHICAL MODEL

One way of reading this story is simply to consider it as one of Bhīma's many adventures, meant to emphasise his prowess, strength, and, in the present case, his attachment and gallantry towards his wife Draupadī. But the story also goes deeper than that. Here I propose to read it as a variant of the age-old myth of the theft of the nectar of immortality (*soma* or *amṛta*).[8] The first model for this narrative would then be the myth of the theft of the *soma* from Indra's heaven by the heavenly bird Garuḍa, which is narrated in Mbh 1.14–30. The second model, which is itself based on the first, would be Hanumān's several jumps, which are described in Rām 7.35–36; 5.1; 6.61 and 6.89. I have discussed these topics elsewhere:[9] what we should note here is that Hanumān's huge leaps to get the sun, or the mountain with the reviving herbs, are in a sense reproductions of the theft of the *soma*. My basic contention is that Bhīma, by getting the golden lotuses, which can on many counts be compared with the nectar of immortality, and also with the sun, fits into a lineage of subversive heroes who go against the established powers to climb to heaven and to obtain immortality in one form or another.

Bhīma, as Hanumān's younger brother, is naturally suited to become heir to some of his elder brother's exploits and functions. This is underscored by the fact that in this episode the two brothers meet for the only time in their life: it is likely that this meeting should take place at a crucial moment, when Bhīma is reduplicating his elder brother's exploits and taking over some of

---

to Allen (2000a: 131), the heroes of these two episodes (dealing with a third-functional quest for beauty – the lotuses or five-coloured flowers) were originally the twin-brothers Nakula and Sahadeva, and not Bhīma. This would explain why the episode is duplicated.

[8] Obviously, this interpretation does not preclude different sorts of readings. For instance, Allen (2000b) draws a parallel between Bhīma's quest for the golden lotuses and Odysseus' home-coming and his reunion with his old dog Argos (*Odyssey* 17.201–327).

[9] See Feller 2004, chapter 4, and Feller 2009.

his functions. Indeed, at this point in his life, having received the boon to live as long as Rāma's story lives on in the world,[10] Hanumān is extremely ancient. Although by no means as decrepit as he pretends to be when he meets Bhīma, he is nevertheless spending his old age in retirement in the Himalayas, and is ready to make over some of his functions to his younger brother. In my opinion these functions are essentially two-fold:

- providing immortality
- overcoming mountains

We shall presently take up these two points in the above order.

PROVIDING IMMORTALITY

Before tackling this point, one problem must be solved: how can we keep the golden lotuses on a par with the *soma-amṛta*, or with any other token or symbol of immortality,[11] such as the sun, which Hanumān tries to eat in his childhood? As a matter of fact, these divine lotuses are described in this passage as a sort of mixture of the *soma* and the sun. Like the *soma*, they have a vegetal nature.[12] Furthermore, the first lotus found by Draupadī is blown into the hermitage by a wind which comes from the north-east (*pūrvottaro vāyuḥ*, 3.146.6).[13] The north-east is traditionally the direction assigned to the god Soma. This is one more way of underscoring the flowers' somic origin. Besides, when Bhīma plucks the lotuses, the verb which is used is *utpāṭya* (3.152.22), exactly the same which is used when Garuḍa tears out the *soma*, or when Hanumān tears out the mountain on

[10] Cf. Mbh 3.147.37: *yāvad rāmakathā vīra bhavel lokeṣu śatruhan* | *tāvaj jīveyam ity evaṃ tathāstv iti ca so 'bravīt* ||

[11] According to Biardeau (1973–74: 140), theses lotuses are 'un motif cosmogonique constant'.

[12] And like the *soma* too, lotuses – or at least their roots – are comestible, but the text does not stress this point. Perhaps because these lotuses are golden, and therefore inedible.

[13] Also in 3.153.15: *prāgudīcī diśam*.

which grow the reviving herbs, which are themselves similar to *soma*. One of the lotuses' most striking characteristic is that they are sweet-smelling,[14] a trait which probably gives the whole mountain its name (Gandhamādana, 'intoxicating with its perfume'), not to speak of the name of the sub-parvan itself (Saugandhikāharaṇaparvan). Now, the *soma*'s fragrance may not exactly be its main characteristic, but its juice is nevertheless sometimes described as sweet-smelling (*surabhi*) in the ninth book of the *Ṛgveda*.[15] And the intoxicating or exhilarating effect of the *soma* is most frequently rendered by the verbal root √*mad* (as in *gandhamādana*).[16] Furthermore, the water in which the lotuses grow is twice likened to *amṛta* (*amṛtarasam*, 3.151.4; *amṛtakalpam ambho*, 3.152.22). Undoubtedly, the pond in which these fragrant golden lotuses grow, which is inaccessible to humans and is the playground of divine beings, is in paradise, and mount Gandhamādana itself is a part of heaven.[17] As we know, the oldest heaven is the mountain-top.[18]

On the other hand, the lotuses are golden (*kāñcana*, 3.150.25; *jātarūpamaya*, 3.151.5) – and the stainless and undecaying gold is itself a symbol of immortality. They are also said to shine like the sun (*arkābha*[19]) and they have a thousand petals (*sahasrapattra*, 3.146.6), which reminds us of the sun's well-known epi-

---

[14] 'Having a divine smell' (*divyagandham*, 3.146.7); 'sweet-smelling' (*saugandhikam*, 3.146.8); 'endowed with perfume' (*gandha* ... *saṃpannam*, 3.146.9), etc ...

[15] *Ṛgveda* 9.97.19 and 9.107.2. See Geldner 1951: 2.

[16] In the Veda and in the *Avesta*. See Geldner 1951: 2.

[17] Biardeau (2002: 546–547) similarly insists on the heavenly location of this episode.

[18] Subsequently, in Mbh 3.160, the Pāṇḍavas' *purohita* Dhaumya describes to Yudhiṣṭhira, from the summit of mount Gandhamādana, the various peaks surrounding them, and reveals that each one of them is the abode of a god.

[19] Cf. also Mbh 3.150.27: *saugandhikavanam* ... *ḇālārkasadṛśadyuti*, 'a forest of perfumed lotuses whose splendour is similar to that of the rising sun'. The shining nature of the lotuses is often mentioned: *surucira* (3.146.9); *virajāṃsi* (3.153.1).

thet *sahasrāṃśu* 'having a thousand rays'. Thus the lotuses simultaneously share in the *soma*'s vegetal nature and the sun's fiery, golden and shining nature. This makes them doubly a symbol of immortality.

When Bhīma climbs the mountain to get the golden lotuses, we are reminded of Garuḍa flying up to heaven to get the *soma*. Bhīma's speed is likened to that of Garuḍa in one verse (*pakṣirāḍ iva vegitaḥ*, 3.146.37). Yudhiṣṭhira makes the same comparison even more emphatically in verse 3.153.19: 'Swift, and, in leaping, Vainateya's equal on earth, he could even jump up to heaven, and jump down again at will' (*tarasvī vainateyasya sadṛśo bhuvi laṅghane* | *utpated api cākāśaṃ nipatec ca yathecchakam*).[20] Moreover, as Hanumān warns Bhīma (3.149.23–25), these lotuses should not be plucked in a hurry, but with devotion, after honouring the gods with '*bali*s, *homa*s, *namaskāra*s and *mantra*s'. This shows that the lotuses are much more worthy of respect than ordinary flowers. However, to be truthful, Bhīma does not seem to follow this piece of advice: quite the contrary, as when Garuḍa carries off the *amṛta* from heaven, or when Hanumān tears off the summit of the mountain on which grow the reviving herbs, the deed is done with a certain ruthlessness. Like Garuḍa, Bhīma encounters a tough resistance in his at-

[20] There are many other points of similarity between Bhīma and Garuḍa. On his way back from heaven, after stealing the *amṛta*, Garuḍa meets Viṣṇu who places him on his flag-staff; Bhīma meets Hanumān who offers to be on Arjuna's standard in the great battle. As a reward for his exploit, Garuḍa gets the boon to have the snakes for his food, and he becomes the master of the antidote against snake-poison. Bhīma, in his youth, when poisoned by Duryodhana, and thrown into a river, is *revived* by the poison of the snakes who bite him at the bottom of the water. Thus the deadly poison of the snakes, far from killing him, functions on the contrary as an antidote for him (see Mbh 1.119). Garuḍa is also mentioned several times in the second episode. The five-coloured flowers fall down after Garuḍa has carried away a great snake. When Bhīma reaches the top of mount Gandhamādana, he has to fight a *rākṣasa* named Maṇimat ('having a jewel'), a name which evokes a snake, like those guarding the *soma*. And indeed, while fighting with Maṇimat, Bhīma is compared to Garutmat fighting a snake (3.157.66).

tempt, having to fight off hordes of *rākṣasas* and *yakṣas* who guard the pond and sacrifice their lives for the sake of the lotuses. The necessity to fight off demons is also seen in Hanumān's jumps.[21] This treasure-like nature of the lotuses, which makes it necessary to guard them closely, is one more trait they share with the nectar of immortality. When Garuḍa steals the *soma*, or when Bhīma plucks the lotuses, ominous portents appear: the earth trembles, meteors fall, the sun is obscured, etc. (see Mbh 1.26.28–32 and 3.153). Such bad omens usually indicate that an event threatening the world-order or *dharma* is taking place. This would surely not be justified if the lotuses were ordinary flowers.

Last but not least, the state of mind in which the deed is performed, and successfully accomplished, is also typical for similar stories of quests for immortality. For *unselfishness* or *selflessness* are necessary prerequisites for the successful completion of Bhīma's feat. Bhīma sets out to climb the mountain and pluck the golden lotuses without even a moment's hesitation, because they are for Draupadī and will please and console her in her exile (*draupadyā vanavāsinyāḥ priyā kartuṃ samudyataḥ*, 3.146.34cd). When he reaches his destination and sees the lotuses, he never thinks of plucking them for himself, but only thinks of the pleasure they will give Draupadī, and 'he went with his mind to his beloved, who was tortured by the stay in the forest' (*vanavāsaparikliṣṭāṃ jagāma manasā priyām*, 3.150.28cd). When Kubera comes to know of Bhīma's deed, he decides not to punish Bhīma nor to prevent him from taking the lotuses, because he knows that they are meant for Draupadī (*kṛṣṇānimittaṃ viditaṃ mamaitat*, 3.152.24d). A corollary of this detachment and selflessness – indeed the quasi-asceticism in which the exploit is performed – is the fact that Bhīma does not eat

[21] He fights against the demon Rāhu when he jumps up to eat the sun in his youth. When he jumps over the ocean to find Sītā, he has to fight against two *rākṣasīs*.

during his quest. He 'has Draupadī's words for sole viaticum' (*draupadīvākyapāthey*[*aḥ*], 3.150.24), and when Hanumān tries to tempt him and make him go back by promising him 'roots and fruit similar to *amṛta*' (*amṛtakalpāni mūlāni ca phalāni ca*, 3.146.81), Bhīma simply ignores his offer.[22] This is of course quite extraordinary behaviour on the part of Bhīma, the notorious glutton.[23] This *tapas*-like selflessness, the concentrated frame of mind and abstinence from food have clear parallels in other stories of *soma*-thefts.[24]

OVERCOMING MOUNTAINS

I now come to the second function which Hanumān makes over to Bhīma, namely, that of overcoming mountains. In the stories which are associated with them, the two brothers turn out to be masters of mountains. Hanumān encounters various mountains at crucial points in his life. In Rām 5.1, when he flies over the ocean, he meets a first obstacle on his way, Mount Maināka, the only mountain who has been able to escape from Indra's wrath when the latter hacked off the wings of the mountains. His wings intact, Maināka took refuge in the ocean.[25] As a matter of fact,

---

[22] This also shows that Bhīma had a different sort of *amṛta* in mind just then.

[23] One of the two reasons why Bhīma 'falls' while climbing up mount Meru at the end of his life is precisely because he has always eaten too much (see Mbh 17.2.23–25).

[24] Thus Garuḍa was rewarded by Viṣṇu precisely because he had not even tasted the *amṛta*. Hanumān, in his childhood, does not succeed in reaching the sun, because he wants to eat it and satisfy his own greed, but he manages to bring back the mountain with the reviving herbs during the war, for he is acting out of devotion for Rāma, and indeed to save Rāma's life, and not for himself.

[25] Maināka and Hanumān share one common experience: both of them have had hostile encounters with Indra. The god wanted to hack off the mountain's wings with his thunderbolt, and he hit the young Hanumān with this weapon when the latter tried to eat the sun (Rām 7.35–36). Maināka managed to escape, but Hanumān did not. In this sense, Maināka may appear as a model for Hanumān.

Maināka is not hostile to Hanumān; he merely rises out of the water to provide him with a resting-place in his flight over the ocean. The obstacle to be overcome by Hanumān in this case is the temptation to rest on Maināka's slopes and eat the fruit and roots offered by him. Hanumān does not yield to the temptation, and pays homage to the mountain before going on his way. We can draw an obvious parallel between this and Hanumān's own offer of fruit and roots to Bhīma, which Bhīma also refuses.

Hanumān's second encounter with mountains takes place during the war against Rāvaṇa. Hanumān twice manages to pluck the mountain-top on which grow the reviving herbs and to bring it to the battlefield in Laṅkā to revive Rāma and Lakṣmaṇa (Rām 6.61 and 6.89). And the third important mountain for the great monkey is Mount Gandhamādana itself, where he is spending the last part of his life when he meets Bhīma.

Bhīma is similarly associated with mountains and with overcoming mountains. Shortly after his birth,[26] Kuntī, frightened by a tiger, dropped him from her lap and he shattered the rock (*śilā*) or mountain (*giri*) on which he fell (Mbh 1.114.11–13). In the episode which concerns us here, Bhīma manages to climb mount Gandhamādana, which is usually out of bounds to humans (3.146.79). When Bhīma starts his ascent of the mountain, Mount Gandhamādana is likened, in an elaborate comparison, to an adorned dancer, 'who is dancing as if with wings, by means of the clouds adhering to its sides' (*sapakṣam iva nṛtyantaṃ pārśvalagnaiḥ payodharaiḥ*, 3.146.25ab). This comparison is certainly an allusion to the ancient myth of the flying mountains, whose wings precisely became clouds after being cut off.[27] In a sense, Mount Gandhamādana is here compared to Maināka, the only mountain which could save its wings. Thus Gandhamādana

---

[26] Bhīma is probably born on mount Śataśṛṅga, in the Himalayas. This is the last mountain mentioned on Pāṇḍu's north-bound journey (Mbh 1.111.5) and presumably where his sons are born.

[27] *Maitrāyaṇīsaṃhitā* 1, p. 152, line 12.

first appears as an obstacle to Bhīma, just as Maināka did to Hanumān.

The third important mountain that Bhīma meets in his life is, strangely, Hanumān himself. Indeed, when Bhīma meets him in Mbh 3.147, Hanumān is lying on the ground, and appears to be nearly fused with the mountain, so that Bhīma, in spite of his proverbial strength, is not even able to lift his tail (3.147.17). In this passage, Hanumān is several times explicitly compared to a mountain, especially when he shows Bhīma the huge form in which he jumped over the ocean. Thus in 3.149.4cd, 'the monkey stood there, having reached the size of a mountain' (*gireś cocchrayam āgamya tasthau tatra sa vānaraḥ*). He is 'like a second mountain' (*dvitīya iva parvataḥ*, 3.149.5b); like a golden mountain (*sauvarṇam iva parvatam*, 3.149.7b); or, most tellingly, like Mount Maināka himself (*mainākam iva parvatam*, 3.149.13d). In a certain sense, Hanumān has 'become' the mountain: he is reduced to a static position, similar to that which Maināka had been reduced to when he hid in the ocean. Hanumān's role here is comparable to Maināka's. Both the mountain and Hanumān first appear as obstacles on the way, when they actually want to help. Both Mount Maināka and Hanumān have to be honoured, by Hanumān and Bhīma respectively, before the latter can proceed on his quest. Here, the now immobilised Hanumān confers the power to move heavenwards in the quest for immortality on his 'successor', his younger brother Bhīma. Thus, the ancestor or model appears at a crucial junction of the story to put his successor to the test and then to help him proceed on his way. This is a way of marking the 'filiations' between these heroes or personages.

Furthermore, Bhīma's adventure and his encounter with Hanumān has the nature of an initiation. Hanumān puts Bhīma to the test: he agrees to let him continue on the way to the golden lotuses not only after testing his courage, but also his respect and devotion. He curbs Bhīma who behaved somewhat boisterously and violently before meeting him – as Hanumān himself was occasionally prone to behave in his younger age, before he was cursed not to know his own strength. Hanumān also gives his younger brother lessons on *dharma*, and when he assumes his terrible form, he reminds us of course of Kṛṣṇa showing his cosmic form to Arjuna in the *Bhagavadgītā* and instructing him before the battle.[28] In his epiphany, he teaches Bhīma on the nature of Time, and describes to him the eons, *yugas*, *kalpas*, etc. (3.148). Such instructions on Time seem to have formed a part of initiation tests, as some other stories in the *Mahābhārata* reveal.[29]

THE FATE OF THE GOLDEN LOTUSES

After this great effort on Bhīma's part to get the lotuses for Draupadī, the end of the story is something of an anti-climax. After climbing up the mountain, meeting his brother Hanumān, fighting off whole armies of *rākṣasas* and finally plucking the lotuses, we would expect Bhīma to give them to Draupadī with some pomp and ceremony. But nothing of this kind happens. When Bhīma's brothers and wife arrive on the scene, alerted by the terrible omens they see, Yudhiṣṭhira, after admonishing Bhīma, simply receives the flowers from him (*anuśāsya ca kaunteyaṃ padmāni pratigṛhya ca*, 3.153.28ab) and these are, to my knowledge, never mentioned again. We do not even know if Draupadī herself, for whom they were meant and who wanted to

[28] As Rāma's companion and vehicle in the *Rāmāyaṇa*, Hanumān may in a sense be considered as a form of the great god Viṣṇu himself.

[29] For instance in Mbh 1.3.19–176. See Feller 2004, chapter 5. Moreover, as Laine (1989: 43) remarks: 'The accounts of Bhīma's and Arjuna's revelation experiences share a number of elements in common with the story of Yudhiṣṭhira's meeting with the god Dharma. All these accounts involve the appearance of a mythic counterpart: Arjuna and Yudhiṣṭhira meet their divine fathers, Bhīma meets his brother son of Vāyu; all involve the appearance of a disguised or unrecognised god; all involve a test, followed by divine self-revelation or didactic discourses and a granting of boons.'

bring them back to her hermitage in the Kāmyakavana, ever receives them. At least, when the Pāṇḍavas eventually go back to the Kāmyakavana (Mbh 3.180), the lotuses are not mentioned again. In the second episode, the situation is even more extreme, because Draupadī does not even make it to the mountain-top, and the flowers are likewise not mentioned again after Bhīma has plucked them.

How to explain this strange state of affairs? As far as I can see, there can be three ways of explaining it:

(1) The flowers are just a pretext for highlighting Bhīma's prowess and strength, and are therefore not in themselves important in the story. This is a somewhat weak way of doing away with the problem, since, as we have seen, the golden lotuses are symbols of immortality and can therefore not be trivial. Besides, the same story is told twice, with some differences. How could this be if the flowers were unimportant?

(2) The second solution would be to draw a parallel between these golden lotuses and other golden lotuses which appear in Mbh 1.189 in the story of the five Indras and of Śrī's tears, which change into golden lotuses as they fall into the water.[30] The story is told to explain and justify the fact that Draupadī will have five husbands. The five Indras and Śrī are cursed to be reborn on earth as the five Pāṇḍavas and Draupadī. Thus Draupadī's getting back the golden lotuses may in a sense be tantamount to saying that the Goddess Śrī, or Fortune herself, is 'swallowing back' her tears.[31] This may be a way of announcing that the wheel of fortune is about to turn and that revenge is at hand. But

---

[30] This parallel is already made by van Buitenen *ad* Mbh 3.146.5. See van Buitenen 1975: 828.

[31] In any case, the text insists on the community of essence, or resemblance, between Draupadī and lotuses. As soon as she is born, she is said to be sweet-smelling like a blue lotus, and her eyes are lotus-like (see Hiltebeitel 2001: 187). At the beginning of the episode of the golden lotuses, we find the following: '[Draupadī] the pure one, having taken the pure [lotus]' (*tac chubhā śubham āsādya*, 3.146.8).

if we accept this attractive solution, we would again expect Draupadī to receive the lotuses, which is not the case.

(3) The third frame of interpretation lies in comparing the fate of the lotuses with the fate of the *amṛta*, or of other symbols of immortality, when they are brought from heaven to earth. Such a comparison reveals a surprising thing, namely, that far from being exceptional, it is on the contrary the usual fate of the *soma* and of its various substitutes, either not to reach the earth at all;[32] or to reach it only very briefly before being immediately taken back to heaven, either without achieving the desired effect – like the *amṛta* stolen by Garuḍa (Mbh 1.30) – or after achieving the desired effect – like the mountain with the reviving herbs brought to the battlefield by Hanumān (Rām 6.61 and 6.89); or even to change into something else, usually of a monstrous nature, on reaching the ground, as when in certain late Vedic versions of the *soma*-theft (e.g. in *Aitareyabrāhmaṇa* 3.26; also in *Suparṇākhyāna* 14.28.1), either the *soma*'s leaf or the Gāyatrī's wing or claw, are cut off by an archer, and change into porcupines or varieties of snakes when they fall on the ground. Thus the fate of the golden lotuses, which are doomed never to come down from the mountain – or from heaven – and never to reach Draupadī's hermitage, far from being exceptional, is on the contrary in keeping with the usual fate of the nectar of immortality.[33]

Thus Bhīma's quest, his search for the golden lotuses, ultimately turns out to be vain since Draupadī never receives the flowers. Although following in the footsteps of his more successful models – Garuḍa, Hanumān, even the flying mountains, all of whom

---

[32] This is for example the case for the sun, a symbol of immortality, in the episode when Hanumān jumps up to heaven and tries to eat the sun (Rām 7.35–36).

[33] The only exception to this rule is the case of the *soma* in the *Ṛgveda*: there the eagle brings it for Indra, and the *soma* is then used by men on the earth, for the purpose of sacrificing. See on this topic Feller 2004: 188.

went against the king of the gods' wishes – Bhīma fails to achieve the desired end. We may notice in this respect that, in the episode which concerns us here, Bhīma has dealings with two different sets of elder brothers: first with Hanumān, his brother on the paternal side, then with Yudhiṣṭhira, his brother on the maternal side. Hanumān first tests him, but finally encourages him and shows him the way to the lotuses. Yudhiṣṭhira accepts the lotuses from him, but scolds him and tells him that his behaviour was unacceptable. Hanumān and Bhīma's father is the wind-god Vāyu, himself a somewhat capricious god, capable of wreaking havoc and going against the set order of things.[34] Yudhiṣṭhira's father, on the other hand, is the god Dharma, Law and Order personified: Yudhiṣṭhira, as Dharma's son, could therefore never condone such a revolutionary and dissident act as stealing the nectar of immortality from heaven, and by making the lotuses over to him, Bhīma as it were seals his defeat.

WOMEN AND THE QUEST FOR IMMORTALITY

While studying the different versions of the myth of the *soma*-theft, we cannot fail to notice that women are often depicted as the original cause for the quest of the *soma*. They are frequently, if not always, the conscious or unconscious instigators of the quest. This theme appears as early as the Ṛgvedic account of the myth, in which Indra is abandoned by his mother at birth: to survive, he has to kill his father and get the soma (*Ṛgveda* 4.18.10–13). Similarly, in his childhood, Hanumān leaps up to the sun to eat it, thinking that it is some kind of fruit, because his mother has left him alone and hungry in a cave (Rām 7.35–36). In these two cases, the abandoned heroes' quest for immortality could be seen as the quest for a substitute of the mother's milk and love.

[34] Thus Vāyu saves mount Maināka from Indra's wrath, and hurls the mountain into the ocean. See Rām 5.1.111–112.

More frequently, conflicting women are the more or less deliberate instigators of the whole enterprise. This trait appears in the late Vedic texts, where the Sky and the Earth, personified as women or female animals, hold a bet which the Sky loses. As a result of her defeat, the Sky has to get the *soma* for the Earth and sends her children, the meters, to get it. This theme is taken up again in the more developed version of the *Mahābhārata*, where Kadrū and Vinatā hold a bet which Vinatā loses. Her son, Garuḍa, then steals the *soma* to redeem himself and his mother from slavery. In these cases, the quarrelling women serve as the instigation for the theft of the *soma*, but they usually do not lay hands on the *soma* themselves.

In other cases, women send men to get the ambrosia, or its substitutes, on a whim, out of a seemingly childish desire,[35] thus being sources of conflict and jealousy. Draupadī twice sends off Bhīma on the quest for the golden lotuses / five-coloured flowers,[36] and in both cases, to goad him on, she either favours another brother over him or seems to compare him unfavourably to one of his brothers. The first time, she tells Bhīma that she must give the lotus she has just found to Yudhiṣṭhira, but that if he, Bhīma, loves her, he should go and get more of the same for her (Mbh 3.146.10–11). The second time (3.157.17), seeing the five-coloured flowers, she tells him that Arjuna has accomplished extraordinary exploits and that the least he, Bhīma, can do, is to

---

[35] Thus Biardeau 2002: 552: 'Quant à Pāñcālī, c'est un pur et simple caprice qui la fait envoyer Bhīma explorer les hauteurs ...'. Also on p. 551, talking of Sītā and Draupadī: 'L'une comme l'autre a ses moments de puérilité, de frivolité ...'.

[36] She sends him on a third similar quest in book 10 of the *Mahābhārata*. To avenge her sons' death she wants to get Aśvatthāman's divine *maṇi* (jewel), which is said to protect its wearer against weapons, sickness, hunger, gods, demons, *nāgas*, *rākṣasas* and thieves (Mbh 10.15.29–30), in other words, which makes him close to immortal. This attempt is crowned with success: the Pāṇḍavas get the jewel from Aśvatthāman and give it to Draupadī, who in turn makes it over to Yudhiṣṭhira, who fixes it on his head (see Mbh 10.16.33–34).

allow them all to go to the mountain-top where these flowers grow. The first instance may be accounted for by the fact that every mark of respect is due to the eldest brother, Yudhiṣṭhira; but the second instance, where Bhīma is compared to his *younger* brother, is rather unaccountable, except if we see it as an attempt on the part of Draupadī to make Bhīma jealous.

From this point of view, we can compare this episode with that of Sītā and the golden deer in the *Rāmāyaṇa* (3.40–3.43). When Sītā sees the fabulous golden deer studded with precious stones – actually the demon Mārīca in disguise – prancing in front of her hermitage, she wants to have it[37] and does not heed Lakṣmaṇa's words, who suspects it may be a demon. To please her, Rāma sets off to hunt the deer. Then, when Rāma fails to return, and Lakṣmaṇa, obeying Rāma's instructions, does not go after him at once for fear of leaving her unprotected, Sītā abuses Lakṣmaṇa and accuses him of wanting her for himself and wishing for his brother's death. We recognise here the motifs of provoking – or evoking – feelings of jealousy and goading one brother on. Sītā's punishment is as swift as it is terrible: she does not get the golden deer, for that deer, as soon as it dies, changes back into the horrible demon it really is.[38] Worse still, Sītā is subsequently kidnapped by the frightful Rāvaṇa, the king of the *rākṣasas*. Draupadī's punishment for her attempts to get immortality are not punished as ruthlessly as Sītā's, but some time after the first episode of the golden lotuses, she and some of her husbands are kidnapped by a *rakṣas* disguised as a Brahmin one day when Bhīma is absent. But Bhīma quickly comes back and

[37] Sītā's reasons for wanting the deer are strikingly similar to Draupadī's: Sītā wants the deer in order to decorate her *antaḥpura* once their exile is over (see Rām 3.41.16), and Draupadī wants the lotuses in order to decorate her hermitage in the Kāmyakavana, once their pilgrimage is over.

[38] This trait reminds us of the *soma*'s leaf changing into poisonous snakes or porcupines when it touches the ground. Besides, the deer is of course golden, and it is also likened to the sun in Rām 3.41.21. As we have seen, both gold and the sun are symbols of immortality.

kills him. This kidnapping appears like an abridged version of the kidnapping of Sītā by Rāvaṇa.[39]

Thus women, seemingly on a whim, send men off on an impossible quest, for ultimately immortality can never be obtained. It either goes back to heaven, transforms into monsters, or simply vanishes, and causes suffering in vain, bringing pain and punishment in its wake. And as a general rule, women can never lay their hands on it. At a human level, therefore, one possible interpretation of the myth could be that the power of immortality is not for women.

But there is another possible interpretation of the myth at a cosmic level. Bhīma unhesitatingly sets off to get the lotuses because he wishes to fulfil his Queen's desires (*abhiprāyaṃ tu vijñāya mahiṣyāḥ*, 3.146.13). Now, any queen, Draupadī as well as Sītā, represents of course the Earth. What we have here is therefore once again the age-old story of the competition between the earth and the sky for the nectar of immortality: here the earth – Draupadī – sends the son of the wind[40] – Bhīma – to get it for her in the form of the golden lotuses. Thus, if we understand that these women ultimately all symbolise the Earth, another interpretation of the myth at a cosmic level could be that, however much we may strive for it, immortality is not for this earth.

---

[39] The *rakṣas* in brahmanical disguise who kidnaps Draupadī is called Jaṭāsura, a name evoking at the same time a Brahmin (the *jaṭā* being a way of reciting the Veda), and a demon (*asura*); this reminds us of Rāvaṇa, who is a *rākṣasa* of brahmanical descent on his father's side. Besides, after this, Bhīma also has to fight against Kubera himself, who is after all Rāvaṇa's brother: one more parallel between the two narratives.

[40] Not a bad choice, considering that the wind can easily go upwards and blow down the flowers.

REFERENCES

Aitareyabrāhmaṇa = *The Aitareya Brahmanam of the Rigveda*. Ed. & trans. by Martin Haug. 2 vols. Delhi: Bharatiya Publishing House, 1976.

ALLEN, N. J. 2000a. *Categories and Classifications. Maussian Reflections on the Social*. New York: Berghahn Books.

——— 2000b. Argos and Hanuman: Odysseus' dog in the light of the Mahābhārata. *Journal of Indo-European Studies* 28.1–2: 3–16.

BIARDEAU, Madeleine 1973–74. Conférence de Mlle Madeleine Biardeau. In: *Annuaire de l'Ecole Pratique des Hautes Etudes* 80–81: 121–141.

——— 2002. *Le Mahābhārata. Un récit fondateur du brahmanisme et son interpretation*, vol. 1. Paris: Editions du Seuil.

BUITENEN, J. A. B. van (trans. & ed.) 1975. *The Mahābhārata*, vol. 2: *The Book of the Assembly Hall*; vol. 3: *The Book of the Forest*. Chicago: The University of Chicago Press.

CHARPENTIER, Jarl 1920. *Die Suparṇasage*. Uppsala: A.-B. Akademiska Bokhandeln.

FELLER, Danielle 2004. *The Sanskrit Epics' Representation of Vedic Myths*. Delhi: Motilal Banarsidass.

——— 2009. Hanumān's jumps and their mythical models. In: Petteri Koskikallio (ed.) *Parallels and Comparisons. Proceedings of the Fourth Dubrovnik International Conference on the Sanskrit Epics and Purāṇas, September 2005*: 193–219. Zagreb: Croatian Academy of Sciences and Arts.

GELDNER, Karl Friedrich 1951. *Der Rig-Veda aus dem Sanskrit ins Deutsche übersetzt und mit einem laufenden Kommentar versehen*. Dritter Teil: neunter bis zehnter Liederkreis. Cambridge, Mass.: Harvard University Press.

GRÜNENDAHL, Reinhold 1993. Zu den beiden Gandhamādhana-Episoden des Āraṇyakaparvan. *Studien zur Indologie und Iranistik* 18: 103–138.

HILTEBEITEL, Alf 2001. *Rethinking the Mahābhārata. A Reader's Guide to the Education of the Dharma King*. Chicago: The University of Chicago Press.

LAINE, James W. 1989. *Visions of God. Narratives of Theophany in the Mahābhārata*. (Publications of the de Nobili Research Library, 16.) Vienna: Gerold.

Mahābhārata, Critical Edition. Ed. by V. S. Sukthankar, S. K. Belvalkar, P. L. Vaidya *et al.* 19 vols. Poona. Bhandarkar Oriental Research Institute, 1933–66.

Maitrāyaṇīsaṃhitā = *Maitrāyaṇī Saṃhitā. Die Saṃhitā der Maitrāyaṇīya-Śākhā.* Ed. by Leopold von Schroeder. 4 vols. Wiesbaden: Franz Steiner Verlag, 1970–72.

Rāmāyaṇa = *The Vālmīki-Rāmāyaṇa.* Ed by Govindlal Hargovind Bhatt *et al.* 7 vols. Baroda: Oriental Institute, 1960–75.

Ṛgveda = *Rig Veda. A Metrically Restored Text with an Introduction and Notes.* Ed. by Barend A. van Nooten & Gary B. Holland. (Harvard Oriental Series, 50.) Cambridge, Mass.: Harvard University Press, 1994.

Uttaṅka's Quest

PAOLO MAGNONE

The theme of the quest is central to a widespread category of myths – indeed, it constitutes the basic pattern of what has been termed by Campbell 'the monomyth' by reason of its universal diffusion (Campbell 1949: 30). Classical mythology is rife with examples of heroic quests: Prometheus, Odysseus, Jason are only some of the most obvious instances that come to mind. The archetypal quest, however, is no doubt embedded in the Sumerian Epic of *Gilgameš*, whose first redaction, during the Ur III period, antedates by some 12 centuries the earliest beginnings of Greek literature. There, the object of the quest is avowedly life immortal, that primal target of human craving which is only dimly adumbrated in the symbolic objects of other quests: often understated or trivialized, as in the magical objects of fairy tales, or else sublimated, as in the spiritual quests of the *buddha*s that be. In brahmanical India, immortality as such is usually not an article much sought after, the ideal seeker being typically set on escaping birth rather than death. Furthermore, immortality is never forever in the frame of the cosmology of the revolving aeons: at the most, one can strive to gain admission to the exclusive club of the *cirajīvin*s, 'the long-lived ones', which are bound to last *ābhūtasamplavam*, 'till the (universal) dissolution of beings' at the end of a *mahākalpa*. And yet in spite of this the overwhelming multiversity of Indian civilization does contemplate in its literary heritage more than one specimen of the mythologeme of the quest for immortality, such as the theft of ambrosia by the solar bird, or the churning of the ocean for the elixir[1] – or the lesser known, enigmatic story of sage Uttaṅka, who seems to conceal under the guise of the perfect brahmanical disciple the earnest seeker for life immortal along the perilous path to the underworld and back, after the prototypical fashion of Gilgameš.

Uttaṅka,[2] one of the minor epic *ṛṣis*, is a sage of the Bhārgava family, unknown to the Vedic literature but present in the *Mahābhārata* and the Purāṇas with a small *corpus* of myths, falling under two main rubrics: 1) the quest for a pair of wondrous earrings (*kuṇḍala*); 2) the killing of *asura* Dhundhu. The quest story, which will detain us for the greater part of this paper, occurs in two different versions in the Mbh, Ādiparvan (3.81–170)[3] [hereafter called version A] and Āśvamedhikaparvan (56–58) [hereafter version B]; as well as in the SkP, Prabhāsakhaṇḍa 3.2 – mainly an abridgement of B, with a few dashes of A. The Dhundhumāra story, on the other hand, is extant in several puranic versions, including the VāP (88.27ff.), BḍP (2.63.28ff.), BrP (5.54ff.), ŚiP (5.37.8ff.) and BhgP (9.6.22) as well as in the Mbh, Āraṇyakaparvan (201ff.) and Hv (1.11.22ff.).[4] Less prominent, yet of no scanty importance for our present study, is a third anecdote, occurring in a single instance in the Mbh, Āśvamedhikaparvan, immediately before B – clearly an aetiological story meant to explain the formation of the so-called 'Uttaṅka's clouds' (*uttaṅkamegha*) in the desert. Other minor feats of the sage will be left unnoticed here as they are of little import to our

[1] For a review of both, as well as of the kindred story of Kaca's apprenticeship to learn the *saṃjīvanī vidyā* from Śukra, see Dange 1997.

[2] Or sometimes Utaṅka.

[3] The reference edition of the *Mahābhārata* is the vulgate, unless specified otherwise. The reasons for the choice will become apparent by the commentaries to individual passages.

[4] On the Dhundhumāra myth, whose discussion is besides the scope of the present paper, see Dange 1996: 296–307 (also footnote 38 below).

discussion.[5] In spite of their seeming diversity, the three myths are not without correlations, as will be shown in the following.

Uttaṅka makes his first appearance as a *brahmacārin* in the Pauṣyaparvan, in a context devoted to initiatory stories[6] illustrating the virtue of obedience (*śuśrūṣā*) as paramount in the master–disciple relationship. Unlike those of his immediate predecessors (Aruṇi, Upamanyu and Veda) however, there seems to be more to Uttaṅka's story than a mere exemplification of the attitude of the ideal disciple, as becomes all the more apparent by a comparison with the alternative version in the Āśvamedhikaparvan (and its *Skandapurāṇa* abridgement). We are now going to review the salient features of both versions before proceeding to a comparison.

---

5 After leaving his master, Uttaṅka, eager to take revenge on Takṣaka, proceeds to Hastināpura and denounces to Janamejaya Takṣaka's responsibility in Parikṣit's death, inducing him to perform the great snake sacrifice (Mbh 1.3.171–188; also DBhgP 2.11.16ff.). Elsewhere Uttaṅka redeems evil Gulika teaching him *viṣṇubhakti* (NāP 1.37–38).

6 Feller (2004: 206–251) analyses the tales of Āyodha Dhaumya's three disciples Aruṇi, Upamanyu and Veda, as well as the latter's own disciple Uttaṅka, as initiation stories where the initiates through trials and ordeals are led on to higher knowledge or even immortality (symbolized in Upamanyu's golden teeth and the cow-dung offered to Uttaṅka) – such immortality, however, being 'nothing but the immortality which they [...] gain through the revelation, knowledge or belief which is imparted to them. Thus most of these passages consciously reuse the age-old initiation pattern, involving death (to the old beliefs) and rebirth (to the new), including the promise of immortality tendered by the new belief' (pp. 250–251). Although this approach is certainly pertinent, it seems to me somewhat reductive, in that it evokes a myth/ritual relationship which is not warranted: for, as Feller herself notes, 'we are here dealing with mythical, prototypical initiations, which are much more drastic and "real" than initiation rites. What is described is not a "dramatic" *mise en scéne*, but the "real thing"' (p. 223). Indeed, initiation myths (and the corresponding rituals) may aptly be viewed as but a special implementation for the needs of social functioning of a more universal psychological pattern – or, to express it the other way round: '[t]he standard path of the mythological adventure of the hero is a magnification of the formula expressed in the rites of passage: *separation – initiation – return* which may be named the nuclear unit of the monomyth' (Campbell 1949: 30). As a matter of fact, more distinctively initiatory features, such as fasting and seclusion, are absent in Uttaṅka's myth.

VERSION A

Uttaṅka is a disciple of Veda, in turn one of Āyoda Dhaumya's disciples. On one occasion, being about to leave to do duty as sacrificer, Veda charged Uttaṅka to see to it that nothing was wanting in the house during his absence. Some time later, the women of the *āśrama* solicited Uttaṅka to do the needful, as the *guru*'s wife was in her fertile period and worried that it might go in vain, but he refused to comply: he would not commit an improper act at the behest of women, for the master had not commanded him to do even what should not be done on any account.[7]

Learning about the incident on his return, Veda was pleased with Uttaṅka and dismissed him with a promise to fulfil all his wishes. Uttaṅka insisted on offering a *dakṣiṇā* according to *dharma* precepts, and after some reluctance Veda finally consented, leaving it to his wife to determine its nature. The latter required Uttaṅka to beg a certain couple of earrings from King Pauṣya's queen, imposing a tight deadline and threatening a curse in case of non-compliance.[8] Soon after his departure, Uttaṅka came across a giant mounted on a huge bull, who bade him eat the animal's excrement. At first Uttaṅka turned down the unseemly request,[9] but finally acquiesced on learning that his master had done the same previously.

---

[7] Mbh 1.3.87 (vulg.): *na hy aham upādhyāyena saṃdiṣṭo 'kāryam api tvayā kāryam iti.*

[8] It is unclear whether we have here a variant of the motive of Potiphar's wife, i.e. whether the *upādhyāyinī* is acting out of spite for having been rejected. As a matter of fact, in version B, where the sexual temptation is absent, the *upādhyāyinī* looks more benevolent: she does not fix a term, and is apparently unaware of the risks involved in her request (in B the king is anthropophagous).

[9] Feller (2004: 235 n.) observes that 'eating cow-dung and urine may not seem as disgusting in the Indian context as it is to our eyes', with reference to the purifying practice of consuming the *pañcagavya* (i.e. milk, curds, clarified butter, dung and urine). In spite of this, the vulgate seems to imply that Uttaṅka did indeed regard it as a loathsome act when it adds *saṃbhramād utthita evāpa upaspṛśya* anticipating the detail of the ritual breach

Upon reaching his destination, Uttaṅka asked the king for the earrings, and was sent to the inner apartments where, however, the queen was not to be found. He blamed the king for speaking untruth, but the king clarified that the queen was invisible to anyone who was rendered impure by the leftovers of food (*ucchiṣṭa*) for lack of proper purification. In fact, Uttaṅka, upset by the uncouth meal, had performed his ablutions standing and on the move (instead of sitting and facing east, as prescribed). After correcting the ritual misconduct, he was able to see the queen, who gave him the earrings with a warning to look after them carefully, for they were coveted by the *nāga* king Takṣaka.

Pauṣya invited Uttaṅka to stay as a guest at a *śrāddha*, but a detail of ritual purity occasioned an exchange of curses (afterwards annulled or mitigated) between the king and the sage.[10] On his way back to the *āśrama*, catching sight of an extra-Vedic naked ascetic,[11] Uttaṅka went in search of some water to perform purifications, leaving the earrings unattended; but alas! the ascetic, who was none else than Takṣaka under disguise, suddenly grabbed hold of them, and resuming his serpentine form sneaked through a crevice in the ground all the way down to his

---

in the manner of the ablutions which will result in Uttaṅka's inability to see the queen (see below).

10 Upon being cursed by Uttaṅka because the food was cold and polluted by a hair, the king at first pronounced a counter-curse, believing the charge to be groundless, but later apologized after verifying its legitimacy. Uttaṅka then toned down (to all practical effects withdrew) his curse, but the king professed himself unable to do the same, for 'brahmans are known to have soft hearts and sharp tongues, but the reverse is true with *kṣatriyas*'; however, Uttaṅka cleverly escaped the king's curse by declaring it void as its precondition (that the food had been blameless) was not fulfilled. In view of this, Uttaṅka's reaction when first receiving the curse – *na yuktaṃ bhavatānnam aśuci dattvā pratiśāpaṃ dātuṃ tasmād annam eva pratyakṣīkuru* – seems indeed apposite, although rejected by the critical edition. This is not the only case where the text of the vulgate recommends itself on internal considerations.

11 *nagnaṃ kṣapaṇakam* (Mbh 1.3.127 vulg.) or *śramaṇam* (Mbh 1.3.137 crit. ed.).

dwelling in the *nāgaloka*. Uttaṅka struggled helplessly to dig a path with his stick, until Indra came to his rescue lending him his *vajra*, which quickly ripped the soil open to the subterranean regions.[12] Here Uttaṅka beheld lofty palaces and stately mansions everywhere, and propitiated the *nāgas* with a hymn of praise, to no avail. Meanwhile, he noticed two women weaving a fabric on a loom with black and white threads, and a twelve-spoked wheel revolved by six youths; furthermore, a man and a horse. Eulogized by Uttaṅka, the man prompted him to blow into the horse's anus. No sooner had he done so than flames and smoke issued forth from the horse's bodily apertures pervading the *nāgaloka* and inducing Takṣaka to hand back the earrings.

The time allowed being about to expire, Uttaṅka mounted the wondrous horse, who swiftly brought him back to his master's house, averting the curse the *guru*'s wife was already on the verge of pronouncing for his default. On being questioned by Veda about the reasons for his delay, Uttaṅka related the facts, asking for an explanation of the mysterious events that had occurred. But we shall revert to this later on.

VERSION B

Uttaṅka is Gautama's favourite disciple. Gautama held him especially dear because of his great devotion, and so one by one he dismissed all his other disciples, but retained Uttaṅka until he was overtaken by old age. One day, back to the *āśrama* from

---

[12] The last sentence summarizes a passage found solely in the vulgate (Mbh 1.3.130b–132), which here exhibits a major discrepancy with respect to the critical edition, to its own advantage, it might be judged, at least going by textual considerations. Expunging the five-liner, whose gist is Indra's coming to the succour of the helpless Uttaṅka, the critical edition has Uttaṅka cleaving through the rift to the *nāgaloka* unaided. However, Indra's intervention at this junction, besides being corroborated by version B, seems all the more plausible in view of the donor/helper role manifoldly played by him (albeit reluctantly on occasions) both in this same myth and elsewhere (see below).

gathering firewood, while throwing down the faggot he noticed a hoary tuft of hair which had got entangled in the sticks. Suddenly realizing that he had grown old, Uttaṅka burst into tears so hot that in vain did the master's daughter strive to collect them in her cupped hands: not even the earth was able to sustain their scorching heat.[13]

Uttaṅka was dejected, for old age had crept unnoticed over him as he was absorbed in single-minded devotion to his master, and would now prevent him from experiencing happiness in married life. For his part, Gautama declared he was ready to let him go without even demanding a *dakṣiṇā*; indeed, he was pleased to such an extent that Uttaṅka would become a sixteen year old youth, and he would give him his own daughter as a bride, for no one else would be able to serve his *tejas*. Uttaṅka rejuvenates then and there without further ado and accepts her,[14] still insisting, however, on paying his fees to the master's wife, protesting he would spare no efforts to satisfy her. Yielding to his pressures, Ahalyā requested King Saudāsa's queen's earrings, apparently unaware of the fact that Saudāsa had taken to anthropophagy in consequence of a curse.[15]

In a secluded forest spot Uttaṅka meets Saudāsa wearing a long beard besprinkled with human blood, who at once sets his mind upon devouring him. Uttaṅka is ready to submit, on condition that he should first be allowed to acquit himself of his obligation towards his master, after which he shall come back to

---

13 The fiery teardrops announce Uttaṅka's supernatural *tejas* in its double value as energy and seed, whereas the master's daughter's attempt to hold them foreshadows her destiny as a fit receptacle for the latter. On the values of *tejas* see Magnone 1993.

14 Mbh 14.56.22–24ab (vulg.): *ittham ca parituṣṭaṃ māṃ vijānīhi bhṛgūdvaha* । *yuvā ṣoḍaśavarṣo hi yad adya bhavitā bhavān* ॥ 22 ॥ *dadāmi patnīṃ kanyāṃ ca svāṃ te duhitaraṃ dvija* । *etām ṛte 'ṅganā nānyā tvattejo 'rhati sevitum* ॥ 23 ॥ *tatas tāṃ pratijagrāha yuvā bhūtvā yaśasvinīm* ।

15 See footnote 19 below.

meet his destiny. Saudāsa agrees and sends Uttaṅka off to his wife, whom he himself cannot approach at meal time.

Madayantī, however, demanded a token of her husband's will to part with the celestial earrings, which were craved by gods, goblins and great serpents[16] on account of their wonderful virtues: they yielded gold day and night, shone brightly in the darkness and could adapt to the size of their wearer, making him immune from hunger and thirst, poison, fire and wild animals.[17] He should beware, however, for they would be grabbed by *pannagas* as soon as cast on the ground.

Uttaṅka went back to Saudāsa to obtain the required token, and was imparted the following shibboleth: 'neither is this way easy, nor is there any other way',[18] which at once convinced the queen. Questioned by Uttaṅka about the meaning of the enigmatic sentence, Saudāsa explained: according to natural disposition, *kṣatriyas* owe respect to *brāhmaṇas*, but the latter are often the cause of great evil. Saudāsa himself, in spite of his submissiveness to *brāhmaṇas*, had come to harm because of one of them.[19] Nevertheless, he did not see any other way of conduct, for

---

[16] Here the vulgate reading (Mbh 14.57.22a) [*devāś ca yakṣāś ca*] *maharṣayaś* [*ca*] (for *mahoragaś* in the critical edition) is evidently incorrect, because it does not tally with the same three categories of prospective thieves as specified subsequently: i.e., serpents would steal the earrings if cast on the ground, goblins if borne by *ucchiṣṭhas* and gods in case their bearer fell asleep.

[17] Mbh 14.57.25–27 (vulg.): *syandete hi divā rukmaṃ rātrau ca dvijasattama | naktaṃ nakṣatratārāṇāṃ prabhām ākṣipya vartataḥ || 25 || ete hy āmucya bhagavan kṣutpipāsābhayaṃ kutaḥ | viṣāgniśvāpadebhyaś ca bhayaṃ jātu na vidyate || 26 || hrasvena caite āmukte bhavato hrasvake tadā | anurūpeṇa cāmukte jāyete tatpramāṇake || 27 ||*

[18] Mbh 14.58.2 (vulg.): *na caivaiṣā gatiḥ kṣemyā na cānyā vidyate gatiḥ*. In the *Skandapurāṇa* version the riddle is differently worded, although the import turns out the same: *yair vinā sugatir nāsti durgatiṃ* [*t*]*e nayanti vai* (SkP 7.3.27). Both refer to brahmans, who are the only way to spiritual welfare but are often hard to get along with and a source of trouble.

[19] Saudāsa, better known as Kalmāṣapāda or Mitrasaha, formerly a glorious king, attained his present condition as a cannibal owing to a *brāhmaṇa*'s curse. The story is narrated differently in Mbh 1.166–167 (crit. ed.) and ViP 4.4.40ff. (as well as elsewhere).

a king adverse to *brāhmaṇas* can neither stand in this world nor enjoy happiness in the other.

Uttaṅka then tricked the king into freeing him of the obligation to return to be eaten and set off for his master's house after wrapping the earrings in a black deerskin, according to Madayantī's instructions. On his way, overcome by hunger, he climbed a *bilva* tree to pluck some fruits, hanging the bundle onto a branch; but alas! the knot being hit by some fruits came undone and the earrings fell down.[20] No sooner had they touched the ground than a snake caught them in his mouth and disappeared into an anthill. Uttaṅka immediately dashed in pursuit, violently overturning the soil for thirty-five days with his stick to no avail, until Indra came there disguised as a brahman and after trying in vain to dissuade him eventually put his *vajra* on top of the stick, which instantly clove the earth to the subterranean regions. There Uttaṅka, beholding the *nāgaloka* stretching out for thousands of leagues, girdled by bulwarks of gold studded with gems and pearls, was filled with despondency – but lo! there appeared a black horse with a white tail and coppery red eyes and mouth, flaming with *tejas*, who prompted him to blow into his anus; he should not shrink in disgust, for he had already done the same formerly, in Gautama's *āśrama*. Uttaṅka was perplexed, until the horse manifested as Agni Jātavedas, assiduously worshipped by Uttaṅka on his master's behalf, and now ready to help him out of his predicament.

Uttaṅka then did as requested, and flames blazed up brightly, while thick smoke exuded from all the horse's pores filling the whole *nāgaloka*. Seared and choked, their eyes red-

---

[20] According to the text of the vulgate (Mbh 14.58.20–24); the text of the critical edition (Mbh 14.57.19–21ab), which is only half in length, appears seriously corrupt (cf. e.g. *sa kasmiṃścit kṣudhāviṣṭaḥ phalabhārasamanvitam* । *bilvaṃ dadarśa viprarṣir āruroha ca taṃ tataḥ* ॥ [20 vulg.] which becomes *sa kasmiṃścit kṣudhāviṣṭaḥ phalabhārasamanvitam* । *bilvaṃ dadarśa kasmiṃścid āruroha kṣudhānvitaḥ* ॥ [19 crit. ed.]).

dened, the serpents handed back the earrings to Uttaṅka who made for Gautama's hermitage with the promised gift for Ahalyā.

COMPARISON[21]

The explicit purpose of version A is clearly to uphold brahmanical values. We are told about the loyalty and obedience of a disciple to his *guru* as well as of his constant concern to fulfil dharmic prescriptions, like the gift of *dakṣiṇā*. Uttaṅka is willing to confront any danger, to submit to any humiliation for his master's sake. After a first refusal, he is finally persuaded to eat the bull's *mūtrapurīṣa* only by the desire to imitate his master. Moreover, issues of ritual purity take up a conspicuous place in the narration. Uttaṅka is unable to see the queen at first, because he has not performed his ablutions according to the prescribed procedure; the impurity of the food offered at the *śrāddha* triggers Uttaṅka's curse against the king. The entire episode of the *śrāddha* brings out the respective values of *brāhmaṇa*s and *kṣatriya*s.[22] Likewise, Takṣaka exploits Uttaṅka's preoccupations with ritual purity, when disguising as a heretic to induce him to part from the earrings to perform his purifications. On the other hand, the earrings as such do not seem to carry any special value, if not as an instrument of Uttaṅka's zeal to please his *guru*'s wife. We are left with an impression of Uttaṅka's paragon virtue which after many trials finally earns him his *guru*'s blessing.

This dominance of brahmanical axiology is hardly surprising, considering the hand brahmans had in shaping the epics in their extant form. Admittedly, in edifying their literary monuments, brahmans have often employed pre-existing narrative materials, through a process of adaptation to make them subservient to

---

[21] See the synoptic table at Appendix 1.

[22] See footnote 10 above.

aims and developments originally alien. Besides these purposeful elaborations, the ancient stories have often incurred alterations of a more haphazard nature over the course of transmission: they have been expanded or condensed, curtailed or supplemented, deliberately or accidentally modified through misunderstandings, errors or memory defects. Thus down the process of their evolution these stories have often ended up looking like the façades of some renovated old *palazzi*, where the incongruous vestiges of medieval trefoils and lancets can still be descried among the orderly rows of modern square windows. Such traces are distinguishable also in our story, and they will stand out all the more conspicuously by comparing the version summarized just now with the alternative version.

In version B, preoccupations of purity that were so pervasive in A are notably absent, as are the relevant episodes of the eating of disgusting food and the ensuing state of *ucchiṣṭa*, as well as the episode of the impure *śrāddha*. Here, furthermore, the motive for Uttaṅka's carelessness in looking after the earrings has undergone an inversion so as to dispense with purity issues: instead of going for lustral water, being repelled by an impure sight, as in A, here Uttaṅka goes for food, being attracted by the sight of a *bilva* tree. On the other hand, the earrings advance prominently to the foreground with a value of their own, no longer merely a pretext for Uttaṅka's display of staunch disciple's devotion. They are newly characterized as magical objects, whose wonderful virtues are praised in detail, so that we now realize that the gist of the story, in its present form, is really about a treasure quest of the type which is so common in folktales.

Indeed, features typical of folktale occur in B that did not figure in the alternative account: besides the magical objects (the earrings), the ogre (King Saudāsa, the owner of the earrings, only here characterized as man-devourer) and the difficult questions (King Saudāsa's riddle). Looking back from this vantage point, we can discover more folktale motifs in A: the animal helper (the nourishing bull and the flaming horse), the swift mount (the horse Agni taking Uttaṅka home just in time), the enigmatic visions (the weavers and the wheel in the *nāgaloka*).

But there is more to it than just the random occurrence of a bunch of folktale motifs: indeed, the very underlying structure of the story evidences the distinctive and specific functions of folktale as identified by Propp's analyses, whose main methodological thrust was indeed to reject what he regarded as the inane fragmentariness of motif classification in favour of the pinpointing of constant and stable elements provided by the functions of the *dramatis personae*, which he believed to be limited in number and arranged in a fixed order. In spite of criticisms[23] and cautionary assessments, Propp's methodology still constitutes a valuable tool for the analysis of folktales, as well as of kindred genres – like quest myths, which share with folktales much of their underlying structure. It may be recalled that the total sequence of 31 functions that make up the model folktale according to Propp need not be completely filled in any given instance, whereas the order of the sequence is supposed to be invariable. The table given at Appendix 2 shows a significant congruence between the development of version A and Propp's sequence. Although a detailed discussion lies beyond the scope of the present paper, a few remarks may be in order. The prologue structure βγδ (absentation/interdiction/violation), highly characteristic of folktales, is present also in our myth, although the couple 'interdiction/violation' has undergone an inversion, being replaced by 'command/fulfilment', in view of the authorial intent to make the story into

[23] Most famously by Lévy-Strauss. His review was prefaced to the 1966 Italian translation of Propp's *Morfologija skazki. Transformacij voléebnych skazok* published by Einaudi, followed by Propp's rejoinder appearing for the first time. For some background information and an appraisal of the controversy (albeit heavily biased against the French anthropologist) see Dundes 1997.

a eulogy of *śuśrūṣā*.[24] Brahmanical preoccupations with ritual purity and paradigms of *brāhmaṇa–kṣatriya* relationship are likewise responsible for the specific embodiment of the 'struggle' as an exchange of curses according to the modalities described above.[25] The next point that deserves mention is the thrice iterated occurrence in version A of the sequence DEF ('donor's intervention / hero's reaction / acquisition of magical agent') with Indra in the capacity as donor, whether openly or under disguise. Now this very sequence is undoubtedly one of the most distinctively characteristic of folktale as a genre.

On the other hand, where exactly lies the demarcation line between myth and folktale as genres is a problem in itself. Propp pointed out that 'a similar construction [is also displayed by] some most archaic myths, some of which present this structure in an amazingly pure form',[26] and again:

> [t]here are myths based on the same morphological and compositional system as the wonder tale [...] At times they correspond, down to minute details, to the compositional system studied in *Morphology of the Folktale*. In some cases myth and wonder tales have the same form. (Propp 1984: 79.)

It has been argued by some that the difference between myth and folktale is not one of structure, but rather of context: myths become fairy tales when transposed into a socio-cultural context where they are no longer believed as true; still others deny this

---

[24] In any case, similar inversions do occur also in folktales, as exemplified by Propp (1966: 40), who remarks that in such instances the fulfilment of the command entails the same (unwished for) consequences as elsewhere the violation of an interdiction. As a matter of fact, Uttaṅka's refusal to oblige the women (which, incidentally, is not without ambiguity, to the extent that he somehow infringes the letter of his master's command while fulfilling the spirit) may lie at the root of the grievous task imposed upon him by the *upādhyāyinī*.

[25] Given the peculiar nature of the 'struggle', the slight alteration in the canonical sequence is not significant.

[26] Propp 1958: 90. (The original translation has, incorrectly: 'displays' instead of the bracketed expression).

genetic relationship between myth and folktale to posit them side by side in the same context, whereupon independently from 'belief' folktales are characterized by the genre-specific quality of the 'fabulous' or 'marvellous', variously defined.[27] What all these different approaches have in common seems to be that folktale emerges as the realm of the unbelievable/unaccountable/unmotivated, as against myth as a story which is eminently 'true' (in some sense). Whereas the folktale hero is not really a hero, actuated upon as he is by incomprehensible agencies exacting of him a usually preposterous task he obtusely submits to, the hero of quest myths is genuinely acting out of his own understanding and will through trials and ordeals in deliberate pursuit of his own purpose. Accordingly, whereas the invariable outcome of folktale is a final reward that comes as totally gratuitous, the quest hero may fail, or else succeed in conquering the well deserved prize of his endeavours.

In addition to 'truth', the 'sacred' is another item often involved in definitions of myth – which is hardly surprising, truth and sacredness being like the two faces of the same coin of Reality:

> [pour les sociétés archaïques] le mythe est censé exprimer la *vérité absolue*, parce qu'il raconte une *histoire sacrée*, c'est-à-dire une révelation trans-humaine qui a eu lieu à l'aube du Grand Temps, dans le temps sacré des commencements (*in illo tempore*). Étant *réel* et *sacré*, le mythe devient *exemplaire* et par conséquant *répétable*, car il sert de modèle, et conjointement de justification, à tous les acts humains.[28]

Accordingly, the quest hero is usually moving in a sacred milieu as he pursues the life-empowering object of his quest – perhaps symbolized by the Woman as the archetypal source of life universal. By contrast, the folktale 'hero', moving about in a secular-

[27] For a survey of these views see Somoff 2002.

[28] Eliade 1957: 17–18 (author's italics).

ized, profane environment, is content with some paltry surrogate – maybe a bride with whom to live 'happily together ever after'.

If it is so, there seems to be no denying that folktales may at times, although they need not necessarily always, represent the outcome of a degeneration or trivialization of myths no longer understood in their own rights. Such would seem to be the case of Uttaṅka's story.

On closer investigation, there gleams through the surface layers of the edifying vicissitudes of the model brahmanical disciple and of the fabulous adventures of the folktale hero an even deeper level, seemingly unfolding before us an Indian version of the universal myth of the hero's descent to the netherworld in quest of immortality. Several clues are there, most of them exclusive to version B, the less thoroughly brahmanised one: a prologue mentioning Uttaṅka's despair upon realizing his old age; an incongruous hint at his rejuvenation; the vivifying properties of the *kuṇḍala*s; Saudāsa's and his queen's characters, displaying features (grim appearance, anthropophagy, invisibility) typical of the guardians of the dead; the ordeal; the theft of the *kuṇḍala*s by a serpent; the time symbolism of the netherworld. We are now going to review each of them in greater detail.

The prologue in B, where Uttaṅka perceives for the first time his hoariness as a sign of impending old age, ostensibly serves the purpose of motivating Uttaṅka's dismissal; the aftermath of Uttaṅka's dejection over his forfeited householder life, however, stretches the boundaries of acceptable absurdities even by epico-puranic standards: by the *guru*'s grace Uttaṅka will become sixteen-year-old then and there, and will obtain the *guru*'s daughter as bride. This is offered by the *guru* as a matter of course, and casually accepted by Uttaṅka in stark contrast with his despondency of a minute before.[29] Nay, the offer comes

---

[29] Dange (2001: 22) too observes that 'the change of age, from old to young, is a typical mytho-folkloristic motif' without, however, remarking the incongruity of this casual rejuvenation; as a matter of fact, in the other ex-

abruptly as an afterthought, while discussing about the *dakṣiṇā* after Uttaṅka has already been given leave. The whole thing takes up no more than three strophes, and never again is there any mention of Uttaṅka's wife in the rest of the story or elsewhere. Needless to say, there is no trace of the whole rejuvenation and marriage affair either in version A or in the *Skanda-purāṇa*. As it stands, the episode is utterly unconvincing; furthermore, it represents the only instance of a glaring violation of the regular sequence of Proppian functions, as is evidenced in the table at Appendix 2: indeed, we know well the regular pattern of so many folktales: the king's daughter is given as bride only in the end, once the hero has successfully gone through all the trials imposed upon him. So, we would expect Uttaṅka to first leave in quest of his lost youth, only eventually to obtain his bride after his triumphant rejuvenation. The incongruous transposition of the rejuvenation and marriage before the journey must in all likelihood be a side effect of brahmanical elaboration, bent on working up the hero into a model of *gurubhakti* and *guruśuśrūṣā*: as a result, the more brahmanized version A has utterly expunged any hint to the theme of old age, rejuvenation and the promised bride, whereas the more conservative version B has preserved some incongruous vestiges, maladroitly trying to make them fit the new context: thus, Uttaṅka's long overlooked aging is made to be the paradoxical outcome of his single-minded devotion to his master, and his unwarranted sudden rejuvenation the effect of the latter's grace.

On the hypothesis that the rejuvenation and marriage, if at all, should occur at the end of the story, according to the time-honoured practice of folktales, it would naturally occur as a consequence of the successful outcome of the quest. Its designated prize, at the surface narrative level, is represented by the

---

amples recalled by him (Yayāti exchanging age with his son Puru, Cyāvana rejuvenated by the Aśvins, the ascetic Vṛddhakanyā conjuring a young body for herself just for one night) the process is not at all so easily accomplished.

queen's *kuṇḍalas*, which, in both versions, are sought after by the selfless disciple not for his own sake, but for his master's wife – yet it is conceivable that they might have been originally meant to restore Uttaṅka's faded youth. In version B, the *kuṇḍalas* are characterized as magical objects of the type commonly found in folktales; however, their specific virtues betray them as symbols of the universal life source: like the objects of other well known quests, they can make their wearer immune from harm; they produce gold, a symbol of the incorruptible reality; their very name of *kuṇḍalas* (not to mention the fact that they are claimed by serpents) recalls that serpentine essence which is a universal as well as genuinely Indian symbol of the vital energy continually regenerating the cosmos. It is true that in the passages being discussed they are not explicitly stated to bestow either youth or immortality; nevertheless, one cannot help recollecting, in this connection, another pair of *kuṇḍalas*, originally belonging to Āditi, stolen by *asura* Naraka and restored by Kṛṣṇa, whose story is narrated in the Mbh, Udyogaparvan (48.80ff.), ViP (5.29), BhgP (10.59) and KāP (40). The latter version specifies that the *kuṇḍalas* poured forth gems and *amṛta*, besides healing pain and removing obstacles. Further ahead, while enumerating Naraka's wrongdoings, the earrings are summarily described by what must amount to their foremost characteristic, as a 'source of *amṛta*'.[30]

As is commonly the case in regeneration myths, Uttaṅka must descend to the nether world to conquer life slipping away. The characters he comes across, indeed, bear the marks of the

[30] Dange (2001: 26–27) sees striking similarities also with some features of the myth of the Syamantaka gem given by Sūrya to Satrājit, likewise endowed with the virtue of producing gold (Hv 1.38.25: *sa maṇiḥ syandate rukmaṃ* – cf. Mbh 14.57.25a (vulg.): [*kuṇḍale*] *syandete hi divā rukmam*) ensuring timely rain and banishing fear of disease. After some vicissitudes the gem is stolen by the king of bears, Jāmbavat, who disappears into a crevice (like the *nāga* Takṣaka). Kṛṣṇa eventually defeats Jāmbavat and wins both the gem and Jāmbavat's daughter as a bride.

sovereigns and guardians of the world of the dead. According to version B, Saudāsa is a man-devourer, even though this property is ascribed to the effects of a brahmanical curse; on the other hand, the more brahmanized version A ignores the king's anthropophagy, but preserves the queen's invisibility – another well-known peculiarity of the dead – albeit attributing it to reasons of ritual impurity. In any case, the *kuṇḍala*s will eventually end up in the *pātāla* through Takṣaka's agency. The connection of the subterranean regions with the realm of the dead is further brought out by a detail in *Skandapurāṇa* version: after grabbing the *kuṇḍala*s Takṣaka heads for the southern direction[31] (i.e., the direction presided over by Yama, the king of the dead). Likewise, in the parallel story of Āditi's *kuṇḍala*s, these are detained by Naraka, clearly a homologue of Yama, as is evinced by his very name (meaning 'hell'), his chthonian parentage (Bhauma is another name of his, through having been begotten by the boar *avatāra* upon the goddess Earth), and the fact of his citadel being protected by a forest of razor-blade-sharp nooses (the *pāśa* being the emblem of Yama).

Again, Uttaṅka must undergo an ordeal in order to obtain the *kuṇḍala*s: in version B it is the threat of being devoured, which he escapes by a quibble. Furthermore, the riddle-like phrase he must communicate to the queen as a token of her husband's will also points to a test, with special developments in the version of the *Skandapurāṇa*[32] and some sort of parallel in the skirmish between the king and Uttaṅka in version A.

Finally, Uttaṅka wins the earrings only to forfeit them again because of his negligence on occasion of a certain incident de-

---

[31] SkP, Prabhāsakhaṇḍa 3.2.38: *etasminn eva kāle tu takṣakaḥ pannagottamaḥ* । *gṛhītvā kuṇḍale tūrṇam agamad dakṣiṇāmukhaḥ* ॥ 38 ॥

[32] In the *Skandapurāṇa* Uttaṅka, baffled by the riddle (SkP, Prabhāsakhaṇḍa 3.27: *yair vinā sugatir nāsti durgatiṃ [t]e nayanti vai*), reverts to the king for an explanation and by the simple fact of doing so releases the king from the curse of anthropophagy, who in turn sets Uttaṅka free from the obligation to serve as food.

scribed differently in either version. The setting of version A, where Uttaṅka, startled by the unseemly sight of a naked heretic, goes in search of lustral water leaving the earrings unattended, which are then taken to the netherworld by a *nāga*, wears a striking resemblance to the circumstances in which Gilgameš forfeits the plant of youth, bathing in a pool at whose bottom lurked a snake, who snatched away the plant to plunge again to his haunt in the depth.[33] On the other hand, version B, where Uttaṅka's negligence is occasioned by the desire of picking fruits, reminds one of the motive of food binding to death, like Persephone's pomegranate[34] (or for that matter Eve's apple[35]).

Undergoing the theft, Gilgameš gives up all hope of ever recovering the plant and his youth, and sadly retraces his way to Uruk empty-handed, submitting to his destiny of death, as befits Mesopotamian pessimism. Uttaṅka, for his part, rushes in pursuit and digs his way to the underworld.

According to version A, his attempts to recover the earrings at first suffer a setback; he is subsequently tested by an allegorical vision featuring – according to the interpretation eventually offered by the *guru* – the two maidens of destiny – Dhātṛ and Vidhātṛ, curiously and perhaps not without reason turned into females – weaving on the loom the black and white threads of day and night (not unlike the Greek *Moira* (or *Moirai*) spinning

---

[33] A remarkable inversion of the theme of the hero being deprived of immortality by a serpent while bathing occurs in the myth of the abduction of *amṛta* by the solar bird. In Mbh, Ādiparvan 34.17–23 it is the snakes who are deprived of *amṛta* by Indra while performing ablutions. The defrauded snakes must content themselves with licking the *kuśa* grass on which *amṛta* had been laid – which further evokes the *plant* of immortality of the Sumerian myth.

[34] Confront, to the contrary, Naciketas' ability to escape seizure by Yama thanks to his fasting during the latter's three day's absence (TaiBr 3.11.8).

[35] The circumstance of Uttaṅka's being deprived of immortal life by a serpent as a consequence of picking fruits from a tree cannot but evoke weird associations with the Judaeo-Christian myth of the original sin.

the thread of life) while six youths representing the six Indian seasons turn the twelve-spoked wheel of twelve-month year. Such deferred interpretation, is, however, redundant, in so far as Uttaṅka had already given proof of penetrating insight in his *stotra* elicited by the vision.[36] He had also unequivocally recognized the attendant man with the horse as Indra and Agni,[37] and it is this successful solution of the enigma that brings about the final *dénouement*.

Version B omits the mysterious vision altogether, only preserving the horse, who of his own accord manifests as Agni and helps Uttaṅka recover the *kuṇḍalas*. The motif of the netherworld and the fiery subterranean stallion (elsewhere the fiery submarine mare) is otherwise of special interest, as it links the quest and the Dhundhumāra myth in elusive yet intriguing ways,[38] while evoking other well known (and not unconnected)

[36] Mbh 1.3.146–147 (vulg.): *trīṇy arpitāny atra śatāni madhye ṣaṣṭiś ca nityaṃ carati dhruve 'smin* | *cakre caturviṃśatiparvayoge ṣaḍ vai kumārāḥ parivartayanti* || 146 || *tantraṃ cedaṃ viśvarūpe yuvatyau vayatas tantūn satataṃ vartayantyau* | *kṛṣṇān sitāṃś caiva vivartayantyau bhūtāny ajasraṃ bhuvanāni caiva* || 147 || Uttaṅka's expressions testify that he has understood the cosmic significance of the loom. According to Nīlakaṇṭha's advaitic interpretation, besides symbolising Time with its joints (360 days-and-nights, 24 lunar nodes, 6 seasons), the wheel symbolises Prajāpati's (i.e. the *anima mundi*'s) cosmic body with its 24 *tattvas*, and 360 (i.e. innumerable) *vāsanās* perpetually activated by the 5 *kleśas* (*avidyā*, *asmitā*, *raga*, *dveṣa*, *abhiniveṣa*; or the 5 *viparyayas*: *tamas*, *moha*, *mahāmoha*, *tāmisra*, *andhatāmisra*) with *māyā* as the sixth. In this context the black and white threads are disconnected by the number 360, and represent *duḥkha* and *sukha* fruitions.

[37] Mbh 1.3.148–149: *vajrasya bhartā bhuvanasya goptā vṛtrasya hantā namucer nihantā* | *kṛṣṇe vasāno vasane mahātmā satyānṛte yo vivinakti loke* || 148 || *yo vājinaṃ garbham apāṃ purāṇaṃ vaiśvānaraṃ vāhanam abhyupaiti* | *namo 'stu tasmai jagadīśvarāya lokatrayeśāya puraṃdarāya* || 149 ||

[38] The Dhundhumāra myth is all about horses: Kuval(ay)āśva, i.e. 'possessing the horse named Kuvalaya' (or 'possessing fodder horses'), son of Bṛhadāśva ('possessing great horses') with Uttaṅka's help succeeds in killing the monster Dhundhu, who is likened to the submarine mare (who, according to PdP, Sṛṣṭikhaṇḍa 18.86 lives in the vicinity of Uttaṅka's *āśrama*, just like Dhundhu). In a variant of the myth told in VmP 78, Dhundhu is bent on gaining power by means of the horse-sacrifice, whereas according to the

destructive underworld stallions, such as Sagara's missing *aśva-medha* horse responsible for the burning of his sixty thousand sons (to mention but the most obvious); but we shall leave that for another occasion.

So, to repeat once again: old age and rejuvenation, wondrous life-saving gold-producing objects guarded by a dreadful guardian, stolen and carried to the underworld by serpents, weird visions of time and fate: all the ingredients are there to substantiate our suspicion that under the guise of the model brahmanical disciple there might lie concealed the onetime hero in quest of the life-quickening elixir. I would like to mention just one last circumstance in support of this hypothesis: although the *amṛta*-producing character of the wondrous earrings can only be a matter of conjecture, yet *amṛta* does play an explicit and all-important role at least in a couple of places in Uttaṅka's mythic *corpus*. One of them we know already. In version A of the quest myth, before meeting the king, Uttaṅka, as we recall, comes across a man astride an enormous bull and obeys the injunction to eat the animal's dung. As his *guru* will later explain, the man was in reality Indra on his elephant mount Airāvata, who, being a friend of the *guru*, had taken pity on Uttaṅka: the excrements were in reality *amṛta*, who had preserved Uttaṅka from death in the underworld. If the divine food preserving from death (the reverse of the demonic food binding to death) is a substitute for the *kuṇḍalas*, which are subsequently stolen by Takṣaka – qualified

---

Mahābhārata version of the regular myth he is said to burn Bṛhadaśva's son just as Kapila had burnt Sagara's sons – who were after the *aśvamedha* horse. Commenting on the *Vāmanapurāṇa* version, Agrawala (1983: 207) remarks that '[i]n the earlier story Dhundhu was an *asura* having the form of a wild horse but now he gives demonstration of his power in the form of the free-roaming horse of the sacrifice.' Furthermore, Dhundhu is said to be the son of Madhu and Kaiṭabha who according to some were killed by a horse-headed incarnation of Viṣṇu. For a ramble through the mind-boggling intricacies of the stallion and mare symbolism in Indian mythology see O'Flaherty 1981: 213ff.

in B as Airāvata's progeny[39] – then we might say that what Indra gives – *amṛta*, through the mediation of Airāvata – the same Indra takes away – the *amṛta*-yielding *kuṇḍalas*, through the mediation of Takṣaka-Airāvata.[40] Now this duplicity of Indra in his *amṛta*-giving attitude is exactly the theme of the aetiological myth about Uttaṅka's clouds immediately preceding version B of the quest myth. According to it, Uttaṅka was granted by Kṛṣṇa the boon of being able to get water at will. Once, as he was suffering from thirst in the desert, he saw a filthy naked outcaste, surrounded by a pack of dogs, who mockingly invited him to drink the liquid he was urinating.[41] When he declined the offer, Kṛṣṇa appeared, revealing that the outcaste had in reality been Indra offering him *amṛta* at Kṛṣṇa's behest. Indra, however, was reluctant to grant immortality to a mortal, and for that reason had obtained permission to camouflage the gift under loathsome appearances. Had Uttaṅka accepted, he would have gained life immortal. So both in the quest and in the *uttaṅkamegha* myth envious Indra offers Uttaṅka ambrosia disguised as *mūtra*[42] and

[39] Elephant Airāvata's camouflage as a bull, and his relationship with serpent Airāvaṭa, favoured by the double meaning of *nāga* (i.e. 'serpent' and 'elephant'), is made the object of much speculation by Dange (2001: 14ff.), along, however, euhemeristic lines irrelevant to the present discourse.

[40] A further *liaison* of Indra with Takṣaka is known from the episode of the burning of the Khāṇḍava forest, Takṣaka's haunt, and on that account under Indra's protection.

[41] The text here is less than explicit: *tasyādhaḥ srotaso 'paśyad vāri bhūri dvijottamaḥ* (Mbh, Āśvamedhikaparvan 55.17cd). Nīlakāṇṭha glosses *śrotas* either as a water-skin or as the (river) Bhogavatī flowing nearby. Nevertheless, it is very tempting to interpret it as the urethra (Feller 2004: 245 n. concurs with this interpretation), also in view of the *functional* parallelism between the outcaste's filthy water/possibly urine and the bull's excrement. This is further corroborated by the rather unusual choice of the word used for outcaste, which is consistently *mātaṃga* (5 times), only finally recapitulated as *cāṇḍāla*; now, it can hardly be a mere chance that the primary meaning of *mātaṃga* is once again 'elephant' – whereby the *mūtra* proffered by the 'elephant' outcaste exactly parallels the *mūtrapurīṣa* elsewhere proffered by the (man on) the 'elephant' bull.

[42] It is worth recalling in this connection Wasson's theory identifying *soma* with the fly agaric (*Amanita muscaria*), which he believes to have

purīṣa, but sees to it that Uttaṅka cannot profit from the gift, and manages in different ways to look as if he were aiding, whereas in reality he is trying to thwart him. Thus both myths are in all likelihood but multiforms of one and the same mythologeme, whose focal point is – the conquest of the life-giving elixir. The outcome, however, is different in either case: for in the quest myth Uttaṅka accepts the ambrosial dung at the outset, and although deprived of the ambrosial *kuṇḍala*s subsequently, still succeeds in winning them back; whereas in the *uttaṅkamegha* myth he refuses the ambrosial urine, and in so doing forfeits life immortal for good: he must go empty-handed, like Gilgameš and all his homologues in universal mythology, and be satisfied with plain old water from the clouds named after him.

REFERENCES

Primary sources

BdP = *Brahmāṇḍamahāpurāṇam.* Vārāṇasī: Kṛṣṇadās Akādamī, 1983.

BhgP = *Śrımadbhāgavatamahāpurāṇam.* Delhi: Nag Publishers, 1987.

BrP = *Śrībrahmamahāpurāṇam.* Delhi: Nag Publishers, n.d.

DBhgP = *Śrīdevībhāgavatapurāṇam.* Delhi: Nag Publishers, 1986.

Hv = *Śrīharivaṃśapurāṇam.* Delhi: Nag Publishers, 1985.

KāP = *The Kālikāpurāṇa. Text, Introduction & Translation in English* by Prof. B. N. Shastri. Delhi: Nag Publishers, 1991.

Mbh = *The Mahābhārata, for the first time critically edited* by V. S. Sukthankar, S. K. Belvalkar, P. L. Vaidya *et al.* 19 vols. Poona. Bhandarkar Oriental Research Institute, 1933–66.

Mbh (vulg.) = *Śrīmanmahābhāratam. Caturdharavaṃśāvataṃsaśrīmannīlakaṇṭhaviracitabhāratabhāvadīpākhyaṭīkayā sametam.* Dillī: Nāg Prakāśak, 1988.

NāP = *Śrīnāradīyamahāpurāṇam.* Delhi: Nag Publishers, n.d.

---

often been consumed in metabolite form through the urine of the officiating priests having previously eaten it. Wasson (1968: 33–34) himself mentions the *uttaṅkamegha* myth in support of his theory (although he somewhat jumbles the characters).

PdP = *Śrīpadmamahāpurāṇam*. Delhi: Nag Publishers, n.d.

ŚiP = *Śrīśivamahāpurāṇam*. Delhi: Nag Publishers, 1986.

SkP = *Śrīskandamahāpurāṇam. Śrīprabhāsakhaṇḍam*. Delhi: Nag Publishers, 1987.

TaiBr = *Taittirīyabrāhmaṇam. Kṛṣṇayajurvedīyam. Sāyanācāryakṛtavedārthaprakāśākhyabhāṣyasahitam*. Calcutta: Baptist Mission Press, 1855

VāP = *Śrīvāyumahāpurāṇam*. Delhi: Nag Publishers, 1983.

ViP = *Śriviṣṇumahāpurāṇam*. Delhi: Nag Publishers, 1985.

VmP = *Śrīvāmanamahāpurāṇam*. Delhi: Nag Publishers, n.d.

Secondary literature

AGRAWALA, Vasudeva S. 1983. *Vāmana Purāṇa – A Study*. 2nd edition. Varanasi: Prithivi Prakashan.

CAMPBELL, Joseph 1949. *The Hero with a Thousand Faces*. Princeton: Princeton University Press (reprint: London: Paladin, 1988).

DANGE, Sadashiv A. 1996. *Towards Understanding Hindu Myths*. New Delhi: Aryan Books International.

——— 1997. *Myths from the Mahābhārata*, vol. 1: *Quest for Immortality*. New Delhi: Aryan Books International.

——— 2001. *Myths from the Mahābhārata*, vol. 2: *Study in Patterns and Symbols*. New Delhi: Aryan Books International.

DUNDES, Alan 1997. Binary opposition in myth: the Propp/Lévi-Strauss debate in retrospect. *Western Folklore* 56.1: 39–50.

ELIADE, Mircea 1957. *Mythes, rêves et mystères*. Paris: Gallimard.

FELLER, Danielle 2004. *The Sanskrit Epics' Representation of Vedic Myths*. New Delhi: Motilal Banarsidass.

MAGNONE, Paolo 1993. The development of 'tejas' from the Vedas to the Purāṇas. *Wiener Zeitschrift für die Kunde Südasiens* 36 (Suppl.): 137–147.

O'FLAHERTY, Wendy Doniger 1981. *Sexual Metaphors and Animal Symbols in Indian Mythology*. New Delhi: Motilal Banarsidass.

PROPP, Vladimir 1958. *Morphology of the Folktale*. Bloomington: Indiana University.

——— 1966. *Morfologia della fiaba*. Con un intervento di Claude Lévi-Strauss e una replica dell'autore. A cura di G. Bravo. Torino: Einaudi. (Original edition: *Morfologija skazki. Transformacij volšebnych skazok*. Leningrad 1928.)

PROPP, Vladimir 1984. *Theory and History of Folklore*. Minneapolis: University of Minnesota Press.

SOMOFF, Victoria 2002. On the metahistorical roots of the fairy tale. *Western Folklore* 61.3–4: 277–294.

WASSON, Robert Gordon 1968. *Soma – Divine Mushroom of Immortality*. New York: Harcourt Brace Jovanovic.

APPENDIX 1

Mbh A	SkP	Mbh B
Guru [G] Veda has 3 disciples including Uttaṅka [U]	*Guru* [G] Gautama has hundreds of disciples including Uttaṅka [U]	
U is tested by sexual temptation		
	U's *śuśrūṣā* is established	
	G pleased keeps him for a long time	
	U discovers his old age by a hoary lock enmeshed in firewood and despairs for having forgone...	
	marriage	happiness
	G's wife relates his sorrow to G	G's daughter tries to collect U's *taijasa* tears
V pleased dismisses him	G regretful dismisses him	
U insists on paying his fees, but G avows himself satisfied with his past services		
		and offers to give him his daughter as wife after his rejuvenation
G's wife asks for a certain pair of earrings worn by...		
King [K] Pauṣya's queen [Q]	King [K] Saudāsa's queen [Q]	
fixing a deadline		
U meets a man riding a bull consents to eat the bull's excrement does not purify properly		
	U meets K	
	who is anthropophagous and threatens to eat U	
	and begs the earrings	
	proposing to postpone the eating	
	K sends U to Q	
Q cannot be seen by U due to his impurity		(Q cannot be seen by K)
U is able to see Q after purifying		
	Q asks for a sign of K's will	
	U relates an enigmatic phrase obtained from K	
U gets the earrings from Q with a warning that they are coveted by		
nāga Takṣaka		*devas*, *yakṣas* and *nāgas* for they have wondrous virtues

U is offered impure food through negligence	K explains the enigmatic phrase	
U curses K	Vasiṣṭha had cursed K	[implicitly the same]
K curses U	[K is to eat U]	
U willingly annuls his curse to K	U has unwittingly annulled Vasiṣṭha's curse to K	
U escapes K's curse by a device	U is no longer to be eaten by K	U escapes being eaten by K by a device
	U sets off for G's house with the earrings	
	wrapped up in black deerskin	
U is repelled by the sight of a naked misbeliever	U is attracted by the sight of a *bilva* tree	
goes in search of water to purify	climbs to pick fruits to eat	
		hanging the deerskin on a branch
laying the earrings on the ground	laying the deerskin on the ground	the earrings fall onto the ground
the misbeliever changes into Takṣaka [N]	Takṣaka [N] appears	A serpent [N] appears
	N steals the earrings	
disappears into a crevice		disappears into an anthill
	U pursues N digging the ground with a stick	
		Indra tries in vain to dissuade U, then
	Indra lends his *vajra*, U reaches the *nāgaloka*; there U sees ...	
two weavers and a loom		
a 12-spoked wheel revolved by 6 youths		
a man and a horse	a white horse endowed with all *guṇas*	white-tailed copper-eyed/mouthed black horse
the man prompts U to blow into the horse's anus	the horse prompts U to touch his private parts	the horse prompts U to blow into his anus
	flames issue forth filling up the *nāgaloka* with smoke	
	N distressed gives up the earrings	
U's time is almost up		U goes back to G
the horse swiftly carries him to G just in time to avoid G's wife's curse		
	U gives the earrings to G's wife	
G explains U's enigmatic visions:		
man on bull = Indra on Airāvata		
bull's dung = *amṛta*		
2 weavers = Dhātṛ and Vidhātṛ		
black & white threads = nights & days		
12-spoked wheel & 6 youths = year & seasons		
man & horse = Parjanya [Indra] & Agni	[horse = Agni]	

APPENDIX 2

<table>
<tr><th></th><th>Fn</th><th>Description</th><th>Version A</th><th>Version B</th></tr>
<tr><td>Prologue</td><td>(α)</td><td>Initial situation</td><td colspan="2">Uttaṅka's [U] discipleship with guru [G]</td></tr>
<tr><td></td><td>β</td><td>Absentation</td><td>G leaves āśrama</td><td></td></tr>
<tr><td></td><td>γ</td><td>Interdiction</td><td>G entrusts U with care of āśrama</td><td></td></tr>
<tr><td></td><td>δ</td><td>Violation</td><td>U behaves correctly</td><td></td></tr>
<tr><td></td><td>ε</td><td>Reconnaissance</td><td></td><td></td></tr>
<tr><td></td><td>ζ</td><td>Delivery</td><td></td><td></td></tr>
<tr><td></td><td>η</td><td>Trickery</td><td></td><td></td></tr>
<tr><td></td><td>θ</td><td>Complicity</td><td></td><td></td></tr>
<tr><td></td><td>(a)</td><td></td><td></td><td>U has grown old unmarried</td></tr>
<tr><td></td><td>(B)</td><td></td><td></td><td>G promises U youth and daughter as /bride</td></tr>
<tr><td></td><td></td><td></td><td></td><td>U is rejuvenated</td></tr>
<tr><td></td><td></td><td></td><td></td><td>U weds G's daughter</td></tr>
<tr><td>Main Plot</td><td>a</td><td>Lack</td><td colspan="2">G's wife desires earrings</td></tr>
<tr><td></td><td>B</td><td>Mediation</td><td colspan="2">U is requested to procure earrings</td></tr>
<tr><td></td><td>C</td><td>Counteraction</td><td colspan="2">U promises to procure earrings</td></tr>
<tr><td></td><td>↑</td><td>Departure</td><td colspan="2">U departs</td></tr>
<tr><td></td><td>D</td><td>Donor</td><td>Indra bids U eat bull's dung</td><td></td></tr>
<tr><td></td><td>E</td><td>Reaction</td><td>U consents</td><td></td></tr>
<tr><td></td><td>F</td><td>Magical agent</td><td>U eats dung (= ambrosia)</td><td></td></tr>
<tr><td></td><td>G</td><td>Transference</td><td></td><td></td></tr>
<tr><td></td><td>H</td><td>Struggle</td><td></td><td>king [K] threatens to eat U</td></tr>
<tr><td></td><td>J</td><td>Branding</td><td></td><td></td></tr>
<tr><td></td><td>I</td><td>Victory</td><td></td><td>U obtains postponement</td></tr>
<tr><td></td><td>K</td><td>Removal of lack</td><td colspan="2">U gets earrings from queen [Q]</td></tr>
<tr><td></td><td></td><td></td><td>mutual curses of U and king [K]</td><td></td></tr>
<tr><td></td><td></td><td></td><td>U escapes K's curse by device</td><td>(U escapes being eaten by device)</td></tr>
<tr><td>Iterations</td><td>A</td><td>Villainy</td><td colspan="2">nāga [N] steals earrings</td></tr>
<tr><td></td><td>(B)C↑</td><td>Counteraction</td><td colspan="2">U rans in pursuit of N</td></tr>
<tr><td></td><td>(DE)F</td><td>Magical agent</td><td colspan="2">Indra lends vajra</td></tr>
<tr><td></td><td>G</td><td>Transference</td><td colspan="2">U cleaves to nāgaloka</td></tr>
<tr><td></td><td>H</td><td>Struggle</td><td>U praises nāga</td><td></td></tr>
<tr><td></td><td>I</td><td>Victory</td><td>U fails to get earrings</td><td></td></tr>
<tr><td></td><td>D</td><td>Donor</td><td>Indra bids U blow into horse's anus</td><td>horse bids U blow into his anus</td></tr>
<tr><td></td><td>E</td><td>Reaction</td><td colspan="2">U consents</td></tr>
<tr><td></td><td>F</td><td>Magical agent</td><td colspan="2">horse emits flames and smoke</td></tr>
<tr><td></td><td>H</td><td>Struggle</td><td colspan="2">N is distressed by fire</td></tr>
<tr><td></td><td>I</td><td>Victory</td><td colspan="2">N surrenders earrings</td></tr>
<tr><td>Epilogue</td><td>↓</td><td>Return</td><td colspan="2">U returns to āśrama</td></tr>
<tr><td></td><td>Pr</td><td>Pursuit</td><td></td><td></td></tr>
<tr><td></td><td>Rs</td><td>Rescue</td><td>(swift horse saves U from curse)</td><td></td></tr>
<tr><td></td><td>o</td><td>Incognito</td><td></td><td></td></tr>
<tr><td></td><td>L</td><td>Unfounded claim</td><td></td><td></td></tr>
<tr><td></td><td>M</td><td>Difficult task</td><td>(U relates enigmatic visions)</td><td></td></tr>
<tr><td></td><td>N</td><td>Solution</td><td>(G solves them for him)</td><td></td></tr>
<tr><td></td><td>Q</td><td>Recognition</td><td></td><td></td></tr>
<tr><td></td><td>Ex</td><td>Exposure</td><td></td><td></td></tr>
<tr><td></td><td>T</td><td>Transfiguration</td><td></td><td></td></tr>
<tr><td></td><td>U</td><td>Punishment</td><td></td><td></td></tr>
<tr><td></td><td>W</td><td>Wedding</td><td>(G dismisses U with blessing)</td><td></td></tr>
</table>

What Need Has He of the Waters of Puṣkara? The Narrative Construction of *tīrtha* in the Sanskrit *Mahābhārata*

JAMES M. HEGARTY

This paper explores the Sanskrit *Mahābhārata*'s conspicuous concern with the *tīrtha*, the fording point or place of pilgrimage. It is my intention to shed light on the way in which the *Mahābhārata* uses the *tīrtha* to address issues of self-presentation, religious authority and its relationship to antecedent, and chiefly Vedic, textual materials. I will suggest that the *Mahābhārata*'s treatment of pilgrimage practice does a very good job of combining exegesis of the old with the elaboration of the new. More than this, I will argue that the *Mahābhārata* uses its discourse on *tīrtha* to position itself as the medium *par excellence* for these ongoing efforts to address issues of continuity and change in early South Asian religious knowledge systems.

The paper is structured as follows: I will first of all consider the way in which the *Mahābhārata* presents various *tīrtha*s. I will show how the *Mahābhārata* uses *tīrtha* to 1) construct the past and 2) contextualise and interpret narrative and ritual activities. I will then consider broader trends in Vedic and post-Vedic exegesis as they relate to the *Mahābhārata*'s approach to *tīrtha*s.

Indological studies of *tīrtha* in the *Mahābhārata* have generally focussed on their being granted a given sacrificial value in the text (a visit to *tīrtha* x is equivalent to Vedic ritual y). This process has been interpreted as part of a broader pattern of the democratisation of religious observance in post-Vedic South Asia. Vassilkov summarises the position rather well when he states,

> In defiance of the Vedic System of Ideas, they proclaim the reward of visiting *tīrtha* to 'surpass even the reward of sacrifices' (3.80.38).[1]

The risk of such an assertion is that it might be guilty of throwing out the inter-textual baby with the historical bathwater. That is to say that there is a danger that a defiant attitude to the minutiae of Vedic observance, and the imposition of vast and exclusivistic sacrificial rites, is mistaken for a fundamental move away from the 'Vedic System of Ideas', which is a system of ideas that I believe is essential for the comprehension of the Sanskrit *Mahābhārata*'s construction of the *tīrtha*.

My arguments, taken together, will demonstrate that *tīrtha*s in the *Mahābhārata* are not natural or neutral descriptions of places of pilgrimage, or part of a move away from core Vedic cosmic and social presuppositions, but are, instead, part of a sustained exegetical agenda in the *Mahābhārata* in which the text seeks to encompass all forms of religious expression and practice which precede it. This will allow us to explore the question 'What need has he of sprinkling with the waters of Puṣkara?' (*kiṃ tasya puṣkarajalair abhiṣecanena*), which occurs toward the very beginning and end of the *Mahābhārata*.[2]

I will begin by exploring the way in which the *Mahābhārata* describes certain prominent *tīrtha*s in relation to four basic headings, the past, narration, ritual, and movement. I will then explore the *Mahābhārata*'s presentation of the relationship between itself and *tīrtha* visitation.

---

[1] Vassilkov 2002: 141. It is important to note that this statement does not negatively impact upon the fascinating analyses, insights and comparisons of this excellent article.

[2] At 1.2.242 and 18.5.54.

TĪRTHA AND THE PAST

In the Āraṇyakaparvan, we find a major description of Kurukṣetra, the location of the main *Mahābhārata* war, as *tīrtha*.[3] The Pāṇḍavas visit the field of Kuru at the commencement of their exile as a result of the disastrous dice-game of the Sabhāparvan (3.6.1). Its status as *tīrtha* is discussed by Pulastya in his enumeration of the *tīrtha*s of the world in the minor book of the tour of *tīrtha*s, the Tīrthayātrāparvan. Pulastya says: 'Then one should go, O Indra of kings, to the celebrated (*abhiṣṭutam*) Kurukṣetra' (3.81.1ab).[4] It is in the course of the Tīrthayātrāparvan that we also find an exploration of the subordinate *tīrtha*s of the Kurukṣetra area: Among the most prominent of these are the lakes that commemorate the slaughter of the *kṣatriya* by the Bhārgava *brāhmaṇa* Rāma Jāmadagnya (Samantapañcaka) and the gate of Kurukṣetra (*dvāraṃ kurukṣetrasya*). We are told in some detail of Rāma Jāmadagnya's annihilation of the *kṣatriya* at Kurukṣetra at various points in the *Mahābhārata*. This narrative 'thread' has been treated in considerable detail by a number of scholars.[5] For the purpose of this paper, it is sufficient to note that this heavy emphasis on Samantapañcaka is an example of the significant past being remembered through the establishment of a *tīrtha*.

The gate of Kurukṣetra is described in detail by Lomaśa, the sage who guides the Pāṇḍavas on their tour of *tīrtha*s in the Āraṇyakaparvan. It is a description that is couched entirely in terms of events in the significant past and more particularly in terms of the actions of ancestors of the Pāṇḍavas:

---

[3] Though it is first mentioned in this way at 1.2.1–2.

[4] *tato gaccheta rājendra kurukṣetram abhiṣṭutam* ।

[5] Mbh 3.117.9 and 3.81.33–34 and also 1.58 for earlier narrations. The *Mahābhārata*'s engagement with the legend of Rāma Jāmadagnya is sustained and elaborate and has been the subject of a number of studies, see Fitzgerald 2002: 89–133. See also the classic study of Sukthankar 1936. A broad study of the Bhārgava *brāhmaṇas* in the *Mahābhārata* has been undertaken by R. P. Goldman (1977).

> Here, sire, King Yayāti Nahuṣa brought up a sacrifice that abounded with many gems, in which Indra delighted. ... In this same spot King Bharata repeatedly sent off his black-dappled sacrificial horse, after he had obtained, by means of *dharma*, all of the earth. (3.129.12, 15)[6]

The layering of narratives concerning the significant past is continuous: we are also told, in the Ādiparvan, of how Kurukṣetra was named after Kuru, the common ancestor of the Pāṇḍavas and Kauravas, who practised austerities there:

> Saṃvaraṇa's wife Tapatī Saurī bore him a son Kuru, and seeing that he was *dharma*-knowing (*dharmajña*) the people chose him for their king. It is his name that made *kurujāṅgala* famous, Kurukṣetra which the ascetic made auspicious with his great austerities (*mahātapāḥ*). (1.89.42–43)[7]

We are further told in the Ādiparvan that Kurukṣetra was one of the dwelling places of Takṣaka, the snake king who slew Parikṣit, the father of Janamejaya (1.3.144). The Kuru king Citrāṅgada is also slain on the field of Kuru by his namesake the *gandharva* king Citrāṅgada (1.95.7). In the story of Sunda and Upasunda, the two *asuras* make their home in Kurukṣetra, having subjected the whole earth to their dominion (1.202.27). In the Udyogaparvan, Bhīṣma does battle with Rāma Jāmadagnya in the field of the Kurus (5.179.2). The war books themselves narrate the chief events of the battle of the Pāṇḍavas and the Kauravas at Kurukṣetra. In the Śalyaparvan, we also hear of Kurukṣetra as the location of the investiture of the war god, Skanda (9.43.48 and 51). In the Anuśāsanaparvan, we hear of the conquering of death by Sudarśana and Oghavatī who dwell, again, on the field of Kuru (13.2.39). In this way, we can begin to

---

6 *atraiva nāhuṣo rājā rājan kratubhir iṣṭavān* । *yayātir bahuratnāḍhyair yatrendro mudam abhyagāt* ॥ 12 ॥ ... *atraiva bharato rājā medhyam aśvam avāsṛjat* । *asakṛt kṛṣṇasāraṅgaṃ dharmeṇāvāpya medinīm* ॥ 15 ॥

7 *tataḥ saṃvaraṇāt saurī suṣuve tapatī kurum* । *rājatve taṃ prajāḥ sarvā dharmajña iti vavrire* ॥ 42 ॥ *tasya nāmnābhivikhyātaṃ pṛthivyāṃ kurujāṅgalam* । *kurukṣetraṃ sa tapasā puṇyaṃ cakre mahātapāḥ* ॥ 43 ॥

discern the extent to which Kurukṣetra is peppered with areas and features that pertain to the past. It is also in the Tīrthayātrāparvan of the Āraṇyakaparvan that we find an assertion of the *tīrtha* status of another location that is of particular significance to the *Mahābhārata*, the Naimiṣa forest:

> Then he should proceed (in his tour of *tīrthas*) to the auspicious Naimiṣa which is attended upon by *siddhas*; there dwells Brahmā amongst the hosts of the *devas* (3.82.53).[8]

The Naimiṣa forest is subject to a similar, if somewhat less elaborate, process of narrative development as Kurukṣetra as the *Mahābhārata* proceeds. Indeed, we find more general references to narratives concerning the Naimiṣa forest all over the *Mahābhārata*: Thus we find that Naimiṣa is mentioned as the site of a sacrificial session of the gods in the opening of the narrative of the five Indras (1.189.1). In the Udyogaparvan, we find further reference to Naimiṣa, where it is described as the location of the sacrifices of the descendants of Yayāti, who is the progenitor of the entire dynasty of Bhārata (5.119.9–10). It is also conspicuous by its presence in the various descriptions of *tīrthayātrās* in the *Mahābhārata* such as those of Pulastya, Dhaumya and Aṅgirasa in the Āraṇyaka and the Anuśāsana Parvans.[9]

TĪRTHA AND NARRATION

All these narratives of the past lead to a close association between narrative activity and *tīrtha* in the *Mahābhārata*: We can see this most clearly by returning to the very beginning of the *Mahābhārata* when the bard Ugraśravas states:

> I was at the *sarpasattra* of the great spirited royal Ṛṣi Janamejaya, son of Parikṣit, where Vaiśampāyana told various auspicious stories (*supuṇyā vividhāḥ kathāḥ*), just as they had happened, in the pres-

---

[8] *tataś ca naimiṣaṃ gacchet puṇyaṃ siddhaniṣevitam* | *tatra nityaṃ nivasati brahmā devagaṇair vṛtaḥ* || 53 ||

[9] To be found, respectively, at 3.80–83, 3.85–88, and 13.26–27.

> ence of the great prince, tales first uttered by Kṛṣṇa Dvaipāyana. I myself, having listened to these stories of diverse meaning (*vicitrārtha*) that are peculiar to the *Mahābhārata* (*mahābhāratasaṃśritāḥ*) and having wandered around far flung *tīrthas*, came to the auspicious *tīrtha*, Samantapañcaka by name, that is attended upon by the twice-born, the country where there took place the battle of the Kurus and the Pāṇḍavas and all the great kings (*sarveṣāṃ ca mahīkṣitām*) ... so what shall I tell you, O twice born? Auspicious tales of the past (*purāṇasaṃśritāḥ*), tales pertaining to *dharma* (*dharmasaṃśritāḥ*), or tales of the events pertaining to great spirited kings and sages? (1.1.8–14)[10]

The main narrative of the *Mahābhārata* is thus told by Ugraśravas while he is on a tour of *tīrthas* in a *tīrtha*! The Naimiṣa forest is thus associated in particular with, not just the events of the past, but acts of narration. So, for example, it is mentioned as a site for the recitation of histories in the alternative provenance Ugraśravas offers for the Āstikaparvan of the Ādiparvan. This text, which investigates the interruption of the *sarpasattra* of king Janamejaya, postdates the main *Mahābhārata* narrative and could not, by its very nature, have formed part of the recitation of Vaiśampāyana. It is credited, however, once again to Vyāsa. Both the original and the subsequent narrations are located in the Naimiṣa forest:

> The elders (*vṛddhāḥ*) used to tell this *itihāsa*, (this) Purāṇa which Kṛṣṇa Dvaipāyana told to the sages of the Naimiṣa forest. My father Lomaharṣaṇa the *sūta*, Vyāsa's wise pupil, was asked by *brāhmaṇas* to tell it. Therefore I shall tell it as I heard it from him, this Āstīka *ākhyāna* requested by you, Śaunaka. (1.13.6–8)[11]

---

[10] *janamejayasya rājarṣeḥ sarpasatre mahātmanaḥ* I *samīpe pārthivendrasya samyak pārikṣitasya ca* II 8 II *kṛṣṇadvaipāyanaproktāḥ supuṇyā vividhāḥ kathāḥ* I *kathitāś cāpi vidhivad yā vaiśaṃpāyanena vai* II 9 II *śrutvāhaṃ tā vicitrārthā mahābhāratasaṃśritāḥ* I *bahūni saṃparikramya tīrthāny āyatanāni ca* II 10 II *samantapañcakaṃ nāma puṇyaṃ dvijaniṣevitam* I *gatavān asmi taṃ deśaṃ yuddhaṃ yatrābhavat purā* I *pāṇḍavānāṃ kurūṇāṃ ca sarveṣāṃ ca mahīkṣitām* II 11 II ... *bravīmi kim ahaṃ dvijāḥ* II 13 II *purāṇasaṃśritāḥ puṇyāḥ kathā vā dharmasaṃśritāḥ* I *itivṛttaṃ narendrāṇām ṛṣīṇāṃ ca mahātmanām* II 14 II

[11] *itihāsam imaṃ vṛddhāḥ purāṇaṃ paricakṣate* I *kṛṣṇadvaipāyanaproktaṃ naimiṣāraṇyavāsinaḥ* II 6 II *pūrvaṃ pracoditaḥ sūtaḥ pitā me*

We see here how a narration that is not integral to the Vaiśampāyana recitation of the *Mahābhārata*, even though it is the work of Vyāsa himself, is explicitly mentioned as having been told, and re-told, in Naimiṣa, that is to say in a *tīrtha*. Narrative activity is, time and time again, located in *tīrtha*s. It is worth noting here that *tīrtha*s, and particularly the Naimiṣa forest and Kurukṣetra, came to be the paradigmatic sites for narration in the Purāṇas (this includes, amongst others, the *Bhāgavata* and *Devībhāgavata*, which are narrated in Naimiṣa, while the *Vāyu* and *Skanda* are narrated in Kurukṣetra, while still others are narrated in the Himālaya, Puṣkara, Kailaśa or other *tīrtha*s).[12]

TĪRTHA AND RITUAL

The foregoing narratives have, amongst other things, emphasised the *tīrtha* as a site where great Vedic rituals are performed. The *tīrtha* is used both to commemorate specific ritual or ascetic acts in the past and, in some cases, to facilitate the performance of a great ritual act. I will take two further, representative, textual examples that illustrate each of these functions.

In the *Māndhātar-upākhyāna*, which is narrated as part of the Tīrthayātrāparvan, there is a fascinating narrative of a literal birth of a king from a king. King Yuvanāśva (an ancestor of

---

lomaharṣaṇaḥ || *śiṣyo vyāsasya medhāvī brāhmaṇair idam uktavān* || 7 || *tasmād aham upaśrutya pravakṣyāmi yathātatham* | *idam āstīkam ākhyānaṃ tubhyaṃ śaunaka pṛcchate* || 8 ||

[12] See Bonazzoli 1981 and 1979. Bonnazzoli comments that 'the places of purāṇic recitation indicated by the Purāṇas are something more than a mere geographical or topographical problem. They tend, rather, along with many other points, to form the structure or inner logic of the Purāṇas themselves. They deserve therefore a thoughtful and careful examination' (Bonazzoli 1981: 48). This paper stands as a response to this injunction. It should also be noted that the descendents of primary characters in the *Mahābhārata* become interlocutors in Purāṇic recitation. Thus Śuka, the son of Vyāsa, narrates to Parikṣit, the son of Abhimanyu and father of Janamejaya, in the *Bhāgavatapurāṇa* (and this narration takes its place within a further narrative frame of a *sūta*, or bard, and Śaunaka in the Naimiṣa forest), see *Bhāgavatapurāṇa*.1.3.40–44, cited in Doniger 1993: 35.

the Pāṇḍavas) gives birth to a king Māndhātar. King Māndhātar is described as performing great ritual acts at Kurukṣetra at the close, and culmination, of the tale:

> This is the place where he offered sacrifice to the *devas*, radiant like the sun. Behold it, in this auspicious place in the middle of Kurukṣetra. (3.126.42)[13]

The *tīrtha* here is, once again, a spatial marker for the events of the past. More than this, Kurukṣetra, in particular, is established as the site *par excellence* for the ritual activities of kings (we need think only of Yayāti and Bharata in this regard). Indeed, the culmination of Lomaśa's description of the narrative geography of the field of Kuru is the assertion that it is Prajāpati's main altar (*vedī prajāpateḥ*, 3.129.22).

My second text deals not just with the commemoration of ritual activity but also with its facilitation: The material is drawn from the Śalyaparvan and is concerned with the Naimiṣa and Kurukṣetra *tīrtha* and provides a foundation narrative for the *tīrtha* status of a certain portion of the Naimiṣa forest (*saptasārasvata*). These narrations form a part of the *tīrthayātrā* of Balarāma (the brother of Kṛṣṇa) and take up about one third of the entire *parvan*. In this text, Balarāma follows the Sarasvatī from estuary to source[14] as part of his journey to the mace duel of Bhīma and Duryodhana, a duel which marks the culmination of the book. The description of the creation of a *tīrtha* at the Naimiṣa forest is based upon the event of the turning of the Sarasvatī river in the first of the ages of humankind, the *kṛtayuga*. King Janamejaya asks:

---

[13] *tasyaitad devayajanaṃ sthānam ādityavarcasaḥ* | *paśya puṇyatame deśe kurukṣetrasya madhyataḥ* || 42 ||

[14] This is a feature of the *tīrthayātrās* that was to become stereotyped in later Māhātmya literature, for example in the *Godāvarī* and the *Kṛṣṇā Māhātmyas*.

> Why, O brāhmaṇa, did the Sarasvatī turn back (*nivṛttā*) east (*prāṅmukhī*)? O best of *adhvaryus*, I wish you to tell me all about this in detail. (9.36.37)[15]

Vaiśampāyana answers that the turning of the Sarasvatī was accomplished in order to provide a *tīrtha* for sages engaged in sacrifice in the Naimiṣa forest (9.36.50–53). There is specific mention of this sacrifice being a twelve-year *sattra*:

> Previously, in the *kṛta* age, O king, the ascetics dwelling in the Naimiṣa forest were engaged in a grand *sattra* lasting twelve years (9.36.39).[16]

This resonates with the twelve-year *sattra* of Śaunaka and even the *sarpasattra* of Janamejaya in which the *Mahābhārata* is itself narrated.[17] The turning of the Sarasvatī is accomplished at the close of this twelve-year sacrifice. Vedic ritual activity is here interrelated with the establishment of *tīrtha*s: *tīrtha*s which are located precisely in relation to the Sarasvatī river, a river which is closely associated with Vāc and Vedic knowledge in general (a connection which has been well established in the work of Hiltebeitel and Oberlies).[18] The narrative of the turning of the Sarasvatī closes with a further fascinating declaration:

> This *tīrtha*, O Indra of kings, is known to belong to Naimiṣa, in Kurukṣetra, O best of the Kurus, perform great ritual actions! (9.36.54)[19]

---

[15] *kasmāt sarasvatī brahman nivṛttā prāṅmukhī tataḥ* I *vyākhyātum etad icchāmi sarvam adhvaryusattama* II 37 II

[16] *pūrvaṃ kṛtayuge rājan naimiṣeyās tapasvinaḥ* I *vartamāne subahule satre dvādaśavārṣike* II 39 II

[17] The *sarpasattra*, or snake sacrifice, is a rite that is wholly unique to the *Mahābhārata*. The *sattra* itself is a Vedic rite in which there are twelve or more days of *soma*-pressing.

[18] Hiltebeitel 1998: 166–167. Hiltebeitel gives credit for the first recognition of this association to Thomas Oberlies; see Hiltebeitel 2001: 121, n. 90 and cf. Oberlies 1995.

[19] *evaṃ sa kuñjo rājendra naimiṣeya iti smṛtaḥ* I *kurukṣetre kuruśreṣṭha kuruṣva mahatīḥ kriyāḥ* II 54 II I take *kuñja* as a reference to the turning of the Sarasvatī the explicit topic of this narration, and the proper

Here it seems that Naimiṣa and Kurukṣetra are closely connected to one another. It would be unwise to suggest that this in any way fixes their locations (the location of the Naimiṣa forest, as Hiltebeitel has demonstrated, is somewhat fluid)[20] but it is of interest that, in the last of the war books of the *Mahābhārata*, there is an assertion that these two prominent locations are connected to a prominent trope of Vedic knowledge, the Sarasvatī. They are, in this way, established as sites which were founded in order to facilitate ritual activity and which are still the sites of great ritual acts. Kurukṣetra is also, of course, the site of the main war of the *Mahābhārata*, while Naimiṣa is the location of its narration. The two areas, by being interrelated, and by being connected to the Sarasvatī, are being granted a subtle sort of parity with each other and a rich sense of connection with the Vedic ritual order. This cannot but subtly aggrandise the *Mahābhārata* itself, both in terms of the events of its main narrative (which culminate at Kurukṣetra) and the act of its narration (which takes place in Naimiṣa). Indeed, as we shall see, *tīrtha* visitation and the recitation of *tīrtha* narratives are being constituted as functionally equivalent to Vedic ritual activity. The importance of *tīrtha* visitation is neatly encapsulated by the following blunt declaration:

---

name of the *tīrtha* there, as simply translating it as 'wilds' of one sort or another would make little sense here. The middle imperative form of √*kṛ*, apart from allowing a pun on Kuru, seems to particularly emphasise this being an exhortation to take on the role of *yajamāna* (patron) and not that of the officiant in a given sacrifice in that it emphasises the completion of the action for the sake of the performer. Indeed, the *sattra* is a ritual act performed for the benefit of the officiant and without *yajamāna*. Further evidence of the closeness of Naimiṣa and Kurukṣetra may be found at line 42 of the same text which places the sacrificing Ṛṣis, the best of the twice born (*dvijasattamāḥ*), on the bank of the river 'up to *samantapañcaka*' (*samantapañcakaṃ yāvat*) and of course in Ugraśravas' introduction to his telling of the *Mahābhārata*, cited above.

[20] See Hiltebeitel 1998 (this article argues for a literary 'convention' of the Naimiṣa forest which in some ways anticipates my wider discussion of *tīrthas*); Bhardwaj 1986: 130–134; Kane 1930–62, vol. 4: 783; and Witzel 1987a: 191.

> Previously in the *devayuga*, the age of the gods, I saw it all, O lord, the *devas* delighted in *dharma* and the *asuras* abandoned it. The *devas* visited *tīrthas*, O Bhārata, the *asuras* did not. (3.92.6–7ab)[21]

Here, what had been a conflict for ritual dominance between two phratries of divine beings is re-cast in terms of *tīrtha* visitation. The relationship between *tīrtha* and ritual is also formalised in the aforementioned equivalencies between a *tīrtha* and the performance of a given Vedic ritual: a visit to the *tīrtha* of Puṣkara, for example, is given as the equal of ten *aśvamedhas* (royal horse sacrifices), while a visit to Kurukṣetra has the equivalent sacrificial value of the two main sacrifices of the *Mahābhārata*'s main narrative, namely the *rājasūya* and the *aśvamedha.*[22]

The texts I have cited show, very clearly, how the *Mahābhārata* projects ritual practice into narrative accounts of the significant past, and in so doing transforms ritual into a form of history and a form of geography. Patton's perspective on the *Bṛhaddevatā*'s narrativisation of *mantra* is instructive here:

> ... the function of narrative as commentary is the opposite of the Eliadean escape from time. In the *itihāsa* explanations, *mantra* is inserted into the progression of events (one might say inserted into time) in order to provide a credible framework for its efficacy. (Patton 1996: xx)

In the *Mahābhārata*, it is chiefly the *yajña* that is 'inserted in time' and also in 'space'. Unlike the *Bṛhaddevatā*, however, the ritual 'framework for efficacy' is subsequently usurped by a range of radically different modes of religious practice, chief amongst these being both pilgrimage and, crucially, narrative activity itself (a point I will develop further below). We can see, very clearly, then, the development of *tīrtha* visitation as a productive form of religious practice. It is important, however, to

---

21 *purā devayuge caiva dṛṣṭaṃ sarvaṃ mayā vibho* | *arocayan surā dharmaṃ dharmaṃ tatyajire 'surāḥ* || 6 || *tīrthāni devā viviśur nāviśan bhāratāsurāḥ* |

22 Given at 3.80.48 and 3.81.6.

note the extent to which the description of *tīrtha*s in the *Mahābhārata* emphasises the movement *between tīrtha*s and not solely their visitation in isolation.

TĪRTHA AND MOVEMENT

The fact that one should always remain conscious of the connection between *tīrtha*s and the movement between them is made clear in the *Mahābhārata* from the following: In the Āraṇyakaparvan, before Yudhiṣṭhira embarks with his brothers on a tour of *tīrtha*s, he asks of the rewards or fruits (*phala*) of the *pṛthivī-pradakṣiṇā*s, the circumambulation of the earth by means of the clockwise tour of the world's *tīrtha*s:

> If a man makes a *pradakṣiṇa* tour of the earth to visit *tīrtha*s, what is his reward (*phala*)? You are obliged to speak of this in its entirety, O *brāhmaṇa*. (3.80.10)[23]

Nārada then cites the discourse of Pulastya and Bhīṣma on the value of a tour of the *tīrtha*s of the world. What is of key significance here is that this entire discourse is predicated not on the description of *tīrtha*s in isolation but on their description in relation to one another. This is something which is prominent throughout the *Mahābhārata*. Whenever *tīrtha*s in the *Mahābhārata* are described or visited it is as part of a tour. Ugraśravas is on such a tour when he meets the Bhārgava *brāhmaṇa*s sacrificing in the Naimiṣāraṇya at the commencement of the Ādiparvan, and still in the Ādi, Arjuna also embarks on a *tīrtha-yātrā* (1.1 and 1.206–210). The remaining Pāṇḍava brothers complete a tour of *tīrtha*s in the Āraṇyakaparvan, a princess Ambā[24] does so in the Udyogaparvan whilst, as we have already mentioned, Balarāma also does so in the Śalyaparvan (3.89–155;

---

[23] *pradakṣiṇaṃ yaḥ kurute pṛthivīṃ tīrthatatparaḥ* | *kiṃ phalaṃ tasya kārtsnyena tad brahman vaktum arhasi* || 10 ||

[24] Herself the subject of complex narrations culminating in her change of gender and slaying of Bhīṣma during the main *Mahābhārata* war.

5.187; 9.34–53). Indeed, the very close of the *Mahābhārata* is taken up with a *tīrthayātrā* that spans the last two books and which culminates in the giving up of life (the great departure, or *mahāprasthāna*) by the Pāṇḍavas and their common wife, Draupadī. In this way, the discussion of the significance of *tīrthas* is inseparable from the movement between *tīrthas*. Dhaumya, the chief religious officiant of the Pāṇḍavas, describes the *tīrthas* of the world in the order of the four directions, or cardinal points, just before the Pāṇḍavas embark on their own tour:

> Listen to me describe, O king, Bull of the Bharatas, the auspicious and *brāhmaṇa*-approved (*brāhmaṇānumatān*) *aśramas*, regions, *tīrthas*, and major mountain ranges (3.85.2).[25]

This parallels the conquest of the directions that is such a prominent feature of the Vedic royal consecration rituals. In the *rājasūya*, the conquest of the earth is expressed explicitly in terms of the four directions. In the *Mahābhārata*'s description of the *rājāsūya* of Yudhiṣṭhira, this conquest of the directions is described in *pradakṣiṇa* order:

> The son of the punisher of Pāka (Arjuna) conquered the direction beloved of the lord of wealth (Kubera the guardian of the north), Bhīma the east (*prācī*), Sahadeva the south (*dakṣiṇā*), Nakula the west (*pratīcī*) while Yudhiṣṭhira, king Dharma (*dharmarāja*), inhabited Khāṇḍavaprastha (2.23.9–10).[26]

The association of movement in relation to *yajña* and *tīrtha* is further developed in the text quoted above that dealt with the attitudes of the gods and *asuras* to *tīrtha* visitation. In the passage that follows this text, the *asuras* are described as perishing due to their failure to visit *tīrthas*. Lomaśa then makes the com-

---

[25] *brāhmaṇānumatān puṇyān āśramān bharatarṣabha* I *diśas tīrthāni śailāṃś ca śṛṇu me gadato nṛpa* II 2 II

[26] *diśaṃ dhanapater iṣṭām ajayat pākaśāsaniḥ* I *bhīmasenas tathā prācīṃ sahadevas tu dakṣiṇām* II 9 II *pratīcīṃ nakulo rājan diśaṃ vyajayad astravit* I *khāṇḍavaprastham adhyāste dharmarājo yudhiṣṭhiraḥ* II 10 II

parison explicit: the gods visited *tīrtha*s and prospered and so too shall the Pāṇḍavas if they likewise embark on a tour of auspicious places. In this way, the *tīrthayātrā* is emphasised as being as essential to the maintenance of prosperity, of a functional social order, as the *yajña* is in other contexts (such as the Brāhmaṇa literature and also, of course, within the *Mahābhārata*), and indeed both are predicated on a core opposition between two groups of empowered beings (be these phratries of gods or men) vying for, or failing to engage with, the means of the creation and maintenance of the cosmos (either *yajña* or *tīrtha*, the latter, of course, being predominantly established precisely to facilitate and commemorate great *yajña*).

There is something of a sequence of development here: It begins with the minutiae of ritual observances; the gestures of the ritual act stand as a micro-instantiation of the concern with empowered, generative, and personally and collectively rewarding movement.[27] This is further developed in the movements across the earth of the grand *śrauta* rituals of royal consecration. The movement between *tīrtha*s is an extension of this: All of these acts are also attached to 'litanies of benefit', a point I will take up and develop below.[28] These movements can, in turn, as

[27] Natalia Lidova (1995) makes a very interesting case for the close interrelationship of the post-Vedic myth of the churning of the elixir of immortality, the *amṛtamanthana*, with the minutae of Vedic ritual activity.

[28] It is also the case, of course, that *pradakṣiṇa* came later to describe the movement round a central shrine in a temple. This movement connects the internal features (which can often include visual narrative data) of the area being circumambulated. For details see Shulman & Handelman 1997: 9–10. The *tīrtha*s of the *Mahābhārata* adumbrate the cosmos much like both the temple and the movement of the devotee. Temples, indeed, are explicitly built on *tīrtha*s (also known as *kṣetra*) and mark a homological recreation, or model, of the cosmos (for details see Kramrisch 1946: 3) that parallels the *Mahābhārata* as something of a narrative model, and the Vedic *yajña* as a ritual model. The connections between these developments and the elaboration of the *yajñakṣetra*, 'the field of the *yajña*', in Vedic literature require further research. For more details regarding the *yajñakṣetra* see Minkowski 1991. It is also interesting to note the extent to which the degree of expansion of the sacrificial arena, such that it becomes co-extensive with the world

Patton puts it 'by a process of abstraction' (Patton 1996: 444), be replaced by alternate forms of religious activity. These may then vie with one another for dominance. This is especially clear in the way in which the *Mahābhārata* positions itself in relation to *tīrtha* visitation.

TĪRTHA AND THE *MAHĀBHĀRATA*

Returning to the title of my paper, 'What need has he of the waters of Puṣkara?', we can, if we translate a little more of the context of this statement, see that it is indeed narrative activity, and specifically the recitation of the *Mahābhārata*, that, in a third order act of revolutionary exegesis, is said to obviate *tīrtha* visitation:

---

itself, parallels the progressive elaboration of the *dharma* concept in post-Vedic discourse. Pollock states: '*dharma* ultimately spilled over the conceptual confines of "sacrificial ritualism" and came to encompass virtually the entire range of activities of Sanskrit society' (Pollock 1997: 398). The post-Vedic elaboration of the *dharma* concept (loosely translated as duty, law or meritorious action) is the ultimate expression of constructivist Vedic logic liberated from the confines of the sacrificial arena. Conformity to *dharma* (which operated on a vast number of hierarchically ordered and mutually resembling levels of being) became constitutive, in the post-Vedic period, of cosmic and social order in a way that precisely paralleled the earlier conception of the profound necessity of the performance of Vedic ritual for the sake of cosmic stability. Thus, by the time of the Dharmaśāstras, the role of empowered religious agent, has been extended (at least ideally) to all beings as an ongoing existential commitment to the stabilising practice of the performance of *dharma*. Hacker defines this conception of *dharma* as '... a concrete, positive ... model of conduct that has already existed before its realization in some way' (Hacker 1965: 103). This description of *dharma* resonates with the notion of *smṛti* as a process of exegetical recovery of *śruti* (this is Pollock's hypothesis, for details see fn. 34 below). The narrativised rituals of the *Mahābhārata* concretise, socialise, and provide an interpretive arena for the progressive 'dharmacisation' of early South Asian religious discourse by situating ritual activity in determinate cosmic, social and *spatial* contexts. For further details of the development, and elaboration, of the *dharma* concept see Olivelle 2004: 491–511, 423–448. It is also worth noting that the conquest of the directions was also to retain a political currency that extends to this day in South Asia (see Sax 2000).

> It once fell from the lips of Dvaipāyana, immeasurable, sanctifying, purifying, and blessing – what need has he of sprinkling with the waters of Puṣkara? (1.2.242 and 18.5.54)

In this way, the *Mahābhārata* presents itself as surpassing *tīrthas* and *tīrtha* visitation. These are of course forms of religious practice which themselves had been granted functional parity with the performance of Vedic ritual activities. The *Mahābhārata* does include within itself, however, *narratives* of *tīrtha* visitation. The value of these is made clear in the Āraṇyakaparvan where we find a text which states that the act of hearing about *tīrtha* is itself productive:

> This truth (*satyaṃ*) one should whisper (*japet*) into the ear of the twice born, of the good, of one's son, one's friends and one's pupil and one's follower. It is dharmic, it is auspicious, it is sacrificially pure (*medhyam*), it is pleasure, it is heavenly, it is lovely, it is the most purifying (*pāvanam uttamam*), it is the mystery (*guhyam*) of the great *ṛṣis*, it frees from all evil (*pāpa*). Having studied (*adhītya*) amidst the twice born one should obtain stainlessness (*nirmalatvam*). *He who should hear of the eternally auspicious tīrtha is always pure* (*śuciḥ*). He recalls for himself his many births and rejoices in the highest heaven (*nākapṛṣṭhe*). (3.83.84–87)[29]

Indeed, even the declaration of an intention to visit a *tīrtha* can be enough:

> All people who go there are freed from their evil (*pāpebhyaḥ*). He or she who should perpetually say (*brūyāt*), 'I will go to Kurukṣetra, I will live in Kurukṣetra' is freed from evil. (3.81.1c–2)[30]

---

[29] *idaṃ satyaṃ dvijātīnāṃ sādhūnām ātmajasya ca* | *suhṛdāṃ ca japet karṇe śiṣyasyānugatasya ca* || 84 || *idaṃ dharmyam idaṃ puṇyam idaṃ medhyam idaṃ sukham* | *idaṃ svargyam idaṃ ramyam idaṃ pāvanam uttamam* || 85 || *maharṣīṇām idaṃ guhyaṃ sarvapāpapramocanam* | *adhītya dvijamadhye ca nirmalatvam avāpnuyāt* || 86 || *yaś cedaṃ śṛṇuyān nityaṃ tīrthapuṇyaṃ sadā śuciḥ* | *jātīḥ sa smarate bahvīr nākapṛṣṭhe ca modate* || 87 ||

[30] *pāpebhyo vipramucyante tadgatāḥ sarvajantavaḥ* || 1 || *kurukṣetraṃ gamiṣyāmi kurukṣetre vasāmy aham* | *ya evaṃ satataṃ brūyāt so 'pi pāpaiḥ pramucyate* || 2 || The emphasis on the efficacy of even the desire to visit the field of Kuru is repeated at 3.81.173–178, where we also find another assertion that Kurukṣetra is Brahmā's main altar.

Thus, merely the expression of *tīrtha* as text, in an oral form, becomes equivalent to its visitation.[31] One should also remember that the Pāṇḍava brothers, in the Tīrthayātrāparvan, *hear* two *tīrthayātrās*, from Pulastya and Dhaumya, before embarking on a tour of their own. This tour is, as we have seen, a further narrative which is, ultimately, as we move through the various nested acts of narration that characterise the self-presentation of the *Mahābhārata*, being narrated in a *tīrtha* by an individual embarked on a tour of *tīrtha*s! Any statement about the benefit of the hearing of *tīrtha* narratives thus resonates across the text in its entirety. In this way, listening to *tīrtha* narratives becomes auspicious and purifying in a way that resonates with the more general declarations of *phalaśruti* ('the consequences of hearing') in both the *Mahābhārata* and subsequent genres of religious narrative.[32] This emphasis, on the impact of the reception of *tīrtha* narrations, can be seen, in addition, to emphasise the act of not just speaking but also listening, i.e. of participating in narrative activity. This itself constitutes a power dynamic: The *Mahābhārata* is privileging its own particular form of discourse whilst simultaneously seeking to take command of space and time.[33] It is exegetically encompassing those forms of religious thought and practice that precede it.

This is founded upon the establishment of an empowering relationship with Vedic ritual activity (the remembrance and facilitation of which, as we have seen, constitutes one of the prime causes of the establishment of a *tīrtha*). This is not only essential to the foundation and status of *tīrtha* but also to the

---

31 And this development once more resonates with Patton's view of the post-Vedic development of Mantra as portable, synedochic, ritual. Here, however, pilgrimage, is the empowered form of religious activity which is condensed and made portable. The interpretive strategy involved is, however, identical.

32 For detailed analysis in a Purāṇic context, see Bailey 1995.

33 As Inden (1990: 246ff.) said of the imperial Rāṣṭrakūtas, it is striving to give 'a better account of the world' than had its predecessors.

status of the *Mahābhārata* which, after all, presents itself as the fifth Veda.[34] This has parallels with wider exegetical developments in early South Asia: Vedic ritual activity, which begins as a form of concrete practice was transformed into more portable and multi-applicable formats, and more often than not by narrative means: this begins with the *arthavāda*s of the Brāhmaṇas, which Malamoud characterises as 'placing the ritual present in relationship to a mythical past' (Malamoud 1996: 29). While it is important to remember that these *arthavāda*s originated as a form of argument in order to encourage observance of the *vidhi*s or *codanā*s (the technical details of the sacrifice: the gestures, postures and formulae), they are, in fact, a form of elaborate justification and provide exaggerated litanies of benefit that resonate strongly with the content of the *Mahābhārata*. For example, from the fifth *adhyāya* of the *Kauṣītakibrāhmaṇa*:

> In that they lay to rest the sacrificer with his own fires, and the fires are the chariot of the gods, thus they lay him to rest on a chariot of the gods; he with this chariot of the gods goes to the world of heaven, which is the world of the good, which is the world of the good (*Kauṣītakibrāhmaṇa* 5.10.30–34).[35]

Compare this with the close of the *Mahābhārata*:

> I have told all that Vaiśampāyana narrated at the command of Vyāsa during the *sarpaṣattra* of King (Janamejaya) ... he who causes

[34] The *Mahābhārata*'s claim to Vedic status must be understood in relation to its understanding, and construction, of that status and as part of a complex strategy of exegetical encompassment. That is to say, a claim to Vedic status here indicates the intention to assume the creative function of these texts as an 'act of recovery'. Similarly, the *Mahābhārata*'s claim to equality, or superiority, to *tīrtha* visitation is dependent on the text's understanding of the nature and function of these sites. It is also a further act of exegetical encompassment. For details of my arguments concerning the nature and function of the *Mahābhārata*'s claim to Vedic status see my paper (in press) and my 2006 article. For details of the conception of *smṛti* as a form of exegetical recovery of *śruti* see Pollock 1997.

[35] *atha yat svair agnibhir yajamānaṃ saṃskurvanti* | *devaratho vā agnayaḥ* | *devaratha evainaṃ tat samāropayanti* | *sa etena devarathena svargaṃl lokam eti* | *sukṛtāṃ yatra lokaḥ sukṛtāṃ yatra lokaḥ* ||

> the learned to hear perpetually, day in and day out, this (narrative) destroys sin, conquers heaven and achieves the status of Brahma ... This *itihāsa* is auspicious and of great significance, the equal of the Veda (*vedasaṃmita*) ... that man (who hears the *Mahābhārata*) is freed of sin and achieves fame, O Śaunaka. (Mbh 18.5.30, 35, 43ab, 44ab.[36]

Malamoud comments: 'In the end, it is this husk of the *arthavāda*s that becomes the flesh and blood of the broader tradition.' This is because they contextualise ritual practice in terms of social and personal benefit and the events of the significant past. It should not be forgotten that any such empowered activity is also the subject of a competition for access to its resources (a fact which is given its most forceful expression in the *asura–deva* opposition). The *Mahābhārata*'s discourse on *tīrtha*s takes its place, then, as a post-Vedic extension of this process in which both pilgrimage, pilgrimage-narrations and, crucially, the *Mahābhārata* itself are constituted as forms of empowered religious practice. It may be as well, at this point, to step back from the text of the *Mahābhārata* and summarise my findings.

CONCLUSION: *TĪRTHAS*, THE *MAHĀBHĀRATA*, AND THE VEDIC CONCEPTUAL ORDER

In order to make my interpretation of the interpretive steps involved in the *Mahābhārata*'s discourse on *tīrtha*s absolutely clear, I will present its development as a logical sequence:

(1) *tīrtha*s are places associated with great, and often ritually focussed, events in the past.

---

[36] These are excerpts from a long sequence of *phalaśruti* given at the close of the *Mahābhārata*. These culminate, of course, in the closing rhetorical question of 'what need has he of the waters of Puṣkara?': *etat te sarvam ākhyātaṃ vaiśaṃpāyanaḳīrtitam* I *vyāsājñayā samākhyātaṃ sarpasatre nṛpasya ha* II 30 II ... *ya idaṃ śrāvayed vidvān sadā parvaṇi parvaṇi* I *dhūtapāpmā jitasvargo brahmabhūyāya gacchati* II 35 II ... *itihāsam imaṃ puṇyaṃ mahārthaṃ vedasaṃmitam* I ... *sa naraḥ pāpanirmuktaḥ kīrtiṃ prāpyeha śaunaka* I

(2) *tīrthas* are also associated with the act of the telling of these 'founding' narratives of the significant past.

(3) *tīrthas*, and the tour of *tīrthas*, are religiously empowered by their range of associations with great ritual activities and empowered heroic and divine agencies.

(4) *tīrtha* visitation is conceptualised as capable of functionally replacing those ritual activities.[37]

---

[37] The notion of functional replacement, of substitution or equivalence, is itself, of course, very much a feature of the philosophy of the Brāhmaṇas. In particular, a concept that is foundational to this sort of exegetical activity is the *bandhu*. Gonda argues that the concept of the *bandhu* encompasses both the knowledge of interconnection and the impact or power of that knowledge: 'Hence also the belief that there may exist a close relationship or correspondence between phenomenal reality or a province of the whole universe or a section made from it, for instance the sacrificial place, a *maṇḍala*, a sacred place or building. All these are centres of power.' (Gonda 1965: 5)

Ritual, in the Vedic conceptual order, is the primary means of ordering a universe which naturally tends toward dysfunction (For details see the work of B. K. Smith and A. Parpola, e.g. Smith 1998 and Parpola 1979). The efficacy of the ritual act is dependent on an interpretive competency typified by an understanding of the *bandhu*, or connection, between different spheres of activity. For example, the year is described as the *pratimā*, copy or image, of Prajāpati in the *Śatapathabrāhmaṇa* (11.1.6.13). Such a description of both connection and equivalency provides a conceptual foundation for symbolic manipulation that could very easily escape the ritual commentary into other forms of religious activity. Parpola emphasises the relation between the technical meaning of *bandhu* and its workaday meaning of 'kin or relation'; he suggests, developing Gonda's perspective, that this term indicates a connection that surpasses even the sense of genealogical relation and incorporates an 'intense consciousness of unity' (Parpola 1979: 151). As early as the *Ṛgveda*, we find evidence of the centrality of these forms of interpretive competence: In one *sukta*, the sages discover the 'secret connections' in their hearts (ṚV 10.129.4). In the *Bṛhadāraṇyaka-Upaniṣad* (3.9.19–20) it is 'by means of the heart that one knows the forms (of things)', *hṛdayena hi rūpāṇi jānāti*. This resonates with the celebrated Vedic conflict of Prajāpati and Mṛtyu in *Jaiminīyabrāhmaṇa* 2.69–70, in which Prajāpati is victorious, not by force of arms, but rather by means of his interpretive competency (which is expressed in terms of superior counting knowledge). This itself parallels a notion of hyper-numeracy in the *Mahābhārata* which is a necessary prerequisite for kingship, this competency is described as the *akṣahṛdaya*, the 'heart of the dice' (perhaps with something of a Vedic pun on both 'eye' and 'syllable', *akṣa* and *akṣara*, see Mbh 2.51.3; 3.70.23, 78.15), and is very clearly related to the dicing component of the Vedic *rājasūya* rite of royal consecration. For an examination of these issues see van Buitenen 1972;

(5) *tīrtha* narratives, and the *Mahābhārata* as a whole, are then conceptualised as being capable of themselves functionally replacing *tīrtha* visitation, which itself was capable of functionally replacing Vedic ritual activity.

Thus we have a sequence of exegetical manoeuvres in the *Mahābhārata* in which Vedic ritual activity is encompassed by pilgrimage practice which is itself encompassed by the narratives of the *Mahābhārata* itself. The *Mahābhārata* projects ritual into a world of narrative action in the significant past. It is on this basis that *tīrtha*s are founded. The text, however, then goes even further; it explicitly connects a form of discourse very like itself to *tīrtha* visitation and suggests the superiority of narrative activity to pilgrimage practice.[38]

Shulman 1993; and my own 2002 article. We can begin to discern a complex network of connections between the Ṛgvedic, Brāhmaṇa and *Mahābhārata* material. Gonda, perspicaciously, captures the constitution of 'alternate centres of power' in the quotation above. However, it is well worth exploring in detail the precise order of the development of the idea of the empowered status of given class of statements, images, practices or locations as well as the forms of interpretive competency associated with them. While the Brāhmaṇas may lay the conceptual foundation and perhaps even provide a framework for these developments, they do not delimit how subsequent genres of religious text manipulated these core ideas.

[38] In terms of granting narrative activity a parity with other forms of religious practice, in this case cosmogonic churning, consider for example the following (drawn from the context of the self-revelation of Nārāyaṇa in the Śāntiparvan):

'This narrative, O king, of the hundreds of other *upākhyāna*s that you have righteously heard from me, is the essence (*sāra*). In the past, O king, the *amṛta* was churned (*nirmathya*) and extracted by the *suras* (*deva*s) and *asuras* and likewise now, here, by the learned (*vipra*) a story-*amṛta* (*kathā-amṛta*) was extracted (*uddhṛta*).' (Mbh 12.326.114–115b) – *matto 'nyāni ca te rājann upākhyānaśatāni vai* | *yāni śrutāni dharmyāṇi teṣāṃ sāro 'yam uddhṛtaḥ* || 114 || *surāsurair yathā rājan nirmathyāmṛtam uddhṛtam* |

From a churning of the ocean to produce a functional cosmos, and an attendant battle between *deva*s and *asura*s, we come to a churning of text, in the here and now, concerned with a vast battle between two opposed groups of cousins. In this way, both in the context of *tīrtha* visitation and narrative activity, new forms of religious practice are related to those of the past and, very often, to the *deva–asura* conflict. This is, of course, the paradig-

In addition, the *Mahābhārata*, to an extent, 'inverts', or at least extends, the exegetic order of practice in the Vedas and Vedāṇgas. Both Jan Houben and Michael Witzel have analysed the extent to which the pragmatics of ritual practice had an impact on the form of their supporting, and even competing, mythologies. Witzel comments, for example:

> The *Brāhmaṇa* authors, indeed, had all the freedom to tell whatever story might appear plausible or appropriate to them in order to explain the problem in question ... One should not forget that they did so in constant competition with other ritual specialists or even with whole schools of specialists. Consequently, they had to come forward with ever new, more ingenious, or simply baffling explanations (Witzel 1987b: 406–407).

While Jan Houben suggests that ritual structure functioned as a 'laboratory of early speculative reflection' (Houben 2000), in the case of the *Mahābhārata* it is no longer the precise details of the ritual forms which stimulate narrative activity, but rather it is the ritually derived mythologies (and geographies) and, very often, further narratives of ritual performances in the human past that themselves provide the basis for further narrative elaborations. Thus we are dealing with second order processes of the exegesis of exegesis, in which earlier forms of narrative elaboration of ritual activity are the point of departure for new interpretations and new forms of practice.[39]

In this way, my analysis of the *Mahābhārata*'s discourse on *tīrthas* is intended to show how 'the destiny of the Veda' is not just, as Patton has termed it, 'a process of abstraction', or indeed

---

matic conflict for the means of ritual, and by extension social and political, re-production.

39 This is, I think, a far more serviceable characterisation of *aitihāsika* practice than Pollock's characterisation of it as 'a mode of explanation that viewed the Vedic texts as what they are, historical-cultural products' (Pollock 1990: 315). This seems far too close to a contemporary scholarly perspective. What we are, in fact, seeing in the *Mahābhārata* (from which we might dare to infer something of the hermeneutic methodologies of the lost *aitihāsikas*) is the treatment of not Vedic text but Vedic *practice* as historical-cultural product.

'internalisation', but also a process of progressive concretisation in alternate modes of religious practice.[40] This is achieved, in the *Mahābhārata*, by means of narratives which historically contextualise Vedic sacrificial activity as part of a deliberate exegetical agenda. The goal of this agenda is to assume both the creative capacities and the exalted status of the Vedas and any and all forms of practice, including *tīrtha* visitation, which are granted equivalent status.[41] It is in this way that the waters of Puṣkara are made redundant.

---

[40] Actually, the matter is more subtle even than this: what we are seeing are processes of abstraction followed by concretisation, sometimes followed by re-abstraction: for example, the Śāntiparvan at 255.39 makes explicit mention of an *ātmatīrtha*, that is to say a *tīrtha* of the self, thus internalising the *tīrtha*! In this way, a series of core exegetical strategies are repeatedly re-applied to their own output in South Asian religious discourse in a fashion which resonates with the explicitly formalistic, recursion-loving and rule-conscious stamp of forms of late Vedic exegesis. For classic expressions of the 'interiorization of the sacrifice' hypothesis in early South Asian religious discourse see chapter two 'Renoncement et intériorisation du sacrifice' in Biardeau & Malamoud 1976: 57–80 and chapter two 'Brahmin, Ritual and Renouncer' of Heesterman 1985, esp. pp. 26–44.

[41] A work to which I did not have access until the completion of this paper was Patton 2005. This work exhibits a wide range of parallels with my own orientation to the transformation and trans-creation of Vedic knowledge in subsequent South Asian religious discourse. The work focuses, in particular, on the development of mantric 'thinking', her central aim, of an examination of the 'wider context of metonymical thinking and the expansion of the Vedic associative imagination' (p. 30) could equally well stand as a summary of my analytic goals in relation to the *Mahābhārata*. The text as a whole seems to see the polysemy of the Vedic *mantras* as the enduring feature, and key practical and cognitive resource, of the Vedic knowledge system. This polysemy acts, for Patton, as a reservoir for metonymic extension. I see, rather, that it is the cross application of core competencies and the fulfilment of core functions that is the enduring feature of Vedic and para-Vedic religious discourse. In this view 'internalisation' and 'externalisation' of the Vedic ritual system (be this in the context of Upaniṣadic speculation or Dharmaśāstric codification) are simply different exegetical manoeuvres on a single hermeneutic continuum. Patton's conclusion, however, concerning 'intertextual metonymy' in which 'Vedic texts show different uses of resemblance for different exegetical purposes ... one text can refer to another, build on another, and yet use the same imagery for very different ends ... the intellectual operations of these kinds of texts thus become of interest in their own right ... making resemblances also involves making claims about the nature, function, and privilege of canonical texts and their authors' (see

REFERENCES

Sanskrit Works

Bṛhadāraṇyaka-Upaniṣad. In: V. Limaye & R. Vadekar (eds.), *Eighteen Principal Upaniṣads*. Poona: Vaidhika Saṃśodhana Maṇḍala, 1958.

Jaiminīyabrāhmaṇa. Ed. by Raghu Vira & Lokesh Chandra. (Sarasvati Vihara Series.) Nagpur: International Academy of Indian Culture, 1954.

Kauṣītakibrāhmaṇa. Ed. by E. R. Sreekrishna Sarma. (Verzeichnis der Orientalischen Handschriften in Deutschland.) Wiesbaden: Franz Steiner Verlag, 1968.

Mahābhārata, Critical Edition. Ed. by V. S. Sukthankar, S. K. Belvalkar, P. L. Vaidya *et al*. 19 vols. plus 6. vols. of indexes. Poona. Bhandarkar Oriental Research Institute, 1933–71.

Ṛgveda. Ed. by Max Müller. 4 vols. Varanasi: Chowkhamba Sanskrit Series, 1966.

Secondary Works

BAILEY, G. 1995. *Gaṇeśapurāṇa*, vol. 1: *Upāsanākhaṇḍa*. (Purāṇa Research Publications, Tübingen, 4.) Wiesbaden: Harrassowitz Verlag.

BHARDWAJ, O. P. 1986. *Studies in the Historical Geography of Ancient India*. Delhi: Sandeep Prakashan.

BIARDEAU, M. & C. MALAMOUD 1976. *Le sacrifice dans l'Inde ancienne* (Bibliothèque de l'École des Hautes Études, Sciences religieuses, 79.) Paris: Presses Universitaire de France.

BONAZZOLI, G. 1979. The dynamic canon of the Purāṇas. *Purāṇa* 21: 116–166.

—— 1981. Places of purāṇic recitation according to the Purāṇas. *Purāṇa* 23: 48–59.

BUITENEN, J. A. B. van 1972. On the structure of the Sabhāparvan of the Mahābhārata. In: J. Ensink & P. Gaffke (eds.), *India Maior*: 68–74. Leiden: E. J. Brill.

Patton 2005: 193–194) is as brilliant as it is programmatic, and my current paper stands as a small contribution to the furtherance of this analytic agenda in a post-Vedic context in the specific domain of *Mahābhārata* studies.

DONIGER, W. 1993. Echoes of the *Mahābhārata*: why is a parrot the narrator of the Bhāgavata Purāṇa and the Devībhāgavata Purāṇa. In: W. Doniger (ed.), *Purāṇa Perennis. Reciprocity and Transformation in Hindu and Jaina Texts*: 31–57, 259–262, 303–304. Albany: University of New York Press.

FITZGERALD, J. 2002. The Rāma Jāmadagnya 'thread' of the *Mahābhārata*: a new survey of Rāma Jāmadagnya in the Pune text. In: Mary Brockington (ed.), *Stages and Transitions: Temporal and Historical Frameworks in Epic and Purāṇic Literature. Proceedings of the Second Dubrovnik International Conference on the Sanskrit Epics and Purāṇas, August 1999*: 89–132. Zagreb: Croatian Academy of Sciences.

GOLDMAN, R. 1977. *Gods, Priests, and Warriors. The Bhṛgus of the Mahābhārata*. New York: Columbia University Press.

GONDA, J. 1965. *Bandhu* in the *Brāhmaṇas*. *Adyar Library Bulletin* 29: 1–29.

HACKER, P. 1965. Dharma im Hinduismus. *Zeitschrift für Missionwissenschaft und Religionswissenschaft* 49: 93–106.

HEESTERMAN, J. 1985. *The Inner Conflict of Tradition. Essays in Indian Ritual, Kingship and Society*. Chicago: University of Chicago Press.

HEGARTY, J. M. 2002. An apprenticeship in attentiveness: narrative patterning in the *Dyūtaparvan* and the *Nalopākhyāna* of the *Mahābhārata*. *Rocznik Orientalistyczny* 54: 33–62.

——— 2006. Extracting the *kathā-amṛta* (elixir of story): creation, ritual, sovereignty and textual structure in the Sanskrit *Mahābhārata*. *Journal of Vaishnava Studies* 14.2: 39–61.

——— (in press). Encompassing the sacrifice: on the narrative construction of the significant past in the Sanskrit *Mahābhārata*. *Acta Orientalia Vilnesia* 8.

HILTEBEITEL, A. 1998. Conventions of the Naimiṣa forest. *Journal of Indian Philosophy* 26: 161–171.

——— 2001. *Rethinking the Mahābhārata. A Reader's Guide to the Education of the Dharma King*. Chicago: University of Chicago Press.

HOUBEN, J. 2000. The ritual pragmatics of a Vedic hymn: the 'riddle hymn' and the Pravargya ritual. *Journal of the American Oriental Society* 120.4: 499–536.

INDEN, D. 1990. *Imagining India*. Oxford: Blackwell.

KANE, P. V. 1930–62. *History of Dharmaśāstra*. 5 vols. Pune: Bhandarkar Oriental Research Institute.

KRAMRISCH S. 1946. *The Hindu Temple*. 2 vols. Calcutta: University of Calcutta (reprint: Delhi: Motilal Barnarsidass, 1986).

LIDOVA, N. 1995. Amṛtamanthana: The Vedic sources of a Hindu creation myth. In: B. Saraswati (ed.), *Prakṛti. The Integral Vision*, vol. 5: *Man in Nature*: 130–139. New Delhi: Indira Gandhi National Centre for the Arts – D.K. Printworld.

MALAMOUD. C. 1996. *Cooking the World. Ritual and Thought in Ancient India*. Trans. by David G. White. Delhi: Oxford University Press.

MINKOWSKI, C. Z. 1991. *Priesthood in Ancient India. A Study of the Maitrāvaruṇa Priest*. Vienna: Sammlung De Nobili.

OBERLIES, T. 1995. Arjuna's Himmelreise und die Tīrthayātrā der Pāṇḍavas: Zur Struktur des Tīrthayātrāparvan des *Mahābhārata*. *Acta Orientalia* 66: 106–124.

OLIVELLE, P. 2004. The semantic history of dharma in the middle and late Vedic periods. *Journal of Indian Philosophy* 32.4: 91–511.

PARPOLA, A. 1979. On the symbol concept of the Vedic ritualists. In: H. Biezais (ed.), *Religious Symbols and their Functions* (Scripta Instituti Donneriani Aboensis, 10): 139–153. Stockholm: Almqvist & Wiksell.

PATTON, L. 1996. *Myth as Argument. The Bṛhaddevatā as Canonical Commentary*. Berlin: Walter de Gruyter.

——— 2005. *Bringing the Gods to Mind. Mantra and Ritual in Early Indian Sacrifice*. Berkeley: University of California Press.

POLLOCK, S. 1990. From discourse of ritual to discourse of power in Sanskrit culture. *Journal of Ritual Studies* 4: 315–345.

——— 1997. Tradition as revelation: *śruti*, *smṛti*, and the Sanskrit discourse of power. In: S. Lienhard & I. Piovano (eds.), *Lex et Litterae. Essays on Ancient Indian Law and Literature in Honour of Oscar Botto*: 395–417. Torino: Edizioni dell'Orso.

SAX, W. 2000. Conquering the quarters: religion and politics in Hinduism. *International Journal of Hindu Studies* 4.1: 39–60.

SHULMAN, D. 1993. Devana and daiva. In: A. W. Van Den Hoek *et al.* (eds.), *Ritual, State and History in South Asia. Essays in Honour of J. C. Heesterman*: 350–365. Leiden: Brill.

SHULMAN, D. & D. HANDELMAN 1997. *God Inside Out. Shiva's Game of Dice*. Oxford: Oxford University Press.

SMITH, B. K. 1998. *Reflections on Resemblance, Ritual and Religion*. Delhi: Motilal Banarsidass.

SUKTHANKAR, V. S. 1936. Epic Studies: VI. The Bhṛgus and the Bhārata: a text historical study. *Annals of the Bhandarkar Oriental Research Institute* 18: 1–76.

VASSILKOV, Y. 2002. Indian practices of pilgrimage and the growth of the *Mahābhārata*. In: Mary Brockington (ed.), *Stages and Transitions: Temporal and Historical Frameworks in Epic and Purāṇic Literature. Proceedings of the Second Dubrovnik International Conference on the Sanskrit Epics and Purāṇas, August 1999*: 133–157. Zagreb: Croatian Academy of Sciences.

WITZEL, M. 1987a. On the localisation of Vedic texts and schools (Materials on Vedic Śakhas, 7). In: G. Pollet (ed.), *India and the Ancient World. History, Trade and Culture Before A.D. 650.* Leuven: Department Orientalistiek.

——— 1987b. On the origin of the 'frame story' in old Indian literature. In: Harry Falk (ed.), *Hinduismus und Buddhismus. Festschrift für Ulrich Schneider*: 380–414. Freiburg: Hedwig Falk.

Mapping *bhakti* through Hospitality and Friendship in the Sanskrit Epics

ALF HILTEBEITEL

Recently, over several publications, James Fitzgerald and I have gone back and forth over the pros and cons of assigning lateness to portions of the *Mahābhārata*, each convinced of what Fitzgerald calls 'the potential value of a careful mapping of the thematic threads of the vast epic' (Fitzgerald 2006: 272), and each offering ways to reorient the discussion principally around the politics of the text. In a nutshell, our two practices can be contrasted as mapping *bhakti* out versus mapping it in. Fitzgerald makes a good point that the *Mahābhārata* poets have a political agenda centered on the vigorous promotion of *varṇadharma*, *svadharma*, a king empowered with the rod of punishment (*daṇḍa*), and the abhorrence of *varṇasaṃkara*, and that this agenda is intelligible in an immediate post-Aśokan setting (Fitzgerald 2006: 275–277). To my mind, however, what such a *bhakti* appendectomy gains in intelligibility it loses in credibility; rather, I believe the Pune Critical Edition offers a largely credible text.[1]

[1] See Fitzgerald 2003: 803, calling me 'A long-time advocate of the intentional unity of the MBh (more or less the Pune text) and, especially, a fierce defender of the divine Kṛṣṇa and Kṛṣṇa-*bhakti* in that text'; Fitzgerald 2006; and below footnote 28 for surgical details; Hiltebeitel 2004; 2005a; 2005b; 2005c; 2006a. See now Brodbeck 2006 on the textual question. I also treat the Baroda Critical Edition of the *Rāmāyaṇa* from a similar perspective while recognizing that it raises different generic issues (see Hiltebeitel

Nonetheless, I believe that Fitzgerald and others, including myself (1989; 2005d) and especially Madeleine Biardeau (2002), have been right to consider the *Mahābhārata*'s 'ideological and narrative grounding' (Fitzgerald 2003: 811), and more to the point, its treatment of *dharma* (including *varṇadharma* and *svadharma* in relation to *daṇḍa*-empowered kings and *varṇasaṃkara*), over and against a post-Aśokan setting in which Brahmanical culture has taken on the task of responding to the about-face provoked by Mauryan imperial rule and the rise of Buddhism and other heterodoxies. The *Rāmāyaṇa*, I believe, fine-tunes *dharma* on these same core *varṇa* and royal issues while projecting an earlier world where *varṇadharma* is more harmonious, where there is far less need to shore up matters of *svadharma* (the *Rāmāyaṇa* has only thirteen usages of the term), and where *varṇasaṃkara* does not really exist. The only two usages of *saṃkara* are in fact preemptive: one states that there was no class mixture during the reign of Rāma's father (1.6.12), and the other has Rāma announce that he will go into exile holding to the truth of his father's word lest 'under the cloak of righteousness (*dharmaveṣeṇa*) it would throw the world into confusion (*lokasaṃkaram*)' (2.101.6ab). Rāma's killing of the upstart Śūdra Śambūka is likewise a preemptive strike against a form of *varṇasaṃkara*: according to Nārada, who puts Rāma to this task, a Śūdra should only be able to perform *tapas* in the *kaliyuga* (7.65.22), an age yet to come. Whatever their differences, the two epics thus take much the same stand on *varṇadharma* and its implications for kings.

If, then, these same problems apply, mutatis mutandis, to both epics, the question to be raised is: what kind of cartography best suits this double terrain? Clearly, it will not just be a matter

---

2005a; 2009), and that it has also been approached from different perspectives on questions of strata and interpolation, positing Rāma's heroism as prior to his divinity (see most recently Brockington & Brockington 2006: ix–xii, 373–380).

of mapping *bhakti* terms (although we will note formations of *ava*-√*tṛ* and the term *bhakti* itself in some passages); nor can it be a matter of passages themselves, or even themes, which can always be set off as pericopes in the name of higher criticism. Though I cannot claim to have done so, one wants to get into the bones of these works. In any case, I have been trying to work this out in a book I am writing on *dharma*, and the first answer is rather obvious: that *bhakti* is to be mapped with *dharma*. I am working from a felicitous contrast made by Biardeau (2002, vol. 1: 85, 87) – that, whereas the *Mahābhārata* marks a *bhakti* 'swerve' (*écart*) in the Brahmanical tradition, Manu '"budges" as little as possible' (p. 96) in its allegiance to the Veda. For present purposes, leaving matters of dating these three texts aside,[2] I take the following view of the relation between them on these issues. First, all three texts express an allegiance to the Veda, but in markedly different ways: the *Mahābhārata* identifying itself as a 'fifth Veda' and filling itself with Vedic allusions; Manu listing Veda as the first and foremost of its four sources of *dharma* (*Mānavadharmaśāstra* 2.6); and the *Rāmāyaṇa* surrounding Rāma with a virtually Vedic world. Second, in contrast to the very ambiguous and capacious treatments of both *dharma* and *bhakti* in the *Mahābhārata*, Manu screens out *bhakti* while putting an orthodox stamp on *dharma*; and the *Rāmāyaṇa* streamlines and straightens out both *bhakti* and *dharma* around its figure of a royal perfect man living in a nearly perfect Vedic time. If so, this offers a way in which we may begin to talk about the politics of *bhakti* while mapping it in each epic.

There is much (though I cannot go into it here) about the politics of *bhakti* in the *Mahābhārata* that is without parallel in the more streamlined *Rāmāyaṇa*. But there are two interrelated

---

[2] See Hiltebeitel 2009 on debates over the priority of the *Mahābhārata* and *Rāmāyaṇa*, siding with those who favor the priority of the *Mahābhārata*.

practices or discourses where *dharma* and *bhakti* coincide that, to my mind, go to the heart (if not the bones) of what they are both about in each epic: hospitality and friendship. With hospitality, if we simply ask who hosts Kṛṣṇa and Rāma in these epics, who do they host in turn, how do they go about it, what are the issues involved, what is the tone or mood created, we get into revealing and, I would argue, indispensable material. And similarly with friendship.

On the last point of tone or mood, I would begin by calling attention to the trope of following or parting from Rāma or Kṛṣṇa and then keeping them in mind even after parting. The epics are sometimes very attentive to such departures, and to the demeanor and moods of those who follow them as far as they can, in ways that are not the case for the comings and goings of other 'characters'. Rāma's prolonged departure from Ayodhyā is perhaps well enough known to be recalled only in passing here, and in a few details below. But the *Mahābhārata* has similar descriptions of Kṛṣṇa's departures, of which it must suffice to cite only the last,[3] which comes after Kṛṣṇa has lingered longer than he wished (see Mbh 14.15.21) to help Arjuna remember the *Gītā* by telling him the *Anugītā*. Having mounted his chariot amid the city folk of Hāstinapura, having said his difficult goodbyes to his sister Subhadrā, his aunt Kuntī, the Pāṇḍavas, and Vidura, and bid them return to the city; and having told his charioteer Dāruka to 'urge the horses to speed'.

> While Vārṣṇeya was proceeding to Dvārakā, O Bharata bull, those foe-scorchers with their retinue, having embraced, turned back. Again and again Phalguna [Arjuna] embraced Vārṣṇeya and as long as he was in eye's range, he saw him again and again. And even so, Pārtha withdrew that sight fixed on Govinda with difficulty, and the unvanquished Kṛṣṇa [withdrew his sight with difficulty from Arjuna] as well. (Mbh 14.52.1–3)

[3] For others, cf. Mbh 2.2.1–23; 2.42.45–59; 5.81.6–57 (cited in part below).

In later texts this is clearly the *bhakti* trope of *viraha* or love in separation (see Hardy 1983), and it is worth pointing out that the first or second century CE *kāvya* poet Aśvaghoṣa, who knew both epics (see Hiltebeitel 2006b), already uses it in that fashion as an established idiom in the *Buddhacarita*'s portrayal of the Buddha's Great Departure. An overriding mood is set when the Brahmanical Ṛṣi Asita [Devala] (1.49–80) predicts this event, and the mood sets in when the prince's father contemplates the Buddha's first outing from the palace:

> Thereon the ruler of men, with tears in his eyes, gazed long at his son and kissed him on his head; and with his voice he bade him set forth, but out of affection he did not let him go in his mind. (3.7; Johnston 1936: 33)

Nor can friendship intervene (4.84–85) as when the prince's friend Udāyin tries to rekindle his interest in beautiful women. Finally, Aśvaghoṣa brings this mood to a climax in a *bhakti* set piece when the Bodhisattva's charioteer Chandaka, about to leave the prince behind in the forest, first recalls the corresponding *Rāmāyaṇa* scene and then behaves like a virtual *gopī*:

> I cannot leave you in the forest, as Sumantra did Rāghava, and go to the city with a burning heart (*Buddhacarita* 6.36; Johnston 1936: 85)

> Looking back once more, he wept aloud and clasped the horse, Kanthaka, with his arms. Then in despair he lamented again and again, and started for the city with his body, but not with his mind. Sometimes he brooded and sometimes he lamented, sometimes he stumbled and fell. So journeying in grief under the force of devotion (*ato vrajan bhaktivaśena duḥkhitaḥ*), he performed many actions on the road in complete abandon. (*Buddhacarita* 6.67–68; Johnston 1936: 91; see Hiltebeitel 2006b: 236 and n. 18)

A. PRACTICES, DISCOURSES, AND IDIOMS OF HOSPITALITY AND FRIENDSHIP

I am aware that with the exception of the friendship of Kṛṣṇa and Arjuna, hospitality and friendship are not the first things one thinks of regarding epic treatments of *dharma* and *bhakti*. But if a post-Mauryan Brahmanical ideology puts kingship front and center in both epics, as is also done in Manu, and may even be said to be what their arguments are ostensibly about, this is not, in any of these texts, the grounds on which the argument was capable of being won. How society is ordered is one thing. How people get along is another. That was encouraged by invoking hospitality and friendship among the more open and flexible civilizational discourses and practices (another is the gift) familiar as custom throughout South Asian Ārya culture under endless local and regional variations – discourses and practices that the Buddha himself relied on rather than criticized (as he did *varṇa*) and could do no better than attempt to refine.[4] In reformulating hospitality and friendship as *dharma*, the Dharmasūtras and Manu both harked back to Vedic practices and discourses, and so did the epics.[5] But the epics could give the same practices and discourses far more complex treatment by telling

[4] See Tsuchida 1991: 67–82 on hospitality practices related to the five daily *mahāyajñas*; *Dīghanikāya* 31 (*Sigālakasutta*) on friendship practices for lay householders (Bodhi 2005: 116–118).

[5] I am in working concurrence with Olivelle's post-Aśokan dating of the Dharmasūtras (Olivelle 2004; 2006), and do not consider the epics' understanding of *dharma* to be older than the Dharmasūtras. The problem arises, for instance, where Fitzgerald says that 'Aśoka preempted the brahmin monopoly on the teaching of *dharma*' (Fitzgerald 2006: 276). As Olivelle's careful reading (2004) of pre-Dharmasūtra Vedic and sūtra texts demonstrates, there is no evidence of Brahmans 'teaching *dharma*' or monopolizing that topic in them. Moreover, even in the Dharmasūtras the monopoly that Brahmans assert is not in teaching *dharma* but in the ritual sphere of teaching Vedic recitation (*adhyāpanam*), performing sacrifices, and receiving gifts – the three 'activities' (*karmas*) reserved for them (e.g., in what seems to be the earliest formulation of this job reservation scheme, *Āpastambadharmasūtra* 2.10.4–7; cf. Mbh 5.29.26; 7.168.22–23; 12.297.15; 13.129.7–8).

stories, among others, about a god among men. In making this 'swerve', they could develop these practices and discourses in narratives that were far more nuanced than incessant top-heavy reminders that the four social classes were created from Puruṣa.

For the most part, the basic vocabularies on friendship and hospitality are shared by the epics and the *dharma* literature. But the epics also innovate and archaize. Several Sanskrit words are often translated by the English word 'friend', three of which, for our purposes, are important to differentiate. I will translate *sakhi*, the most prominent narrative term, as 'friend', and its derivative *sakhya* as friendship. But it is important to note that *Mahābhārata* usages, especially, draw on Vedic precedents in giving this term two senses, which I call 'pact friendship' as when Indra makes pacts with middle Vedic demons such as Vṛtra and Namuci, and 'intimate friendship', as when Indra is helped to defeat Vṛtra by his 'intimate friend' Viṣṇu (*índrasya yújya sákhā*) in *Ṛgveda* 1.22.19. *Mitra*, another Vedic term, has a primary meaning of 'ally', but can also sometimes be translated more usefully as 'friend'. And *suhṛd*, which seems to be an epic coinage, literally means 'good-hearted', 'one with a good heart', and will be translated as 'wellwisher', with all the ambiguity that the term can suggest, while its derivative *sauharda / sauhārda* will be translated 'goodhearted'.[6] Meanwhile, as to hospitality terms, *atithi* is the main old word for guest, and *ātithyam* for hospitality, while there is no consistent term for host, that concept being more contextual. A host may be found in a house, a sacrifice, a performance, perhaps on a chariot, in a heart, etc. A few pre-epic usages are also interesting here. I believe there is something to be made of the fact that Viṣṇu is 'the "guest" par excellence' in the *ātithyeṣṭi*, the *ātithya* or 'guest offering rites' mentioned in the *Śatapathabrāhmaṇa* (3.4.1.1) and *Āpastamba-śrautasūtra* (10.30.1–31, 31.6–7), which 'call the gods to mind',

[6] For discussion of these terms, see Hiltebeitel 2010.

and to which Viṣṇu is 'invited' by the *acchāvāka* or 'inviter' priest, a deputy of the *hotṛ*, by reciting *Ṛgveda* 1.154.1–3, praising Viṣṇu for 'traversing three times', he whose 'power ... is like a terrifying, hungry, wild animal who dwells in the mountains (or in speech), the one of many hymns' in whose 'three steps, all worlds abide' (trans. Patton 2005: 151). Used in simple *iṣṭi* or cake-offering rites, the *ātithyeṣṭi* can also be used 'for the reception of Soma stalks, which are considered kingly, or a royal guest'. Laurie Patton notes that the verses, according to the *Śaṅkhyāyanaśrautasūtra*, are also used during the cutting of the victim for Viṣṇu in the animal sacrifice (*paśuyajña*), while according to the *Ṛgvidhāna* (1.136–137) they are to be recited for the adept 'to attain *dharma*, intelligence (*jñānam*), sons, the increase of Brahmā, and the highest abode of eternal light' (Patton 2005: 9, 171–173). Also on guests, and suggestive in the same vein, is *Āpastambadharmasūtra* 2.7.5–10:

> 'Whether you hold them dear or not', it is stated, 'guests lead you to heaven'. When a man gives food in the morning, at noon, and in the evening, they constitute the three pressings of Soma; when he rises as his guest gets up to leave, it constitutes the final rite of the Soma sacrifice; when he addresses the guest with kind words, it constitutes the praise of the priestly fee; when he follows the guest as he leaves, it constitutes the Viṣṇu steps; and when he returns, it constitutes the final bath. (Trans. Olivelle 1999: 50).

It is interesting, in the light of the passages cited earlier on Kṛṣṇa's and Rāma's departures, that following the guest until he leaves should 'constitute the Viṣṇu steps'. On epic hospitality more widely, however, I would like to cite as a sort of touchstone-text a surprising unit from the *Mahābhārata*'s 13th Book, the Anuśāsanaparvan.

B. *ṚṢIDHARMA*

This is the rather expansive 'Dialogue Between Umā and Maheśvara' (*Umāmaheśvarasaṃvāda*, Mbh 13.126–134 [henceforth UMS]) that comes at one of those points of increasing 'bonhomie' (as Adam Bowles [2004: 294, 296] has put it) that can be traced through the entire arc of Yudhiṣṭhira's postwar instruction, and also one that could have something to do with a staggered but cumulatively increasing emphasis on devotion through that four-part instruction.[7] The UMS begins with Yudhiṣṭhira wanting to know more about Kṛṣṇa who has up to now been standing by, just listening and witnessing, ever since Bhīṣma began his teachings on *rājadharma*. Says Yudhiṣṭhira:

> If I am favored by you along with my brothers, O sinless one, you can speak to our question that I ask you, O king. This Nārāyaṇa, possessed of *śrī*, is esteemed by all kings, and even he waits upon you with much respect and deference (*bahumānena praśrayeṇa ca*). In his presence, you can speak out of affection for me and for the benefit of all these lords of the earth, my brothers. (13.126.4–6)

Bhīṣma is overjoyed to speak on this subject, and tells one story about Kṛṣṇa that leads him to a second in which he quotes Nārada to narrate the actual UMS. As one would expect, there are threads that link the two stories. The first (13.126.10–50) tells how Kṛṣṇa once undertook a twelve-year vow at some mountain to obtain a son who would be his equal in vigor (*vīrya*) and would have half the splendor (*tejas*) of Śiva.[8] Upon completing the con-

---

[7] A possibility Zaehner (1963) and Brockington (2000) have raised for the Mokṣadharma, with its *Nārāyaṇīya* toward the end of that third of four Śāntiparvan anthologies. Indeed, soon after the UMS, Bhīṣma, claiming that he is losing strength, directs Yudhiṣṭhira to ask a question about Brahmans to Kṛṣṇa himself, which turns into further questions about the irascible Ṛṣi Durvāsas that lead Kṛṣṇa to speak also about the greatness of Śiva (13.143–146) – before the dialogue returns for its wind-up with Bhīṣma.

[8] Mbh 13.126.33 and 45. It is not immediately clear who this son would be. As one learns in the *Upamanyu-Upākhyāna* (13.14–18), Jāmbavatī once reminded Kṛṣṇa at the end of twelve years that she wished him to obtain an heroic son for her 'like himself' (*ātmatulya*) by doing something similar to the twelve year vow to Śiva he had done to obtain the eight sons of Rukmiṇī

secration (*dīkṣā*), he was visited by Nārada, Vyāsa and other important sages – a veritable *ṛṣisaṅgha* (49a), in fact – and received them with rites of hospitality (*atithisatkāram*) like those offered to the gods (13). Once he had seated them on gorgeous seats, they chatted amiably on matters of *dharma* until Kṛṣṇa released from his mouth a fire that burnt the entire mountain with its living creatures and came back to him like a docile disciple, after which he brought everything back to life. Noticing that his well-travelled visitors were surprised (*vismito*) and given to dismay (*vismaya*) at this inconceivable wonder (*adbhutam acintyam*), Kṛṣṇa explained that the fire contained his wrath and was his Vaiṣṇava *tejas* or 'energy of Viṣṇu'. Then he asked them to tell him something highly wonderful (*āścaryam paramaṃ kiṃcit*) that they had seen or heard on earth or in heaven, even while admitting that every such wonder is known to him, since what is 'told by the pious and heard by the good is surely beneficial and lasts long on earth, like writing engraved on a mountain' (43)! Once the *ṛṣisaṅgha* is finished glorifying (*vardhayantas*) and worshiping (*pūjayantas*) Madhusūdana with hymns of praise (46), the sages appoint Nārada to tell the sequel.

Nārada, now called 'the well-wisher of Nārāyaṇa' (*nārāyaṇa-suhṛd*, 13.127.1a), then leads us into an idyll of Śiva and Umā on Mount Himavat (1–39) that is interrupted when Umā covers Mahādeva's eyes in jest (26), leaving the world in darkness and distress until Śiva's third eye (30) opens and burns the mountain – which Umā then gets him to restore to its natural beauty. This inspires Umā to ask a series of questions on topics that intrigue her: first Śiva's third eye, then his four faces, his bull, and his preference for the crematorium, which he explains by the delight he takes there with his ghostly host (*bhūtasaṅgha*, *bhūtagaṇa*,

(13.14.12–16). Jāmbavatī, a relative of Jāmbavat, the king of bears in the *Rāmāyaṇa*, is here 'the daughter of the Indra of monkeys' (*kapīndraputrī*, 13.14.24b), and their son, will be Sāmba, who would seem to be the likeliest candidate. Rukmiṇī's eight sons with Kṛṣṇa are listed here at verse 17.

128.18) – that leads Umā to ask next, for the benefit of the hosts of Ṛṣis and Munis (*ṛṣisaṅgha, munigaṇa*) who attend them on Mount Himavat, what *dharma* is and how men learn of it (128.21–23), and then to ask about the *dharma* of the four social classes, and particularly that of Brahmans (128.29a). Śiva, who has just stated his rather capacious view of *dharma*'s many branches (*dharmo bahuśākhaḥ*; 128.27), focuses his reply on householder Brahmans as 'gods on earth' (*bhūmidevāḥ*), for whom several of his prescriptions involve hospitality practices – honoring one's gurus and gods, keeping houses properly rubbed (with cow-dung and water) and fumigated with the smoke of clarified butter (128.30–34, 35–45) – related to the five daily *mahāyajñas* underscored in the Gṛhyasūtras and further developed especially as *dharma* in the Dharmasūtras and Manu. After hearing more about householders, however, Umā wants to know about the *dharma* of Ṛsis or Munis themselves (using the compounds *ṛṣidharma* and *munidharma* at 129.32 and 34). Śiva, who delights not only in his hosts of ghosts but praises outlandish ascetics, now uses the verbal root √*uñch*, 'to glean', several times to describe different types of 'gleaners': *phenapas*, 'foam-drinkers', who as it were 'glean' or perhaps better 'skim' their nourishment from the foam of waters left over from sacrifices (129.36–38b); *cakracaras* who practice the *dharma* of compassion (*dayādharma*), rove in the world of the gods (*devaloka*), and 'glean' their food from moonbeams (129.43–44, 48); others who live with their wives, practice the five *mahāyajñas*, and 'glean' their nourishment from the flames of their fires; and thumb-sized *vālakhilyas* who live in the solar disc, wear deer skins or tree bark, follow the actual *uñcha* or 'gleaning' mode of life by subsisting like birds on grains left in the fields (*uñcham uñchanti dharmajñāḥ śākunīṃ vṛttim āsthitāḥ*), and 'attain equity with the gods in accomplishing the purpose of the gods' work' (*te suraiḥ samatāṃ yānti surakāryārthasiddhaye*) (129.39c–42b). It is with reference to this varied 'gleaning' ideal that Śiva then summarizes the *dharma* of Ṛṣis and Munis in the passage that should most interest us:

> In all the *ṛṣidharmas*, selves are to be conquered, sense faculties are conquered. Afterwards, desire and wrath are then to be conquered, so I think. Maintaining the *agnihotra*, sitting for the *dharmarātri*,[9] Soma sacrifice, assenting to approval, and [offering] honoraria at sacrifices as the fifth, [plus] the daily *dharma* of sacrificial acts: pleasure in the worship of Pitṛs and gods and hospitality to all – [these] are to be done with food acquired by gleaning. Turning away from food prepared with cows' milk and taking pleasure in lying on the bare ground, yoga, enjoying vegetables and leaves, eating fruits and roots, partaking of wind, water, and duck weed – these are some observances of the Ṛṣis by which they conquer the way of the unsubjugated. When there is no more smoke, when the pestle is set down, when there are no more coals, when the people have eaten their meal, when the handing around of vessels is over, when the time for asking alms has passed by, surely [it is then, still] longing for a guest, [that] one eats the food left over. Delighted by the *dharma* of truth, patient, he is yoked to the *munidharma*. Not arrogant or proud, the one who is neither heedless nor surprised (*vismita*), a friend alike to friend and foe, he is the foremost knower of *dharma*. (Mbh 13.129.48–55)[10]

---

[9] I do not know what this is. Ganguli (1970, vol. 11: 296) has: 'occupying a fixed seat employing oneself the while in the sacrifice called Dharmaratri', which hardly tells us what it is. A nightlong vow is probably suggested. Though a Bengali, he seems not to have been tempted to relate it to a Dharmagajan. For recent discussion of this Dharma cult, with bibliography, see Korom 2004.

[10] *sarveṣu evarṣidharmeṣu jeyātmā jitendriyaḥ* I *kāmakrodhau tataḥ paścāj jetavyāv iti me matiḥ* II 48 II *agnihotraparispando dharmarātrisamāsanam* I *somayajñābhyanujñānaṃ pañcamī yajñadakṣiṇā* II 49 II *nityaṃ yajñakriyādharmaḥ pitṛdevārcane ratiḥ* I *sarvātithyaṃ ca kartavyam annenoñchārjitena vai* II 50 II *nivṛttir upabhogasya gorasānāṃ ca vai ratiḥ* I *sthaṇḍile śayanaṃ yogaḥ śākaparṇaniṣevaṇam* II 51 II *phalamūlāśanaṃ vāyur āpaḥ śaivalabhakṣaṇam* I *ṛṣīṇāṃ niyamā hy ete yair jayanty ajitāṃ gatim* II 52 II *vidhūme nyastamusale vyaṅgāre bhuktavajjane* I *atītapātrasaṃcāre kāle vigatabhaikṣake* II 53 II *atithiṃ kāṅkṣamāṇo vai śeṣānnakṛtabhojanaḥ* I *satyadharmaratiḥ kṣanto munidharmeṇa yujyate* II 54 II *na stambhī na ca mānī yo na pramatto na vismitaḥ* I *mitrāmitra samo maitro yaḥ sa dharmavid uttamaḥ* II 55 II

For the most evocative lines in verses 53–54b, Ganguli (1970, vol. 11: 297) has this lovely evocation of a life he surely knew better than I: 'When the smoke has ceased to curl upwards from a house, when the husking machine has ceased to ply, when all the inmates have taken their food, when

The Himalayan setting, where Umā and Śiva are surrounded not only by their own 'hosts' of Bhūtas but – among other celestials – by hosts or crowds of sages, would make them likely to have a shared interest in what they call *ṛṣidharma*. But the term was a surprise to me, as I had never even considered that there was such a thing. One does not hear about *ṛṣidharma* or *munidharma* from Manu, even though the celestial sages form the first audience of Manu's Dharmaśāstra. Nor is it the sort of thing one would expect to hear about from Kṛṣṇa, and even if Yudhiṣṭhira's next questions are about the One God, whom Bhīṣma identifies as Vāsudeva, and about the highest *dharma* (13.135.2–3), which turns out to be the recitation of Viṣṇu's thousand names (the *Viṣṇusahasranāmastotra*), it shows the value of having more than one god's opinion about *dharma*. Indeed, Śiva will round off his dialogue with Umā by getting her reply to *his* question about the *dharma* of women (134.1–29).

Yet even though the term is unusual in the *Mahābhārata*[11] and not found in the *Rāmāyaṇa*, *ṛṣidharma* is all over the place in both epics, and has much to do in them not only with *dharma* but *bhakti* – and, as the quote from Śiva would suggest, with both hospitality and friendship. Think of all the Ṛṣis the heroes meet. Let us begin by noting some of the ways that this can be appreciated straight from this passage. Most crucial is the very idea that there is a *ṛṣidharma* or *munidharma* that is of interest to such a grand audience as a *ṛṣisaṅgha* or *munigaṇa*. I will first

---

dishes are no longer carried from room to room, when mendicants have ceased to walk the streets, it is then that a man desiring to have a guest (but finding his desire ungratified), should eat what remnant of food may still occur in his house.' Verse 53 may partly echo *Baudhāyanadharmasūtra* 2.11.22.

[11] Of nine usages of *ṛṣi-* (5) and *munidharma* (4), all but two are in our current dialogue, and all occur in Bhīṣma's postwar instructions – the other two at 12.61.10, where Bhīṣma speaks of 'the very difficult householder *dharma* seen as *munidharma*', and at 12.185.1–2, where he describes 'forest dwellers (*vānaprasthas*) who follow the *ṛṣidharma*' (1[1], in prose) as living in the wilds while still carrying on their daily rites.

deal with practitioners of *ṛṣidharma* – at least as Śiva describes them – and then turn to the topic of *ṛṣisaṅgha* audiences.

It seems that despite their elemental modes of subsistence and access to higher worlds, the gleaners of the various types who practice *ṛṣi-* or *munidharma* still lead earthly lives, and would *not* be among those in the *ṛṣisaṅgha*s or *munigaṇa*s who attend Umā and Śiva's conversation or come in conclave to honor Kṛṣṇa. As one can gather even from Manu, the gleaner's mode of life in some sense surpasses the categories of *varṇa-* and *āśrama-dharma*. For one thing, it has nothing to do with *varṇasaṃkara*. Gleaners do not belong in a statement about Brahmans 'forced' by *varṇasaṃkara*

> to subsist by performing military roles, shopkeeping or even engaging in agriculture – or they are reduced to 'gleaning' and often starve. The king who allows such degradations to occur is said to be seriously deficient' (Fitzgerald 2006: 275).

As Manu says, although a king 'may eke out a living by gleaning (*śīloñchena*), his fame spreads in the world like a drop of oil on water' (7.33). But as Manu indicates,[12] it is mainly Brahmans who would have this high ideal: that of a sort of liminal householder life somewhere between the second and third *āśrama*, enjoined to perform the five *mahāyajña*s but with what seems the absolute minimum that its practitioners might offer to their own families, not to mention guests. Among those who are mentioned by Śiva, the foam-, moonbeam-, and fire-gleaners could be forest-dwellers, but the actual grain-gleaners would have to live near agricultural fields even if they are dressed in leaves and tree-bark. But they all 'attain equity with the gods in accomplishing the purpose of the gods' work'. It is indeed a high ideal

---

[12] See *Mānavadharmaśāstra* 4.7 ('the length of time one stores up grain in advance shows the superiority of those who store the least'); cf. 3.100; 4.4–10 (limiting the daily sacrifices which, according to 6.5, the forest hermit should perform).

in the *Mahābhārata*, where it is exalted in such terms as these to Yudhiṣṭhira in several *upākhyāna*s or 'substories', where a god in disguise (Dharma), or a saint (that irascible clone of Śiva, Durvāsas), might come as a hungry guest to test its practitioner. Yet one notices that Yudhiṣṭhira and his companions only hear about such gleaners and never meet one, especially in the forest. Clearly the ideal is precisely this: to live out daily, as Śiva says, the challenges of hospitality and friendship to all with patience in a world of marvels, indeed, where one is oneself one of the marvels, but as one who is surprised by nothing.[13] It remains a living ideal in India, for as I learned from my friend T. P. Mahadevan, who grew up in Kerala, he remembers that, whether his mother had ever seen one or not, she thought there was no one finer than an *uñchavṛtti* Brahman who (quoting Thennilapuram Mahadevan, e-mail, July 7, 2006) lived a life beyond 'the taint of the *dakṣiṇā* that regular Vaidika Brahmans had to endure' and could rise 'above it all, a liberated soul squaring the circle, as it were, in the world but out of it'.

As to the audiences of more celestial Ṛṣis and Munis such as the two *ṛṣisaṅgha*s that attend to Kṛṣṇa and to Umā and Śiva, here we meet variants of the widest audiences of the two epics and Manu: in the *Mahābhārata*, the Naimiṣa Forest Ṛṣis who hear the epic from the bard Ugraśravas in its outer frame; the celestial Ṛṣis who come *to* the Naimiṣa Forest from Brahmaloka with Brahmā's permission to hear the rest of the *Rāmāyaṇa* from Kuśa and Lava once Sītā has descended into the earth;[14] and the celestial seers who listen to Manu's son Bhṛgu recite the *Manusmṛti*. We also find another variation in the last verses of the

---

[13] At Upamanyu's *āśrama*, for instance, 'Mongooses sported with snakes, and tigers with the deer, like friends' (Mbh 13.14.42ab).

[14] Rām 7, Appendix I, no. 13, lines 21–49. Although the Baroda Critical Edition rejects this *sarga*, it does so only on the grounds that without it 'the continuity of the narration ... is not hampered and appears in a better order' (Shah 1975: 29).

Śvetāśvatara-Upaniṣad, a text whose chronological relation to the epics and Manu is perhaps too close to call:[15]

> By the power of his austerities and by the grace of God (*devaprasādād*), the wise Śvetāśvatara first came to know *brahman* and then proclaimed it to those who had passed beyond their order of life (*atyāśramibhyaḥ*) as the highest means of purification that brings delight (*juṣṭam*) to the company of seers (*ṛṣisaṃgha*).
>
> This supreme secret was proclaimed during a former age (*purākalpe*) in the Vedānta. One should not disclose it to a person who is not of a tranquil disposition, or who is not one's son or pupil.
>
> Only in a man who has the deepest love for god (*yasya deve parā bhaktir*), and who shows the same love for his teacher as towards God, do these points (*arthāḥ*) declared by the Noble One shine forth (*prakaśante mahātmanaḥ*, which is repeated). (*Śvetāśvatara-Upaniṣad* 6.21–23; Olivelle 1998: 422–423)

Here we meet the term *ṛṣisaṅgha* explicitly, and again with some differentiation made between this 'company of seers' and those in whom they delight who have 'passed beyond their order of life as the highest means of purification'.[16]

So what are these *ṛṣisaṅgha* audiences that take such delights: in this Upaniṣad, a delight in those who have 'passed beyond their order[s] of life as the highest means of purification'

---

[15] I would not go as far as Oberlies, who argues that the *Śvetāśvatara* is from around 0–200 AD (Oberlies 1997: 86; 1988: 57–59), though he may be right (even if I doubt it) that it is younger than the *Bhagavadgītā*. Cf. Olivelle 1996: 252: 'Its thought and vocabulary are close to that other theistic document, the *Bhagavad Gītā*.'

[16] Given the liminal state of certain highly purified gleaners, I am not sure that Olivelle is right to discount the assumption of 'many scholars ... that the expression *atyāśramibhyaj* ... refers to those who had passed beyond the four stages of life known as the *āśramas*', maintaining himself, on the contrary, that '*āśrama* here means just the householder life and the expression refers to ascetics who have moved beyond the householder life' (Olivelle 1998: 628 n.). As Olivelle indicates, there *are* instances where term *atyāśramin* does hold 'the possibility of living beyond the *āśramas*' in later texts. In any case, the closing *Śvetāśvatara-Upaniṣad* passage makes the differentiation mentioned, and, I would suggest, might provide an early instance of this later *āśrama*-transcending meaning by its *bhakti* context, which is similar to that in the UMS – for one thing, in stressing 'the highest means of purification'.

from receiving a 'supreme secret proclaimed during a former *kalpa*' that now 'shines forth' to them after it was disclosed as *brahman* to Śvetāśvatara by the grace of God; in our epic examples and Manu, a delight in hearing and talking about *dharma*; and in our current example from the *Mahābhārata*, a delight in stories that provoke wonder and surprise? What is interesting is how *bhakti* texts do something *bhakti* is famous for: constructing community, and with it audience. I think Manu shows familiarity with this novel textual practice, without mentioning the term *saṅgha*, to give his unbudging orthodoxy some similar kind of constituency. But the term has this cachet elsewhere. I need hardly mention that the use of the term *ṛṣisaṅgha* would have counterparts in the wider sense of the Buddhist second refuge 'in the Sangha', which can also be an audience for texts, and in the use of the same term for the first audiences of Tamil literature.

These universal listeners are not only the greatest of sages, enough of them Vedic to make that part of their aura, but a model audience that listens for and with a carefully constructed 'us'. Indeed, in very broad terms, we can say that they are the hosts of the texts, and in the *Mahābhārata*, sometimes the hosts of subunits such as the UMS, and of course not just hosts of the texts but of their readers since they set the model for receiving the texts into human lives and homes and open questions of interpretation just by their overhearing presence, not to mention specific interventions where they 'do the gods' work'.[17] In the UMS, the first thing to notice is that first Kṛṣṇa and then Śiva and Umā are hosts to two successive *ṛṣisaṅghas*. In the first case, Kṛṣṇa registers the *ṛṣisaṅgha*'s surprise at the miracle of the fire from his mouth, and encourages their telling him the

[17] See notably the intervention of the Ṛṣis (some of them, at least, celestial) of Mount Śataśṛṅga in Mbh 1.110.45–117.33, who clearly know about the secret mantra Kuntī has gotten from Durvāsas, and do the gods' work in facilitating the Pāṇḍavas' births and introducing them to the Kurus of Hāstinapura.

most wondrous thing they have seen, since a story 'told by the pious and heard by the good is surely beneficial and lasts long on earth, like writing engraved on a mountain' (*śaile lekhyam ivārpitam*; 43d) – an interesting juxtaposition of oral and written modes of communication.[18] Kṛṣṇa and Rāma are both hosts to stories: in Kṛṣṇa's case, one about Umā and Śiva; in Rāma's, his own. But Kṛṣṇa is also standing by here, listening and witnessing as both guest and host to stories about himself, as he does in the Mārkaṇḍeyasamasyaparvan of *Mahābhārata* Book 3, where Mārkaṇḍeya springs the revelation that the child he has just described surviving the cosmic dissolution is none other than this Kṛṣṇa in the Pāṇḍavas and Draupadī's very company,

> your ally Janārdana, ... the primal person, the ubiquitous lord, ... who sits here as though at play, ... the Placer and Disposer and Destroyer, this Govinda, ... the unborn God of the beginning, Viṣṇu the Person of the yellow robe,

and that the Pāṇḍavas and Draupadī[19] should 'go to him for refuge, he will grant it' (187.50–53) – which, from the standpoint of the UMS they have now finally already done by becoming devoted to Nārāyaṇa after hearing the White Island story in the *Nārāyaṇīya*,[20] again, with Kṛṣṇa-Janārdana standing by. Meanwhile, in the second case, it is Umā and Śiva's solicitation of their own attendant *ṛṣisaṅgha* that spurs Umā, once she has satisfied her curiosity about Śiva's third eye, four faces, Nandin, and Bhūtas, to ask what *dharma* is and how men learn of it (Mbh 13.128.21–23), which leads on to her question about *ṛṣidharma*.

---

[18] Recall Rāma's parting words to Hanumān, stressing orality: 'As long as I hear *Rāmakathā* on the face of the earth, so long will my breaths reside in my body' (Rām 7.39.16). On these matters, see Hiltebeitel 2005c.

[19] She is mentioned with the rest as bowing to Kṛṣṇa in the next verse (3.188.1).

[20] 'Having heard this best of narratives, O Janamejaya, King Dharma and all his brothers became devoted to Nārāyaṇa ...' (12.326.121).

These highly reflexive and nearly metatextual considerations may seem only to be connecting the dots between passages that have long been considered late, and moreover, pointing us away from the two Sanskrit epics' main stories. But as I have said, *ṛṣidharma* is all over the place in both texts, where it has very much to do not only with hospitality and friendship but with a kind of intimacy and complicity in these matters that is fostered by the texts themselves, and indeed by their poets under the *noms de plume* of their Ṛṣi authors. Vālmīki composes the *Rāmāyaṇa* and imparts it to Kuśa and Lava at his *āśrama* as nothing less than the host of Sītā, and he comes with the twins to have them sing the *Rāmāyaṇa* at Rāma's *aśvamedha* as a guest of Rāma, and then arrives there again with Sītā for her final ordeal. Vyāsa is time and again a surprise visitor and occasional host, and a nearly but not entirely silent (see Hiltebeitel 2006a: 245–249) and definitely honored guest as a *sadasya* at Janamejaya's snake sacrifice where he is present as a great-great-great grandfather of that host for the inner frame recital of his own composition. And if Vālmīki is intimate with his characters not only through his proximity to Sītā but through a boon from Brahmā, Vyāsa is still more intimate – literally so with two Kuru queens and one of their maidservants, but also with his characters' thoughts, beginning with those of his mother, and including those of Kṛṣṇa (see Hiltebeitel 2001: 32–91).

C. RĀMA AND KṚṢṆA AS GUESTS, HOSTS, AND FRIENDS

With such matters in mind, I close with a couple of examples from each epic that show the work of the poets in creating this kind of intimacy around hospitality and friendship. I will focus on two examples of each from each text, noting that hospitality and friendship also often imply each other. Getting back to where we started, each also implies a politics of *bhakti* that maps differently in each epic: as what I have elsewhere (Hiltebeitel 2010: 94) called a master–slave dialectic in the *Rāmāyaṇa* centered on an exaltation of the figure of the divine king; and, in the *Mahābhārata*, a politics of friendship and the enemy. Again, while never forgetting *varṇāśramadharma* and indeed reinforcing it, these instances also open on to matters that extend well beyond its limits.

If one looks at the *Rāmāyaṇa* in terms of hospitality for and by Rāma, there are two broad phases. In the first five books one meets Rāma mostly as a guest of various forest sages, and as one who refuses to be a guest in the monkey capital of Kiṣkindhā. But the tables begin to turn in Book 6, where he gives refuge to Vibhīṣaṇa, and are fully turned in Book 7, where, now as king, he gets to host just about everybody – Rākṣasas, monkeys, Ṛṣis, and kings – more than once, including, as noted, Vālmīki among the Ṛṣis, and ultimately even that intemperate hungry Sage Durvāsas whose demand for food signals Rāma and his brothers' deaths. The first phase is quite interesting because it brings an explicitly Vedic variety of *ṛṣidharma* right into the heart of the story to mark Rāma's path and to provide him and his brothers[21] with mentoring or guidance that always has something to do with *dharma*.

Vālmīki presents himself and Rāma as contemporaries of the eponymous ancestors of five, and the near descendent of one, of the poets of the six 'family books' of the *Ṛgveda*: Rāma Jāmadagnya, descended from Bhṛgu and the Bhṛgu poets of Book 2; Viśvāmitra as the lead family poet of Book 3; Gautama of Book 4;

[21] Lakṣmaṇa is usually with Rāma when Rāma meets them; Bharata – once Rāma has gone into exile with Lakṣmaṇa – meets Bharadvāja (ca. 2.66–80) independently, and so too deals independently with Vasiṣṭha (2.107.6a–c; see also 2.96.24–29). Śatrughna also meets with Vālmīki in a passage that the Critical Edition rejects even though it appears in all the manuscripts collated, when he stops at Vālmīki's hermitage, hears the twins' elegant recitals, and promises that he and his army will keep their birth secret from Rāma (Rām 7, App. I, no. 9; Shah 1975: 26–27).

Atri of Book 5; Bharadvāja of Book 6; and Vasiṣṭha of Book 7. Rāma is in one way or another hosted by all of these except, unless one stretches the point, Rāma Jāmadagnya. In addition, before any of these meetings, Rāma's birth is made possible by his father's hosting of the Sage Ṛśyaśṛṅga, a descendant of the Ṛṣi Kaśyapa whose family is associated with many of the Soma hymns of *Ṛgveda* Book 9; and the culmination of all these meetings is another one with Agastya, less prominent in the *Ṛgveda* but celebrated along with these seven other eponyms as ancestors of eight great Brahman *gotras* or lineages. Moreover, these eight have also, by the time in which the main story of the *Mahābhārata* is set, passed on to become important stars: with Jamadagni instead of his son Rāma Jāmadagnya, the first seven (listed as the composers of *Ṛgveda* 9.67 and 10.137), make up the northern Big Dipper, and Agastya is the southern star Canopus (see Hiltebeitel 1977). This does not stop these seven from having generated descendants in the story and occasionally dropping into it.

Rāma is guided by these Vedic Ṛṣis from his conception to the point where Agastya directs him further south to Pañcavaṭī, where Sītā will be abducted.[22] Gautama's wife Ahalyā explicitly provides Rāma and company hospitality (*ātithyam*, 1.47.18ab), as do Bharadvāja (2.48.30b), Agastya (3.11.24b), and another forest Ṛṣi named Śarabhaṅga (3.4.25d), each calling him their 'dear' or 'beloved guest' (*priyātithi*). One might not want to make too much of this phrase, since Daśaratha uses it to welcome Yudhājit, Bharata's maternal uncle (1.72.6), and again to describe how the boy he killed by accident would come 'like a welcome guest' (*priyam ivātithim*) to feed his parents (2.58.28), though this actually reflects back on how Daśaratha yearns for his own son, the departed Rāma. But *priyātithi* is also used when

[22] See Hiltebeitel 2009, and for some discussion of these scenes and *Rāmāyaṇa* hospitality practices, Brockington & Brockington 2006: xvi, 381–383.

the residents of Ayodhyā return to the city, lamenting after following Rāma's departure as far as they could, and say, 'Every hill and grove Rāma visits will treat him as a beloved guest and not fail to accord him hospitality' (2.42.34). The most telling instance, however, comes in the words of Agastya, the last of the Ṛṣis to provide signposts along Rāma's trail, who says to Rāma with joined hands:

> King of the whole world, one who fares in *dharma* (*dharmacarī*), a great chariot warrior, a man offered reverence and esteem, you have come as my beloved guest (*pūjanīyaś ca mānyaś ca bhavān prāptaḥ priyātithiḥ*). (Rām 3.11.27).

As with the earthly gleaners described by Śiva, these *Rāmāyaṇa* sages do the gods' work and would seem to have more than an inkling of what it is (cf. footnote 17 above). Agastya recognizes his 'beloved guest' even in exile as the king of the world, and after saying these words, gives Rāma a bow of Viṣṇu (3.11.29–32), perhaps to make up for the one Rāma broke in his showdown with Rāma Jāmadagnya. When Sītā is soon abducted from Pañcavaṭī, Brahmā, seeing this outrage with his divine eye, says, 'What is done was to be done', and the Daṇḍaka Forest Ṛṣis are 'thrilled' (*prahṛṣṭā*) at the same sight (3.50.10–11).

As to friendship in the *Rāmāyaṇa*, again, in broad strokes, it is the second Book that first brings the subject to the fore as regards Rāma. Here we meet the first articulation of *sakhi* as a 'friend to master' (*sakhi* to *bhartṛ*) relation specifically focussed on Rāma in his dealings with the low caste Niṣāda Guha.[23] Otherwise, as Rāma emerges into view in his court and palace life as he is about to be made heir apparent, *sakhi* is never used for any relation he has with anyone else in this Book. In departing Ayodhyā, Rāma leaves no friends behind – that is, *sakhi* friends. What he leaves behind are *suhṛd*s. While *suhṛd* is used

[23] See 2.44.9b and 14a; 2.78.5ab and 11cd, 79.2b, 80.7a, 81.16b: Guha is friend and Rāma specifically master in the boldfaced verses.

fairly frequently, what is noteworthy is that although Rāma has *suhṛds*, when they are around him they are impersonal, never named.[24] For example, 2.36.16:

> As for Rāma's wellwishers, they were all bewildered: crushed by the weight of their grief, they could not rise from where they had fallen.

The *suhṛds* are last in the run of those bidding Rāma adieu, mentioned just before 'Ayodhyā, with all its hosts of soldiers and herds and horses and elephants ...' (36.17) – after which Rāma disappears from the sight (37.1–2) of those remaining in the capital, leaving those left behind to think how every hill and grove he visits will treat him as 'a beloved guest' (2.42.34, as just cited). It is not until Sītā is abducted, that is, after Rāma has had his rather fateful advice to go to Pañcavaṭī from Agastya, the last of the Vedic Ṛṣis he meets, that Rāma needs to make some new friends, which he begins to do at the end of Book 3 and then mainly in Book 4 by widening a strange network of low status *sakhi* friends among animals. Here *sakhi* at least initially means 'pact friend', beginning with one of his father's former *sakhis*, Jaṭāyus (3.13.4a; also a *vayasya*: 4.13.3d), a vulture, and soon including Jaṭāyus's brother Sampāti, likewise a former *sakhi* of Daśaratha (4.55.20, 56.9 and 12). But in making his own friendship with Sugrīva, the pact friendship becomes an intimate friendship as well.

This process is set in motion toward the end of Book 3 when Kabandha (a headless Dānava-turned-Rākṣasa who guards the

---

[24] See 2.3.29ab; 2.5.12ab; 2.15.1, 4ab and 9ab; 2.16.55 and 61; 2.31.4 (he has taken leave here of 'all his *suhṛds*'. Lakṣmaṇa likewise has his *suhṛj-jana*, a depersonalized throng of wellwishers (28.11cd and 15ab), whom Rāma bids him say adieu to once Lakṣmaṇa has pleased Rāma with his detailed promise of services to him and Sītā in the forest. With Bharata, however, things are a little different. At least in Rājagṛha among his mother's people of Kekaya he has *sakhis* who can speak to him as his *suhṛds* (2.63.5–7). This is no doubt to nuance the fact that these 'friends of Bharata' are potential rivals of Rāma, for Bharata's own loyalty to (and *saubhrātra* with) Rāma is being at this point still held up to question.

way into the Krauñca Forest) reveals that for Rāma to find Rāvaṇa, and thus the whereabouts of Sītā, he must form a friendship with Sugrīva. Let us note the order of terms by which Kabandha instructs Rāma as to how this should come about.

First, before his cremation, Kabandha says:

(1) 'Now I will give you some good advice, bull among men, and make known to you an ally (*mitra*) once the two of you have purified me with fire' (67.17).

(2) 'You will do well to form a friendship (*sakhya*) with him, Rāghava. He adheres to established usage, and he will be pleased and straightaway render you assistance' (67.30).

Then, once Kabandha is cremated and now a redeemed being, he continues, giving the further instruction that gives this budding friendship its special flavor, at least in the *Vālmīkirāmāyaṇa*:[25]

(3) 'Quickly make him a *vayasya*, having gone there now, Rāghava, sealing your compact in the presence of blazing fire to shun all trickery' (68.13; again at 68.17d). Kabandha also brings in the term *sahāya*, 'companion' (64.18); apparently Sugrīva and Rāma cannot yet be *suhṛds* before they make each other's acquaintance. Soon, Rāma begins to pick up on Kabandha's cue that he should *quickly* make Sugrīva a *vayasya*, and as Kabandha had specifically instructed, Sugrīva and Rāma 'entered into *vayasya* (*vayasyatvam upāgatau*) by reverently circling the blazing fire' (4.5.16cd). That is, their bonding is reminiscent of a marriage. What begins to come across, as the two draw from their parallel or similar experiences, is that *vayasya*, built on the term *vayas*, 'age' – sometimes translated as comradeship or contemporaneity – here suggests 'sympathizer', or, better, 'comforter' or 'commiserator'. I need not discuss four other usages of *vayasya* that soon follow here, as I have done so elsewhere (Hiltebeitel 2009). Suf-

---

[25] See Hiltebeitel 2009, contrasting this emphasis with the term's absence in the *Mahābhārata*'s *Rāmopākhyāna*.

fice it to say that they all call on Sugrīva's parallel experiences to Rāma's of exile and wife-abduction. Yet it remains *sakhi/sakhya* friendship that drives the narrative, with Rāma not only addressing Sugrīva in the vocative '*sakhe*' but Sugrīva grounding what follows in the same term (4.8.2d and 39cd). What is important for present purposes is that, while this relationship is not one that is directly characterized by *bhakti*, it is forged by Hanumān who imparts *bhakti* to all the monkeys' services to Rāma and eventually Sītā. It is Hanumān who next, after Kabandha, tells Rāma about Sugrīva. Speaking as a proto-*bhakta* rather than a *sakhi* or any other type of friend, he defines Rāma and Sugrīva's friendship from the beginning as emerging out of a mutual desire for *sakhya*: to Rāma he says, 'Righteous Sugrīva desires friendship (*sakhyam icchati*) with you' (4.3.20); and to Sugrīva he says, 'These brothers, Rāma and Lakṣmaṇa, are desirous of friendship (*sakhyakāmau*) with you' (4.5.7). Sugrīva then confirms Hanumān's pact-making with the words,

> For me it is indeed an honor and also a great advantage, lord, that you desire goodheartedness (*sauhārdam*) with me, a monkey. And if such a friendship (*sakhya*) is agreeable to you, here is my outstretched arm. Take my hand in yours and let us make a firm pact (*maryādā*) (11cd–12)

– to which Rāma responds by embracing Sugrīva tightly (*paryaṣvajata pīḍitam*), 'having accepted cordial goodheartedness' (*hṛdyaṃ sauhṛdam ālambya*) (13). This embrace sets the conditions under which Rāma and Sugrīva will do each others' will. We shall meet its counterpart in the *Mahābhārata*. As befits the *Rāmāyaṇa*'s master-servant politics, it restricts the role of friend to pacted subordinates, and separates the roles of friend and *bhakta*, leaving the one no less a servant than the other.

As to the *Mahābhārata*, scenes of friendship and hospitality are far more recurrent and also harder to disentangle the one from the other. Things are also more complex than the *Rāmāyaṇa* in other ways: unlike Rāma, Kṛṣṇa is never a king with need for subordinates or grandiose hosting obligations;[26] he is not the center of the story; and he tends not to play things straight. As with Rāma and the Ṛṣis of the *Rāmāyaṇa*, ultimately one would need to track the hosting and guesting relation between Kṛṣṇa and the Ṛṣis and the semantics of Kṛṣṇa's 'friendships' throughout the *Mahābhārata*: much too large a task for this article.[27] Suffice it to say that if one looks at hostings of or by Kṛṣṇa, one can rarely forget that he is *the* special *sakhi* of both Arjuna (as lies behind the instance cited above [Mbh 14.52.1–3]) and Draupadī (see Hiltebeitel 2007b: 126–133). Often when he comes into the *Mahābhārata* action or leaves it to return to Dvārakā, there are shows of hospitality to Kṛṣṇa as a guest, usually by the Pāṇḍavas. And then there are some highly dramatic scenes that clearly relate hospitality and friendship to *dharma*, such as Kṛṣṇa's welcoming Arjuna at Dvārakā by helping him abduct his sister Subhadrā toward the end of Book 1, and his breach of Jarāsandha's hospitality at Magadha by coming along with Bhīma and Arjuna disguised as Snātaka guests only to kill their host near the beginning of Book 2. Two more such dramatic scenes[28] are the *Śiśupālavadha*, a little later in

[26] At Mbh 13.16.8cd, however, Kṛṣṇa gets an eighth boon from Umā that he will be able to feed seven thousand guests daily at his palace.

[27] Nārada notably leads the Ṛṣis to Yudhiṣṭhira's *rājasūya* in time for the guest-gift (2.33.1–21) and reflects that Hari Nārāyaṇa who has 'become man' is the lord praised by sacrifices (2.33.20). Many Ṛṣis could also be tracked in the *Mahābhārata*'s citation apparatus, particularly in the four anthologies with Kṛṣṇa standing by; and in the *Mahābhārata*'s substories (see Hiltebeitel 2005a: 475, n. 43 and *passim*).

[28] These two have been singled out for rejection by Fitzgerald from his 'main *Mahābhārata*', for which he posits a first written redaction composed fairly early in post-Mauryan times that was 'concerned to provide ideological and narrative grounding for a Brahmanical conception of kingly rule and hierarchical society in the wake of the Mauryan Empire and that government's cosmopolitanism and its insufficient recognition of the uniqueness of Brahminic authority' (Fitzgerald 2003: 811) – to which *bhakti* runs were 'probably added some time between 50 CE and 350 CE', including 'the essential portions of the *Bhagavad Gītā*, all episodes that elaborate some theme of devotion to Viṣṇu, Śiva, or Kṛṣṇa' including these two and 'several

Book 2, where it is a question of Śiśūpāla's rage at Kṛṣṇa's being selected by Bhīṣma to receive the 'guest gift' (*arghya*) at Yudhiṣṭhira's *rājasūya* (see footnote 27 above); and the scene where Kṛṣṇa awakens to find Arjuna and Duryodhana seeking boons from him at his bedside in Dvārakā, with Duryodhana claiming, it would seem misleadingly, that Kṛṣṇa has an equal friendship (*samam ... sakhyam*) and relationship (*sambandhakam*) to both himself and Arjuna (5.7.10), which Kṛṣṇa, however (that 'friend alike to friend and foe'), does not deny, but rather devises the order of choosing so that Arjuna gets the first choice between Kṛṣṇa as a noncombatant charioteer (which of course Arjuna chooses) and a whole army division of Kṛṣṇa's Nārāyaṇa Gopā warriors (which seems to content Duryodhana). This episode comes at the beginning of Book 5, the Udyogaparvan, about preparations (*udyoga*) for war, in which Kṛṣṇa truly swings into action, and I will choose my two *Mahābhārata* examples from this Book – both from the heart of the Udyogaparvan: Kṛṣṇa's embassy to the Kuru court.

Here, for once, Kṛṣṇa is being hosted not by the Pāṇḍavas but by the Kauravas. The long sequence – the Bhagavadyānaparvan – is filled with moments that could illustrate our points, including Dhṛtarāṣṭra's raptures at the thought of giving Kṛṣṇa fabulous gifts and displays of welcome (5.84); Vidura's advice that it would be enough for Dhṛtarāṣṭra to give Kṛṣṇa *priyam ātithyam* (5.85.14a) as he deserves – just a jar of water to drink, water to wash his feet, an inquiry into his health, and what he (supposedly) really wants, peace (*śamam*) – rather than planning a costly welcome (85.13–16); Duryodhana's plans to capture Kṛṣṇa (86.13); and Kṛṣṇa's decision at Duryodhana's palace not to accept food, bedding and other hospitality from Duryodhana but to take food and residence only at Vidura's (89.1–34).

---

highly polished expressions of Kṛṣṇa *bhakti* in the narrative wake of Yudhiṣṭhira's *abhiṣeka*' between 12.40 and 12.56 (Fitzgerald 2006: 272–273).

Let us first focus in here. It is at Vidura's, at night, after Kṛṣṇa has relaxed, that Vidura makes a most interesting statement while telling Kṛṣṇa he thinks it was a mistake for him to have come and discouraging him from returning to talk to Duryodhana:

> When all those villains are huddled together, your descent (*avataraṇam*) in their midst does not please me, Kṛṣṇa (*tava madhyāvataraṇaṃ mama kṛṣṇa na rocaye*). (5.90.15cd)

Vidura's words in this line are identical, but for a slight tweaking of the pronouns and syntax, with a line where Yudhiṣṭhira has made the same recommendation *before* Kṛṣṇa's departure on this very mission:

> I do not agree, Kṛṣṇa, that you should go to the Kurus. Suyodhana will not accept your advice. The earth's assembled warrior class comes under the sway of Suyodhana. Your descent (*avataraṇam*) in their midst does not please me, Kṛṣṇa (*teṣāṃ madhyāvataraṇaṃ tava kṛṣṇa na rocaye*). Surely, not a thing would comfort us, not divinity – let alone happiness – not the overlordship of all immortals, with your obstruction, Mādhava. (5.70.82–84)

It can be no coincidence that first the son of Dharma, Yudhiṣṭhira, and then the incarnation of Dharma, Vidura, both discourage Kṛṣṇa from 'descending' into what, in the epic's genealogical and incarnational terms, amounts to the camp of demons. This double usage of *avataraṇa* surely means 'descent' here (as van Buitenen translates it both times) and not 'a "taking down", a relieving of the burden that oppressed the earth'.[29] But the most pertinent passages concern Kṛṣṇa's interaction with the Ṛṣis.

Now as Kṛṣṇa was about to leave the Pāṇḍavas' camp, various birds had circled auspiciously above him (5.81.24–25), and then some of the great Ṛṣis (*brahmadevarṣayaḥ*) – Vasiṣṭha, Nārada, Vālmīkā (sic), and Bhṛgu among them – circumambu-

---

[29] Fitzgerald (2004: 55) says the term *avatāra* has this 'original sense', and '[o]nly later ... came to signify the "descent" of a deity for such a rescue mission'. For him this would be a 'later' *bhakti* usage.

lated him and performed smokeless rites with mantras on his behalf (5.81.26–29). We do not hear where they have come from until he has set out:

> Along the path (*adhvani*) Keśava of strong arms saw the Ṛṣis blazing with brahmic luster, who stood on both sides of the road (*sthitān ubhayataḥ pathi*). Quickly descending from the chariot (*avatīrya rathāt*), Janārdana, having bowed, honoring all these Ṛṣis duly, said, 'Is good health in the worlds, is *dharma* observed, do the other three *varṇas* abide under the Brahmans' rule (*śāsane*)?' Paying them homage (*pūjā*), Madhusūdana proclaimed, 'Where have you reached perfection, your worships (*bhagavantaḥ*)? What path has brought you here? What is to be done for your worships? What can I do for you? For what purpose have your worships come down to earth?' (5.81.61–64)

This show of roadside hospitality to the celestial Ṛṣis now turns, in the narrator's words, into a remarkable scene of friendship: Rāma Jāmadagnya, 'embracing Govinda as an old friend in good deeds' (*purā sucarite sakhā*) (65cd), speaks for them. Having witnessed the old battles of the Gods and the Asuras, the Ṛṣis are

> everywhere (*sarvatas*) wishing to see the gathering of the royal *kṣatra*, the kings sitting in the hall, and yourself speaking the truth, Janārdana. We are coming to watch this grand spectacle, Keśava (67–68).

Jāmadagnya in particular wishes to hear what Kṛṣṇa, Bhīṣma, Droṇa, Vidura and others will say, and closes, 'Go unhindered, hero, we shall see you in the hall' (72cd). Note the depth of precedent behind this exchange: not only have the Ṛṣis witnessed the old Deva–Asura wars; there is the veiled reference to an old *sakhi* 'friendship' between Kṛṣṇa and Rāma Jāmadagnya. This is probably an affinity in their tasks, since this Rāma brought about the 'destruction of the Kṣatriyas', as Kṛṣṇa will now do as well. Or did they meet before when Kṛṣṇa was Rāma?

As we have noted, Kṛṣṇa's welcome by the Kurus is no simple matter: he rejects Duryodhana's hospitality and spends the night at Vidura's. But finally Kṛṣṇa enters the hall:

> There standing amidst the kings, the foe-scorcher Dāśārha, conqueror of enemy cities, saw the Ṛṣis hovering in the sky (*apaśyad antarikṣasthān ṛṣīn*). While watching them, headed by Nārada, Dāśārha said softly to Bhīṣma, 'King, the Ṛṣis have come to watch the earthly assembly, and should be invited and honored with seats and full hospitality (*satkāreṇa ca bhūyasā*). No one can sit before they are seated. Let homage (*pūjā*) be paid to these Munis whose souls have been perfected.' Śaṃtanava, seeing the Ṛṣis arrived at the gate of the hall, hurriedly ordered the servants: 'Seats!' and they brought large and wide smooth seats that sparkled with gold and gems. (5.92.40–45)

Clearly the poets leave questions that an attentive reader or *ṛṣisaṅgha* might ask: was the hall open to the sky? Did Bhīṣma only see the Ṛṣis when they got to the gate? How many seats did the servants provide? How many angels can dance on the head of a pin? But this is serious business. Once everyone falls silent looking at the elixir of the dark yellow-robed Lord (51–53), Kṛṣṇa makes his first speech, beginning, 'Let there be peace ...' (5.93.3ab). The Ṛṣis will not be silent in this assembly. Jāmadagnya will warn Duryodhana with the story of the ancient king Dambhodbhava who foolishly challenged the Ṛṣis Nara and Nārāyaṇa at their Badarī hermitage (5.94). And other stories are told by Kaṇva and Nārada to urge Duryodhana to make peace and not to be obdurate (5.95–121). Then Kṛṣṇa will address Duryodhana directly, as will others, including his mother, all to no avail, before Duryodhana tries to capture Kṛṣṇa. Kṛṣṇa then gives the Ṛṣis, along with only Droṇa, Bhīṣma, Vidura, and Saṃjaya, the divine eye to see his theophany (5.129.13–14), and quits the court. Having met with the Pāṇḍavas' mother Kuntī, his last business before going back to the Pāṇḍavas is with Karṇa.

This scene at the end of Kṛṣṇa's embassy to the Kauravas deals mostly with friendship, but illustrates our whole nexus. Dhṛtarāṣṭra said, 'Before he rode out amidst princes and councilors, O Saṃjaya, Madhusūdana had Karṇa mount his chariot' (5.138.1) – mount Kṛṣṇa's chariot, that is: a kind of hospitality, I believe, for a man away from home. Dhṛtaraṣṭra asks, 'What consolations' (*kāni sāntvāni*)[30] did Kṛṣṇa-Govinda offer with his 'voice roaring like a flood or a cloud' to the Sūta's son, 'whether gently or sharply' (*mṛdu vā yadi vā tīkṣṇam*) (3). Saṃjaya replies,

> Hear from me, Bhārata, what Madhusūdana of boundless self said to Rādheya in the course of their conversing, in words that were smooth (*ślakṣṇāni*) and gentle (*mṛdūni*), dear (*priyāṇi*), joined with *dharma* (*dharmayuktāni*), truthful (*satyāni*), helpful (*hitāni*), and *to be cherished in the heart* (*hṛdayagrahaṇīyāni*) (5.138.4–5).

To be cherished in *whose* heart? Karṇa's? Dhṛtarāṣṭra's? The hearts of Janameyaya and the *sadasyas* at Janamejaya's snake sacrifice? The hearts of the Ṛṣis of the Naimiṣa Forest? All listeners and readers? The phrase puts a big asterisk to these words of Kṛṣṇa's, which provide one of those points where we can feel that the notion of the *suhṛd* as 'wellwisher' ultimately extends to the *bhakti* community of readers that the *Mahābhārata* seeks to create as its audience of aficionados: that is, its *sahṛdayas*.[31]

Recounting Karṇa's true origins, Kṛṣṇa tells him he is legally a Pāṇḍava – on his father's side related to the Pāṇḍavas, on his mother's to the Andhakas and Vṛṣṇis – and offers him the kingship (5.138.9–10). The Pāṇḍavas and the Andhakas and Vṛṣṇis will clasp his feet (12). Draupadī will make him her sixth (15cd). Hospitality indeed! Moreover, Kṛṣṇa himself will consecrate him; as Karṇa rides the royal chariot, Yudhiṣṭhira as *yuvarāja* will fan him (18–19). Arjuna will drive his chariot (21). Karṇa will have a whole new nexus of allies, about whom Kṛṣṇa concludes:

> Your allies (*mitrāṇi*) shall shudder with joy, your enemies (*ripavas*) with fear. Today let there be good brotherhood (*saubhrātram*) between you and your Pāṇḍava brothers! (5.138.28).

---

[30] Conciliations, mild words? Van Buitenen has 'blandishments' (van Buitenen 1978: 444).

[31] For a slightly fuller treatment of these matters, see Hiltebeitel 2010: 107–108.

Both Kṛṣṇa and Karṇa would know how these changes for Karṇa would also change things entirely for Karṇa's real friend (*sakhi*), Duryodhana, and for the larger Pāṇḍava–Kaurava *saubhrātra*-'brotherhood', of whom Kṛṣṇa says nothing here.[32]

Karṇa shows that he takes these words as coming from the heart:

> I have no doubt at all, Keśava, that you are speaking to me out of goodheartedness (*sauhṛdāt*) and affection (*praṇayāt*) and with friendship (*sakhyena*), Vārṣṇeya, as to my best interests' (5.139.1).

Karṇa does not doubt that Kṛṣṇa is a *suhṛd* speaking from love or affection, and further, that he speaks 'with *sakhya*' to Karṇa's best interests. I believe Karṇa is registering here the maternal ties that link him with Kṛṣṇa, which Kṛṣṇa has just spoken of to him: No less than Kṛṣṇa and Arjuna, Kṛṣṇa and Karṇa are cousins through Kuntī (see Hiltebeitel 2007a: 27–28).

The theme of these words being heartfelt is then carried through to the parting words of their meeting, where Kṛṣṇa responds to Karṇa's explanation of why he must refuse.

> Kṛṣṇa said, 'Of a certainty the destruction of the earth is now near, for my words do not touch your heart, Karṇa (*tathā hi me vacaḥ karṇa nopaiti hṛdayaṃ tava*). When the destruction of all creatures is at hand, bad policy disguised as good does not crawl off from the heart' (*anayo nayasaṃkāśo hṛdayān nāpasarpati*). (141.43–44)[33]

As they part, Karṇa 'clasped Mādhava tightly' (*pariṣvajya ca pīḍitam*) (47b).

As I said, we would meet again with something close, indeed, with the same two verbs, to Rāma and Sugrīva's close embrace (*paryaṣvajata pīḍitam*) cited above. I would not argue that these

---

[32] Kuntī too says to Karṇa, just after this: 'Let the Kurus today witness the meeting of Karṇa and Arjuna in a spirit of brotherhood (*saubhrātreṇa*)' (5.143.9).

[33] Kṛṣṇa seems to commend Karṇa for holding to good policy from his heart, despite rejecting Kṛṣṇa's offer.

somewhat conventional though not that common phrases[34] indicate any dependency or even self-conscious relation between the texts. But they are surely parallel iconic moments that allow for some reflection. In each epic, this is an embrace between Viṣṇu's incarnation and a son of the Sun god. As with Rāma and Sugrīva, the embrace of Kṛṣṇa and Karṇa sets the conditions under which each will do the other's will. And in each epic there is a *bhakti* relation that defines this embrace, for just as Hanumān stands behind the embrace of Rāma and Sugrīva, Arjuna stands behind that of Kṛṣṇa and Karṇa. As befits the *Mahābhārata*'s *bhakti* politics of friendship and the enemy, Kṛṣṇa and Karṇa's embrace extends the circle of the Lord's men and women friends to include even the foe, and combines the roles of friend and *bhakta*.

In conclusion, we come back to the necessities and niceties of textual discretion in portraying the hiddenness of gods among men, which I believe Sheldon Pollock (1984) correctly saw in Rāma's case to be tacitly structured into the *Rāmāyaṇa* around the boon that Rāvaṇa can be slain only by a man,[35] and that Bimal K. Matilal broached well for Kṛṣṇa in his 'defence of a devious divinity' (Matilal 2002: 91–108). There is a politics of *bhakti* in these texts, *as we have them*, that can only be obscured by positing prior ideologies or class interests that would relieve us of this feature. For the *Mahābhārata*, I believe that more than a 'rage' at the loss of Brahman privilege under Aśoka and the Mauryas,[36] its *bhakti* politics reflected a probably later, post-

---

[34] The only other such usages are Mbh 2.2.18; 8.50.29; Rām 7.79.6.

[35] See now Goldman & Goldman 1996: 19, 43, holding that in the Sundarakāṇḍa, *bhakti* is often 'assured' by 'conventions' even without being 'foregrounded'. Although González-Reimann (2006) makes some useful criticism of a few of Pollock's supporting arguments, they do not derive from a close reading of the text.

[36] Fitzgerald's 'main Mbh' centers on Dharmarāja Yudhiṣṭhira as a somber extension of Dharmarāja Yama, god of death. Yudhiṣṭhira must preside over a divine raiding party of the gods that descends to earth to restore Brahmans to privileges denied them by Aśoka, projecting 'a deep and bitter

Mauryan, sly and confident sense of taking over the game. For the *Rāmāyaṇa*, it would appear to be the same game disambiguated – a game that Aśvaghoṣa could respond to from a familiarity with both epics, linking his own *bhakti* overtures to the Buddha's quest for the 'true *dharma*'.[37] In both epics, *bhakti* is a trump card played discretely and not that often (though certainly more often in the *Mahābhārata*). And it is played with a deck stacked with Vedic-sounding Ṛṣis who, by 'doing the gods' work', support a *new* Brahmanical *dharma* (see footnote 5 above), one they now make familiar in every sense. It is this combination that could give life to a king who has listened well enough to God (Kṛṣṇa) and even better to the Ṛṣis, as is the case with Yudhiṣṭhira son of Dharma; or to a king who, as a hidden God himself, could be the Ṛṣis 'beloved guest'.

REFERENCES

BIARDEAU, Madeleine 2002. *Le Mahābhārata. Un récit fondateur du brahmanisme et son interprétation*. 2 vols. Paris: Éditions du Seuil.

BODHI, Bhikkhu (ed.) 2005. *In the Buddha's Words. An Anthology of Discourses from the Pāli Canon*. Somerville MA: Wisdom Books.

BOWLES, Adam 2004. Dharma, Disorder and the Political in Ancient India: The Āpaddharmaparvan of the Mahābḥārata. Ph.D. dissertation, La Trobe University.

BROCKINGTON, John 2000. The structure of the Mokṣa-dharma-parvan of the Mahābhārata. In: Piotr Balcerowicz & Marek Mejor (eds.), *On the Understanding of Other Cultures. Proceedings of the International Conference on Sanskrit and Related Studies to*

political rage at the center of the *Māhabhārata*' (Fitzgerald 2001: 85; cf. 2004; and footnote 27 above).

[37] See *Buddhacarita* 1.49 and 74 where Asita comes 'thirsting for the *saddharma*' (1.49) and predicts that the Buddha will deliver it; 6.31 where Chandaka makes one of his ironic last appeals: 'You should not desert, as a nihilist the good law (*saddharmam iva nāstikaḥ*), your loving father, who yearns for his son'; 13.1: Māra is 'the enemy of the true *dharma*'; 13.31: the divine sages in their pure abodes are devoted to it. See Hiltebeitel 2006b: 237, n. 22, 251, 255.

Commemorate the Centenary of the Birth of Stanislaw Schayer (1899–1941): 71–83. Warsaw: Oriental Institute, Warsaw University.

BROCKINGTON, John & Mary BROCKINGTON (trans.) 2006. *Rāma the Steadfast. An Early Form of the Rāmāyaṇa*. London: Penguin.

BRODBECK, Simon. 2006. Ekalavya and Mahābhārata 1.121–128. *Journal of Hindu Studies* 10: 1–34.

BUITENEN, J. A. B. van 1978. *The Mahābhārata*, vol. 3: *Book 4. The Book of Virāṭa*; *Book 5. The Book of the Effort*. Chicago: University of Chicago Press.

FITZGERALD, James L. 2001. Making Yudhiṣṭhira the king: the dialectics and the politics of violence in the Mahābhārata. *Rocznik Orientalistyczny* 54.1: 63–92.

——— 2003. The many voices of the Mahābhārata [Review of Hiltebeitel 2001]. *Journal of the American Oriental Society* 123.4: 803–818.

——— 2004. Mahābhārata. In: Sushil Mittal & Gene Thursby (eds.), *The Hindu World*: 52–74. New York: Routledge.

——— 2006. Negotiating the shape of 'scripture': new perspectives on the development and growth of the Mahābhārata between the empires. In: Patrick Olivelle (ed.), *Between the Empires. Society in India 300 BCE to 400 CE*: 257–286. Oxford: Oxford University Press.

GANGULI, Kisari Mohan (trans.) 1970. *The Mahabharata of Krishna-Dwaipayana Vyasa*. 4 vols. 3rd edition. Delhi: Munshiram Manoharlal (first published by Pratap Chandra Roy & Sundari Bala Roy; Calcutta: Bharata Press, 1883–96).

GOLDMAN, Robert P. & Sally J. Sutherland GOLDMAN (trans.) 1996. *The Rāmāyaṇa of Vālmīki*, vol. 5: *Sundarakāṇḍa*. Princeton: Princeton University Press.

GONZÁLEZ-REIMANN, Luis 2006. The divinity of Rāma in the Rāmāyaṇa of Vālmīki. *Journal of Indian Philosophy* 34: 203–220.

HARDY, Friedhelm 1983. *Viraha Bhakti*. Delhi: Oxford University Press.

HILTEBEITEL, Alf 1977. Nahuṣa in the skies: a human king of heaven. *History of Religions* 16.4: 329–350.

——— 1989. Kṛṣṇa at Mathurā. In: Doris Srinivasan (ed.), *Mathurā. A Cultural Heritage*: 93–102. New Delhi: Manohar – American Institute of Indian Studies.

——— 2001. *Rethinking the Mahābhārata. A Reader's Guide to the Education of the Dharma King*. Chicago: University of Chicago Press.

HILTEBEITEL, Alf 2004. More rethinking the Mahābhārata: toward a politics of bhakti. *Indo-Iranian Journal* 47: 203–227.

——— 2005a. Not without subtales: telling laws and truths in the Sanskrit Epics. *Journal of Indian Philosophy* 33: 455–511.

——— 2005b. Weighting orality and writing in the Sanskrit Epics. In: Petteri Koskikallio (ed.), *Epics, Khilas, and Purāṇas. Continuities and Ruptures. Proceedings of the Third Dubrovnik International Conference on the Sanskrit Epics and Purāṇas, September 2002*: 81–111. Zagreb: Croatian Academy of Sciences and Arts.

——— 2005c. On reading Fitzgerald's Vyāsa. *Journal of the American Oriental Society* 125.2: 241–261.

——— 2005d. Buddhism and the Mahābhārata: boundary dynamics and textual practice. In: Federico Squarcini (ed.), *Boundary Dynamics and Construction of Traditions in South Asia*: 107–131. Florence: University of Florence Press – New Delhi: Munshiram Manoharlal.

——— 2006a The Nārāyaṇīya and the early reading communities of the Mahābhārata. In: Patrick Olivelle (ed.), *Between the Empires. Society in India 300 BCE to 400 CE*: 227–255. Oxford: Oxford University Press.

——— 2006b. Aśvaghoṣa's Buddhacarita: the first known close and critical reading of the Brahmanical Sanskrit epics. *Journal of Indian Philosophy* 34: 229–286.

——— 2007a. Krishna in the Mahābhārata: the death of Karna. In: Edwin Bryant (ed.), *Krishna. A Sourcebook*: 23–76. New York: Oxford University Press.

——— 2007b. Among friends: marriage, women, and some little birds. In: Simon Brodbeck & Brian Black (eds.), *Gender and Narrative in the Mahābhārata*: 110–143. London: Routledge.

——— 2009. Authorial paths through the two Sanskrit Epics, via the *Rāmopākhyāna*. In: Robert P. Goldman & Muneo Tokunaga (eds.), *Epic Undertakings* (Papers of the 12th World Sanskrit Conference, 2): 169–214. Delhi: Motilal Banarsidass.

——— 2010. Mapping *bhakti* through friendship in the Sanskrit Epics. In: Sheldon Pollock (ed.), *Epic and Argument in Sanskrit Literary History: Essays in Honor of Robert P. Goldman*: 91–116. Delhi: Manohar.

JOHNSTON, E. H. 1936. *Aśvaghoṣa's Buddhacarita or Acts of the Buddha*, Part 1: Sanskrit Text, Sargas 1–14; Part 2: Introduction and Translation; Part 3: Translation of Cantos 15–28 from Tibetan and Chinese versions. Lahore: Baptist Mission Press for University of the Panjab (reprint: Delhi: Motilal Banarsidass).

KOROM, Frank J. 2004. The Bengali dharmarāj in text and context: some parallels. *Journal of Indian Philosophy* 32: 843–870.

MATILAL, Bimal Krishna 2002. *Ethics and Epics. The Collected Essays of Bimal Krishna Matilal*. Ed. by Jonardon Ganeri. New Delhi: Oxford University Press.

OBERLIES, Thomas 1988. Die Śvetāśvatara-Upaniṣad: eine Studie ihrer Gotteslehre (Studien zu den 'mittleren' Upaniṣads I). *Wiener Zeitschrift für die Kunde Südasiens* 32: 57–59.

——— 1997. Die Textgeschichte der Śvetadvīpa-Episode des Nārāyaṇīya (Mbh. 12.321–326). In: Peter Schreiner (ed.), *Nārāyaṇīya-Studien* (Purāṇa Research Publications Tübingen, 6): 75–118. Wiesbaden: Harrassowitz Verlag.

OLIVELLE, Patrick. 1996. *Upaniṣads* (World Classics). Oxford: Oxford University Press.

——— 1998. *The Upaniṣads. Annotated Text and Translation*. New York: Oxford University Press.

——— (trans.) 1999. *Dharmasūtras. The Law Codes of Ancient India*. (Oxford World's Classics.) New York: OUP

——— 2004. The semantic history of dharma: the middle and late Vedic periods. *Journal of Indian Philosophy* 32.5–6: 491–511.

——— 2006. Explorations in the early history of the Dharmaśāstra. In: Patrick Olivelle (ed.), *Between the Empires. Society in India 300 BCE to 400 CE*: 169–190. Oxford: Oxford University Press.

PATTON, Laurie L. 2005. *Bringing the Gods to Mind. Mantra and Ritual in Early Indian Sacrifice*. Berkeley: University of California Press.

POLLOCK, Sheldon 1984. Ātmānam mānuṣam manye: Dharmākūtam on the divinity of Rāma. *Journal of the Oriental Institute, Baroda* 33: 505–528.

SHAH, Umakant Premanand 1975. *The Vālmīki-Rāmāyaṇa*, vol. 7: *The Uttarakāṇḍa*. Critical edition. General eds. G. H. Bhatt & U. P. Shah. Baroda: University of Baroda.

TSUCHIDA, R. 1991. Two categories of Brahmins in the early Buddhist period. *The Memoirs of the Tōyō Bunko* 49: 51–95.

ZAEHNER, R. C. 1963. Utopia and beyond: some Indian views. *Eranos-Jahrbuch* 32: 281–309.

Magic Cows and Cannibal Kings: The Textual Performance of the Viśvāmitra Legends in the *Mahābhārata*

ADHEESH SATHAYE

Indologists like to debate nearly every aspect of the Sanskrit *Mahābhārata*, but there is one point that is rarely disputed: the epic is an extraordinarily complex piece of literature. Its complexity is not simply due to its copious plot twists, settings, and characters – features that ordinarily produce literary sophistication. Instead, what makes the *Mahābhārata* particularly labyrinthine is its deeply heterogeneous text. The sustained intermixture of myths, legends, discourses, and trivia within its primary narrative has led many Indologists to regard the extant *Mahābhārata* as at best an encyclopedia and at worst a 'monstrous chaos'.[1] Since the time of Joseph Dahlmann, synchronistic studies have sought to make holistic sense of this complexity, while most diachronists have abided by E. Washburn Hopkins's insistence upon a philological differentiation between the 'original' bardic poems and later 'intrusion of masses of didactic matter [and] addition of Puranic material old and new' (Hopkins 1901: 397–399, as cited by Laine 1989: 25). As James Laine points out, the arguments between the analysts and synthesists

[1] Oldenberg 1922: 1. For further discussion on the *Mahābhārata*'s encyclopedic nature, see Hiltebeitel 2001: 14–15, 161–163; Proudfoot 1979: 45–49.

expose two central questions of epic studies: 'Does the epic have a fundamental unity or not? How can we understand the process which produced the extant text?' (Laine 1989: 26). To begin addressing these questions, this essay takes a performance-centered approach to the interpolation of one specific subnarrative into the *Mahābhārata*'s primary narrative – a case of what Hopkins might well have deemed 'all is fish that comes to the net' (Hopkins 1901: 369–370, as cited in Hiltebeitel 2001: 15).

In the epic's lengthy first book, the Ādiparvan, a Gandharva prince named Citraratha ambushes the Pāṇḍava heroes during their first forest exile.[2] After he is soundly defeated in combat, the Gandharva tells them a number of stories to illustrate the importance of having a wise family priest (*purohita*), especially someone like the Brahman sage Vasiṣṭha.[3] Among the narratives he tells is the account of Vasiṣṭha's conflict with the king Viśvāmitra over a wish-giving cow, a *kāmadhenu* (Mbh 1.164–165). Viśvāmitra's utter defeat impels the Kṣatriya to turn himself into a Brahman. In ancient Indian literature, this is a unique self-transformation: though the epic does attest to others who become Brahmans without shedding their non-Brahman bodies, Viśvāmitra is the only one said to have done so of his own accord, through the determined acquisition of ascetic power (*tapas*).[4]

---

[2] Mbh 1.158.22. All references to the two Sanskrit epics in this essay are to their respective Critical Editions, henceforth abbreviated as Mbh and Rām, respectively. Their use has been greatly facilitated by Muneo Tokunaga, *Machine-Readable Text of the Mahaabhaarata based on the Poona Critical Edition*, first revision (Tokyo, [1991] 1994), along with Tokunaga's electronic Sanskrit text of the critical edition of the *Rāmāyaṇa*; both electronic texts have been revised by John Smith (see his Home Page, http://bombay.indology.info/index.html [accessed March 24, 2007]).

[3] Citraratha, who gives up his name in defeat, explains why he had assaulted the party: 'You were without fire, without oblations, and without a Brahman to guide you' (Mbh 1.159.2).

[4] E.g., Vītahavya's conversion by the sage Bhṛgu (Mbh 13.31); see Goldman 1977: 111–112. Sindhudvīpa and Devāpi are two additional Kṣatriyas mentioned in the *Mahābhārata* to have become Brahmans by bathing in the same *tīrtha* as Viśvāmitra (Mbh 9.39.10).

Outside of the *Mahābhārata*, Viśvāmitra has a long and well-documented presence within the world of ancient Indian literature.[5] A major Vedic sage, Viśvāmitra is inextricable from the earliest Vedic texts as the seer of a large number of *Ṛgveda* hymns and progenitor to an important family of Vedic Brahmans.[6] He is a fixture within later Vedic 'ancillaries' like the *Bṛhaddevatā*, and an elaborate cycle of legends about him – including the *kāmadhenu* conflict – is woven into the first book of the other Sanskrit epic, the *Vālmīkirāmāyaṇa* (Rām 1.50–64). Viśvāmitra is crucial within this epic's early plot as a Proppian donor figure – not only does he provide Rāma with magic weapons (*astras*) that will enable him to slay Rāvaṇa, he also brings Rāma to Mithilā for Sītā's marriage contest.[7] Though Viśvāmitra does not play an equally significant role within the *Mahābhārata*'s primary narrative, the epic is quite aware of his presence. The *kāmadhenu* narrative is repeated in the Śalyaparvan, and scattered throughout the epic is a number of Viśvāmitra narratives: during the Śakuntalā narrative in the Ādiparvan, as a part of Bhīṣma's discourse on the laws of emergency (*āpaddharma*) in the Śāntiparvan, not to mention the *Viśvāmitropākhyāna* (The Viśvāmitra Subnarrative) in the Anuśāsanaparvan.[8] The *Viśvāmitropākhyāna* is delivered by the Kuru patriarch while lying on his deathbed in the aftermath of the war as a response to a query

[5] Early studies of Viśvāmitra include Muir 1868; Roth 1846; Müller 1860; Weber 1854; and particularly Pargiter 1913; 1917; 1922. For more recent studies, see Hariyappa 1953; Rahurkar 1964; Kapadia 1971; Sharma 1975; Chaubey 1987.

[6] A complete list of references to Viśvāmitra in the Vedas is found in Sharma 1975: 39–69.

[7] Propp 1968: 39–50. Robert Goldman (1984: 41, n. 79) notes that Rāma is given magic weapons by Vasiṣṭha and Agastya as well, and that the hero's encounter with Agastya 'is the source for the much more elaborate episode of Viśvāmitra's gift of divine weapons'.

[8] The Śalyaparvan's *kāmadhenu* legend: Mbh 9.39; Viśvāmitra's tryst with Menakā in the Śakuntalā narrative: Mbh 1.65.20–1.66.8; Viśvāmitra's eating dogmeat during famine: Mbh 12.139; the genealogy of Viśvāmitra and Rāma Jāmadagnya: Mbh 12.49 and 13.4.

from Yudhiṣṭhira – a question that reveals the *Mahābhārata*'s fundamental interest in Viśvāmitra.

> 'Mighty King, Lord of men', he asks, 'if Brahmanhood is so difficult to attain by members of the other three *varṇas*, then how did that mighty Kṣatriya, Viśvāmitra, attain Brahmanhood? This is what I wish to hear, righteous King of men, please tell it to me accurately, Grandfather.' (Mbh 13.3.1–2ab)

The nature of *varṇa* as a social ideology appears to be fundamentally at stake in the representation of Viśvāmitra, and more specifically, the boundary between Brahman and Kṣatriya *varṇa* that Viśvāmitra is so uniquely able to cross.

Viśvāmitra is by no means the only legendary figure to have his exploits injected into the Sanskrit *Mahābhārata*. Populating what are considered the 'later' strata of the great epic is a tremendous number of what Alf Hiltebeitel terms 'subtales' about sages, kings, gods, demons, and others.[9] Along with Hiltebeitel, a number of epic scholars have taken subnarratives seriously in theorizing how and why these monumental Sanskrit literary texts were produced.[10] In his comparative study of two versions of the Rāma Jāmadagnya legend, for example, Robert Goldman argues that differences in the presentation of each narrative point to two distinct ideological 'provenances' of the *Mahābhārata*'s composers (Goldman 1972). Accordingly, by identifying distinct interpretive strategies in the epic versions of Viśvāmitra legends, we will be better positioned to assess the cultural value of this character in ancient India.

One immediate observation arises from the fact that Viśvāmitra is a prominent Vedic personality: the *Mahābhārata* does not regard these narratives as fictional. Viśvāmitra would hardly

---

[9] Hiltebeitel 2005. Hiltebeitel translates *upākhyāna* as 'subtale', though *upākhyānas* are not precisely 'tales' in the folkloristic sense of 'prose narratives which are regarded as fiction' (Bascom 1965: 4).

[10] Some examples include Gombach 2000; Reich 2001; Feller 2004.

have been an obscure name among the epic's potential audiences: communities that were deeply immersed in Vedic culture (Minkowski 1989). Stories about him and his accomplishments must therefore have been common knowledge at the time. In fact, narratives explicitly involving a conflict between Viśvāmitra and Vasiṣṭha appear in the *Bṛhaddevatā* and other late Vedic texts (Bhattacharya 1987). Therefore, regardless of whether one takes the position that the *Mahābhārata* developed over a long period of time, or whether one believes it to be the work of a single editor or committee, we are faced with two possibilities of how these subnarratives were inserted into its corpus. Due to the preexistence of a Viśvāmitra narrative tradition, the narratives found in the epics either must be 'fixed' versions of otherwise 'fluid' legends, or they must have been offered as authoritative narratives *in spite of* the existence of other fluid versions. The emphasis on fluidity rather than 'orality' builds upon A. K. Ramanujan's critique of the assumed one-to-one correspondence between oral vs. written texts and folk vs. classical texts, alongside Velcheru Narayana Rao's insistence on the 'oral literacy' or oral scholasticism of the purāṇic composer (Ramanujan 1999: 538; Narayana Rao 1993: 95). It also helps to move the discussion past the largely unanswerable question of whether the epic was composed orally or in written form. What is important to determine is not the extent to which the epic was written, I suggest, but the extent to which it was a fixed text (see Doniger 1991). Assuming that it is indeed a fixed text, I argue that the *Mahābhārata* employs Viśvāmitra as a way of engaging with a fluid storytelling tradition lying outside its domain, a narrative culture that we might in this sense productively regard as 'folk'.

If we begin to regard epic subnarratives as participating in folk narrative culture, it becomes evident that the texts we now have are a product of a negotiation between epic composers and epic audiences within this culture. Alongside Laine's questions of textual process (noted above) appears a deeper question of authorial (and audience) intent: what did these stories mean to the ancient users of the text? In other words, what was the cultural 'work' of the Sanskrit *Mahābhārata* within its ancient social and religious context?

Standing in the way of any cultural analysis of the Sanskrit epics, of course, is the obstacle of history. The physical texts that we find today are but fragmentary copies of an irretrievable original, if such an original ever existed. We can interview neither the composers nor the audiences of the epics, and every theory we have about how, where and when the epics were composed necessarily arrives at broadly speculative conclusions. As Ian Proudfoot observes,

> We have no means of knowing the historical reality [of the *Mahābhārata*'s composition] except through the analysis of the text (Proudfoot 1979: 58).

While this is undeniable, I argue that we may recover a historical dynamic between composer and audience if we approach epic subnarratives not as inert insertions into the text, but as 'textual performances' – that is, as active storytelling events taking place within the fixed context of the epic, parallel to how they would ordinarily happen within a purely fluid discourse. As Stuart Blackburn has done for medieval literary story collections, we would like to situate subnarratives within their larger 'narrative context' (Blackburn 1996). However, before embarking upon a performance analysis of the *kāmadhenu* legend, it is first necessary to develop the concept of textual performance in greater detail.

TEXTUAL PERFORMANCE

The performance-centered approach to cultural expression within folklore studies and sociolinguistics does not ordinarily concern itself with how literary texts are produced. Instead, performance theorists investigate how forms of what Richard Bauman terms 'verbal art' – oral or expressive forms of folklore such as proverbs, riddles, or tales – are enacted by a performer for an audience within situated, lived contexts (Bauman 1975). At the heart of the theoretical model is what Bauman has called the 'emergent quality of performance': the observation that when a performer tells an item of folklore, it is neither simply a *verbatim* parroting of a text nor a purely novel invention. Instead, the text emerges through a negotiation between performer and audience, involving both the audience's evaluation of the competence and cultural relevance of the performer and his material, along with the performer's exercise of temporary social power over his captive audience (Bauman 1975: 293).

This emergence is demonstrable on two levels – text and event. Textual scholars recover the emergent quality of oral texts through a generative model, such as that developed by Milman Parry and Albert Lord towards the Homeric epics.[11] With varying degrees of success, a handful of Indologists have in fact applied the Oral Formulaic Theory to the Sanskrit epics.[12] Perhaps more germane to our concerns, however, is the interest of performance theorists in theorizing folklore as *event* – as Bauman explains, we would like to know how an item emerges through 'a product of the interplay of many factors, including setting, act sequence, and ground rules of performance'.[13] Rather than a written tran-

---

[11] Lord 1960. For a comprehensive survey of the development of Oral Formulaic Theory, see Foley 1988.

[12] Grintser 1974; summarized in de Jong 1975; Vassilkov 1995; Brockington 1998: 103–116.

[13] Bauman 1975: 299. For performance in South Asia, see Blackburn 1988; Narayan 1989; Flueckiger 1996.

scription of an oral tradition, we treat the epic as an event, as a literary setting within which pre-existing narrative material is presented through a particular sequence and with certain implicit social 'ground rules'. The epic, in other words, is designed to be a fixed 'stage' upon which fluid stories are performed.

How do these performances take place? Bauman and Charles Briggs describe performance as the *entextualization* of an abstract form that the performer had previously heard within the regular flow of discourse (Bauman & Briggs 1990). Lauri Honko deems this abstraction a 'mental text' – the constituted text of an item of folklore as it exists in the mind the performer (Honko 1998). A mental text is first extracted from the undifferentiated discourse in which the performer originally hears it (i.e., 'decontextualized'), and then embedded within a new discursive context each time the teller tells it to a new audience (i.e., 'recontextualized'). I argue that the Viśvāmitra subnarratives have in fact been entextualized, the key difference being that the decontextualization-recontextualization process transpires within the fixed confines of the literary text. Proving this to be the case requires some speculation, in the absence of living informants. I suggest that one indicator that an epic subnarrative has been decontextualized from oral tradition is its *intertextuality* – that is, its commonalities with versions found in other texts. Close parallels between the *Mahābhārata* and *Rāmāyaṇa* versions of the *kāmadhenu* narrative, for example, allow us to postulate a mental text – a common archetype of the boundary-crossing narrative that the epic audience would have expected to hear. Recontextualization of an epic subnarrative, on the other hand, is accomplished through a process of *supplementation* – a story's relationship to other narratives and expository matter that surround its appearance in the epic. Supplementation is thus a discursive move. By juxtaposing the *kāmadhenu* narrative with another, Brahman-centric narrative, the *Mahābhārata* is able to retain ideological control over its audience's interpretation of

Viśvāmitra's boundary crossing. An inquiry into textual performance thus fulfills what Proudfoot considers to be the 'two requirements for a fruitful exploitation of the *Mahābhārata*'s wealth of information' – 'a sound delineation of the component passages ... and a reasoned account of their association in the context where they are found' (Proudfoot 1979: 44). With this theoretical picture in mind, we now turn to the intertextuality and supplementation of the *kāmadhenu* legend in the *Mahābhārata*'s Ādiparvan.

MAGIC COWS AND CANNIBAL KINGS

The *kāmadhenu* legend is presented three times in the Sanskrit epics – in the first books of both the *Rāmāyaṇa* and the *Mahābhārata* (Rām 1.50.20–1.55; Mbh 1.164–165) and once more in the Śalyaparvan of the latter epic, where it appears to be derivative of the earlier *Mahābhārata* telling (Mbh 9.39). The *Rāmāyaṇa* version is lengthier and more detailed than the Ādiparvan's version, but it is difficult to determine conclusively which version came first. Robert Goldman notes that 'despite differences in size and in detail, the versions are fundamentally the same' (Goldman 1978: 351), while Madeleine Biardeau also points out that all versions of this story share a common underlying ideological message (Biardeau 1999). Both epic versions do in fact share a common plot: Viśvāmitra, the Kṣatriya ruler, arrives in the hermitage of the Brahman sage Vasiṣṭha, where the king and his troops are served a delicious feast through the services of the sage's *kāmadhenu*.[14] Viśvāmitra tries first to purchase the *kāmadhenu* from the sage, and then tries to take her by force, but is defeated by hordes of barbarian armies emitted from every ori-

[14] There is variation in the presentation of these motifs. In the *Rāmāyaṇa*, Viśvāmitra the king arrives at Vasiṣṭha's *āśrama* while on a tour of his dominions (Rām 1.50.20–21); the *Mahābhārata* specifies that he is 'hunting deer and boar in the charming woods' (Mbh 1.165.5). The *Rāmāyaṇa* names the cow 'Śabalā' while the *Mahābhārata* calls her 'Nandinī'.

fice of the magic cow.[15] Defeated by Vasiṣṭha's Brahmanic power, Viśvāmitra becomes disenchanted with his Kṣatriya status and decides to become a Brahman himself, which he eventually accomplishes through a great amount of *tapas*.[16]

The nature of Viśvāmitra's dissatisfaction with Kṣatriya power has been encapsulated in what is perhaps the most famous Sanskrit *dictum* associated with Viśvāmitra. This half-verse also happens to be the only text these two epic versions have exactly in common. In the *Mahābhārata*, we find:

> *dhig balaṃ kṣatriyabalaṃ brahmatejobalaṃ balam* I
> *balābalaṃ viniścitya tapa eva paraṃ balam* II 42 II
>
> Damn this Kṣatriya force! The force of Brahmanic power is truly the greater force. I have understood which is more powerful and which is less, and clearly, tapas is the highest force. (Mbh 1.165.42)

The *Rāmāyaṇa* gives precisely the same first half, but with a different second half:

> *dhig balaṃ kṣatriyabalaṃ brahmatejobalaṃ balam* I
> *ekena brahmadaṇḍena sarvāstrāṇi hatāni me* II 23 II

---

[15] There are certain key differences in the details of Viśvāmitra's defeat. First, while the *Rāmāyaṇa* has Vasiṣṭha actively order his cow to produce the barbarian armies (Rām 1.53.17), in the *Mahābhārata*, the *kāmadhenu* acts on her own (Mbh 1.165.31). In the *Rāmāyaṇa*, the *kāmadhenu*'s barbarian armies slay Viśvāmitra's hundred sons (Rām 1.53.18–54.7), while in the *Mahābhārata*, only his armies are defeated. Furthermore, the latter specifies that 'none of Viśvāmitra's soldiers was deprived of life by the angered sons of Vasiṣṭha' (Mbh 1.165.39). The loss of his sons in the *Rāmāyaṇa* instigates a mini-quest within the story, as Viśvāmitra leaves the hermitage in dejection, and travels to the Himalayas to procure the divine *astras* from Śiva (Rām 1.54.8–20). His final defeat comes through one-on-one battle with Vasiṣṭha (Rām 1.54.21–55.21).

[16] In the *Mahābhārata*, this transformation is described in one and a half verses (Mbh 1.165.44), while the *Rāmāyaṇa*'s telling will extend this narration to nearly nine chapters (Rām 1.56–64), incorporating his encounters with Triśaṅku (Rām 1.56–59), Śunaḥśepa (Rām 1.60–61), Menakā (Rām 1.62), and Rambhā (Rām 1.63), until Vasiṣṭha eventually is forced to acknowledge him to be 'the greatest knower of the Kṣatriya Veda as well as the Brahman Veda' (Rām 1.64.15).

> Damn this Kṣatriya force! The force of Brahmanic power is truly the greater force. With merely one Brahman's staff, all my magic weapons have been vanquished. (Rām 1.55.23)

Given the word-for-word correspondence between these two passages, it is quite surprising that there are no other shared verses in the two epic versions. There are certainly parallel tropes and plot events – for example, the descriptions of the *kāmadhenu*'s opulent feast, Vasiṣṭha's refusal to sell his cow, the cow's pleas to her Brahman master for protection, and the cow's emission of barbarian armies: the Śakas, Yavanas, Huṇas, Kirātas.[17] But *verbatim* parallels are not to be found.

It is a revealing absence. One epic is not simply copying the *kāmadhenu* story from the other; on the contrary, these are two distinct performances of the same traditional narrative. Because the two versions are virtually identical in plot and motifs, it is clear that both epics are working with the same – or nearly the same – narrative archetype, a common 'mental text' that is being articulated into epic literary verse. Here, the *dhig balaṃ* verse is a 'key to performance', an indexical device demarcating the entire narrative as a distinct performance object (Bauman 1975: 295). The use of an explicit intertext suggests that the epic composer is intentionally demonstrating scholastic competence to his audience – that is, he is trying to show his awareness of the existence of this famous verse. Furthermore, within the verse itself, we find a process of decontextualization and recontextualization that unlocks the interpretive strategy of each epic. The first half, *dhig balaṃ kṣatriyabalaṃ brahmatejobalaṃ balam*, decontextualizes the *kāmadhenu* narrative, condensing it into the essential social ideology that is at stake: the relation of power between Brahman and Kṣatriya *varṇas*. On the other hand, the

---

17 Compare the descriptions of the feast (Mbh 1.165.10–12 and Rām 1.52.1–5); Viśvāmitra's attempts to bargain for the cow (Mbh 1.165.16–19 and Rām 1.52.8–24); and the *kāmadhenu*'s emission of barbarian armies (Mbh 1.165.35–36 and Rām 1.53.18–21, 54.2–3).

second half of the verse is a recontextualization, and the variation between them reflects the basic interpretive difference between the two epics. The *Mahābhārata* emphasizes the superior force of Vasiṣṭha's *tapas* (*tapa eva paraṃ balam*), while the *Rāmāyaṇa* is concerned with the power of Viśvāmitra's magic *astra* weapons (*sarvāstrāṇi hatāni me*).

Understanding the significance of the *astra* motif in the *Rāmāyaṇa* requires a consideration of the 'mini-epic' of Viśvāmitra legends that dominate the *Bālakāṇḍa*. Upon arrival at the kingdom of Mithilā, Janaka's minister Śatānanda tells Rāma a lengthy history of Viśvāmitra in order to 'describe the nature of his power and what happened to him' (Rām 1.50.16). Through its lengthy chronicles of how 'doing the unimaginable, and with unrivalled splendor, the very radiant Viśvāmitra has become a Brahman-sage through *tapas*' (Rām 1.50.14),[18] the *Rāmāyaṇa* glorifies Rāma's guide, but also testifies to the repeated loss of his *tapas* due to *kāma* and *krodha*, 'lust' and 'anger'. As pure distillations of his power (*tejas*), the *astras* serve as symbolic representations of Viśvāmitra's *tapas*, untainted by *kāma* and *krodha*.[19] On the level of narrative, the *astras* are instrumental to Rāma's glorification as the slayer of Rāvaṇa, just as much as his marriage to Sītā will act as an efficient cause for his meeting the demon-king. On the level of discourse, the *astras* are part of the epic's representation of Rāma as an ideal king, possessing *tejas* that is maintained by his righteousness (*dharma*). Jarrod Whitaker notes that in general 'the relationship between *tapas* and *tejas* is unclear, although neither one is given precedence', but in the *Rāmāyaṇa* it appears that the *astras* do produce a distinction between the two categories (Whitaker 2000: 107; see also

---

[18] *acintyakarmā tapasā brahmarṣir amitaprabhaḥ* । *viśvāmitro mahātejā vetsyenaṃ paramāṅ gatim* ॥ 14 ॥

[19] For a comprehensive discussion of the interworkings between *astras* and the Brahman and Kṣatriya *tejas* infused within them, see Whitaker 2000.

Whitaker 2002). While Viśvāmitra's *tapas* is repeatedly lost through his temper, Rāma's *tejas* is tempered by *dharma*.[20]

Though it tells practically the same story, the *Mahābhārata* focalizes its version around Vasiṣṭha rather than Viśvāmitra. Interestingly, the *kāmadhenu* narrative is part of a larger *Vasiṣṭhopākhyāna* (The Subnarrative of Vasiṣṭha) detailing how Vasiṣṭha defeats both *kāma* and *krodha* through his *tapas*.[21] Continuing the contrast between Vasiṣṭha's Brahmanic self-control and Viśvāmitra's Kṣatriya passion, the epic follows the *kāmadhenu* story with a legend about a cannibal king named Kalmāṣapāda – a narrative in which Viśvāmitra becomes the villain and which turns the *kāmadhenu* legend's storyworld upside-down.

The story begins as the Ikṣvāku king Kalmāṣapāda, hunting in the forest, gets into an argument with Vasiṣṭha's son Śakti about who has the right of way on a narrow forest path (Mbh 1.166.1–6). Overhearing this squabble between Brahman and Kṣatriya, Viśvāmitra causes the king to be possessed by a demon (*rākṣasa*) so that he may eat up Vasiṣṭha's son (Mbh 1.166.11–16), while Śakti also curses the king to become a cannibal after Kalmāṣapāda strikes the Brahman with a whip (Mbh 1.166.7–10). Later that evening, back at Kalmāṣapāda's palace, an old, unnamed Brahman comes begging for a meal of meat, but unfortunately, there is no meat left in the kitchen (Mbh 1.166.20–26). The king tells his cook to use human meat from the gallows, and presents the meal to the starving Brahman (Mbh 1.166.27–29).

---

[20] I thank Paolo Magnone for pointing out the significance of *tejas* as a symbolic category during a discussion at the 13th World Sanskrit Conference, Edinburgh, July 10–14, 2006. For a historical discussion of *tejas*, see Magnone 1993.

[21] Mbh 1.165.5: 'Lust and anger, which even the immortals find difficult to overcome, have both been conquered through Vasiṣṭha's *tapas*, and they now wash his feet.' (*tapasā nirjitau śaśvad ajeyāv amarair api* I *kāma-krodhāv ubhau yasya caraṇau saṃvavāhatuḥ* II 5 II)

The sage immediately recognizes what's really in his food and also curses the king to become a cannibal (Mbh 1.166.30–32).

Cursed twice and possessed by Viśvāmitra's demon, King Kalmāṣapāda becomes a man-eater and begins to roam madly in the forest. He finds Śakti and devours him, along with the rest of Vasiṣṭha's one hundred sons, as Viśvāmitra urges him on (Mbh 1.166.33–38). Vasiṣṭha despairs at his sons' death, but instead of vengeance he tries repeatedly to commit suicide.[22] He jumps off a mountain, but the rocks below turn to pillows. He jumps into a fire, but the fire becomes cool. He jumps into the ocean, but the waves return him unharmed (Mbh 1.166.41–45). Finally, Vasiṣṭha laments *martuṃ na śakyam*, 'it is impossible to die', (Mbh 1.167.10c) and then heads home, where he hears the sounds of Vedic chanting coming from his daughter-in-law's womb (Mbh 1.167.11). It turns out to be Śakti's unborn son, Parāśara, later to become the father of Vyāsa. With this, Vasiṣṭha's spirits are revived (Mbh 1.167.15). After some time, Vasiṣṭha encounters Kalmāṣapāda – still a cannibal – in the forest. Vasiṣṭha, using water blessed with *mantras*, frees the king from his curse (Mbh 1.168.4), and tells him: 'Go back to your kingdom and rule it. And do not ever disrespect Brahmans' (Mbh 1.168.9).[23]

The Kalmāṣapāda story thus makes no secret of its allegiances. Clearly, Viśvāmitra is now the villain and Vasiṣṭha is the hero, and this is brought about through the direct structural inversion of the preceding legend. Instead of a Brahman's *kāmadhenu* providing a delicious feast for a king, the king feeds the Brahman a revolting plate of human flesh. Instead of vengeance, Vasiṣṭha seeks suicide. Instead of abandoning Kṣatriya status, Kalmāṣapāda is restored to his Kṣatriya status, upon his accep-

---

[22] The *Mahābhārata* specifies here that his sons' deaths were due to Viśvāmitra: 'But Vasiṣṭha, hearing that his sons had been destroyed by Viśvāmitra, held up his grief, like the great mountains do the earth' (Mbh 1.166.39).

[23] *brāhmaṇāṃś ca manuṣyendra māvamaṃsthāḥ kadācana* ‖ 9 ‖

tance of the Brahman's superiority.[24] While the Kalmāṣapāda story has been subjected to diverse historical, psychological, and structural analyses, scholars have failed to ask why this narrative is embedded immediately after the *kāmadhenu* legend in the *Mahābhārata* – and *only* in the *Mahābhārata*.[25] By presenting the Kalmāṣapāda legend immediately after the *kāmadhenu* legend, the epic recontextualizes Viśvāmitra's extraordinary feat of crossing *varṇa* barriers, and reinterprets the religious force that enables this rupture, *tapas*. Both the social ideology of *varṇa* and the religious force of *tapas* are thus recast within the *Mahābhārata*'s decidedly normative vision. Yes, Viśvāmitra is able to cross social boundaries that are otherwise uncrossable, and yes, *tapas* is a transcendent force; after all, '*tapa eva paraṃ balam*'. But this is only half the story, says the *Mahābhārata*, and in the end, the epic confirms the hierarchical rigidity of *varṇa* and glorifies Vasiṣṭha's *tapas* contoured by Brahmanic self-control and compassion, rather than Viśvāmitra's Kṣatriya *tapas*, fueled by lust and anger.

CONCLUSIONS: SOCIAL HISTORY AND TEXTUAL PERFORMANCE

What, then, is the purpose of the *Mahābhārata*'s textual performance of the *kāmadhenu* legend? Why does this epic tell a seemingly counter-normative story that questions the rigidity of *varṇa* boundaries, only to supplement it with one that reaffirms them? Recent scholarship has located the production of the epics within the historical context of post-Mauryan North India (Fitz-

---

[24] Goldman argues that Vasiṣṭha's impregnation of Kalmāṣapāda's wife through *niyoga* at the narrative's conclusion, serves to 'further punish him by yielding his own wife to an avenging father figure' (Goldman 1978: 357).

[25] Goldman offers a detailed psychological analysis of Kalmāṣapāda as an 'extension of the Oedipal rivalry between Viśvāmitra and Vasiṣṭha' (Goldman 1978: 355). For Kalmāṣapāda, see also Lommel 1965–66; Hariyappa 1953: 322.

gerald 2006; Sutton 1997; Hiltebeitel 2001: 18), arguing that the epics served to create both a Brahman-centered notion of polity (Fitzgerald 2003: 811), as well as a new imagination of non-cruel kingship.[26] Furthermore, scholars are virtually unanimous in regarding the *Mahābhārata* – at least at some point in its textual history – as being infused with a normative Vaiṣṇava religiosity. Laine argues, for example, that the narratives of theophany that 'first found expression in the *Mahābhārata*' may be associated with 'the visually-oriented piety of *bhakti* and the new religious pluralism which emerged concurrently in the post-Aśokan age' (Laine 1989: 273). James Fitzgerald, combining both political and religious concerns, suggests that the epic provides 'an early Brahmanic Vaiṣṇava ideological grounding for an empire' (Fitzgerald 1983: 625). However, in constructing the epic's storyworld, this new religious culture was forced to engage with antinomian ascetic practices, not to mention Buddhism, Jainism, and other non-Vedic traditions. As David Gordon White shows, the legends of Viśvāmitra – the '"renegade" renunciant Kṣatriya' – appear to function as 'chronicles of a changing relationship between sanctioned Brahmanic authority and unsanctioned non-Brahman renunciant power' (White 1993: 55).

Against this political and religious backdrop, we are able to make some tentative observations on the social role of the *Mahābhārata*. Rather than an encyclopedia or a 'monstrous chaos', we have seen that the hybridity of the epic is sometimes a direct result of its cultural function as an authoritative textual stage upon which pre-existing narrative traditions were performed, where fluid texts were made fixed. Particularly useful here is Fitzgerald's provocative observation of the structural similarities between the epic and large-scale Hindu temples, which both appear to have developed contemporaneously (Fitzgerald 1983:

---

[26] See Hiltebeitel's discussion of non-cruelty (*ānṛśaṃsyā*) as an epic reconfiguration of nonviolence (*ahiṃsā*) (Hiltebeitel 2001: 202–214).

611). Within the context of a post-Mauryan political world in which *varṇa* boundaries were dangerously flexible – witness the rise of the Brahman king Puṣyamitra in the second century BCE – the epic's textual performances managed to naturalize both *varṇa* rigidity and Brahman supremacy by presenting stories of these concepts gone wrong.[27] Within a post-Upaniṣadic religious world, the *Mahābhārata* employs Viśvāmitra, 'the stock mythic representative of "non-Brahmanic asceticism", of an individual who exercises a Brahmanic prerogative of renunciation which is properly not his own', as a cautionary example testifying to the ultimate superiority of Brahman orthodoxy (White 1993: 61).

Nevertheless, the epic composers are still obligated to tell the story of a king who became a Brahman through *tapas* – a legend already known to the epic audience. As a result, the *Mahābhārata*'s composers are forced to come to terms with the boundary-breaking force of *tapas*. *Tapas* is, after all, the central puzzle within the *kāmadhenu* narrative. The *Rāmāyaṇa* conceives of *tapas* as a counter-normative yet productive force – a force that Rāma will eventually utilize for heroic and theistic purposes. The *Mahābhārata*, on the other hand, constructs two distinct modes or 'flavors' of *tapas*: Viśvāmitra's Kṣatriya *tapas* and Vasiṣṭha's Brahman *tapas*. The latter is evaluated as superior and naturalizes a larger discourse within the *Mahābhārata*: Brahmans are ultimately 'never to be disrespected'.

It is of course tempting to see the presence of these stories as a tampering of later Brahman redactors, and this argument is commonplace in 'Bhṛguization' or 'Kṣatriya-core' approaches to the question of epic textual production.[28] On the other hand, a performance-centered approach suggests that as a normative

---

[27] For the connection between Puṣyamitra and the *Mahābhārata*, see Fitzgerald 2006: 274.

[28] For Bhṛguization, see Sukthankar 1936; Goldman 1977: 81–147. Critiques of Bhṛguization are found in Minkowski 1991; Hiltebeitel 1999; Fitzgerald 2002; Hiltebeitel 2001: 105–118.

stage upon which folk narratives were presented, the epic was designed to incorporate and indeed to *perform* them in such a way that both constructed authoritative versions of these narratives and controlled the insurgent discourses contained within them. Ultimately, the *Mahābhārata* might indeed have neutralized the socially and religiously dangerous implications of Viśvāmitra's *tapas* through supplementation, but the fact that it needed to do so at all highlights the critical power of folk narratives in ancient India: to provide a site within normative epic culture in which counter-normative voices can be heard, and where dominant social ideology is questioned, if only for a brief, fleeting moment.

REFERENCES

BASCOM, William 1965. The forms of folklore: prose narratives. *Journal of American Folklore* 78: 3–20.

BAUMAN, Richard 1975. Verbal art as performance. *American Anthropologist* (N.S.) 77: 290–311.

BAUMAN, Richard & Charles BRIGGS 1990. Poetics and performance as critical perspectives on language and social Life. *Annual Review of Anthropology* 19: 59–88.

BHATTACHARYA, Deepak 1987. Viśvāmitra in later Vedic literature. In: Chaubey 1987: 36–45.

BIARDEAU, Madeleine 1999. Kāmadhenu: La vache mythique, symbole de prospérité. In: Yves Bonnefoy (ed.), *Dictionnaire des mythologies et des religions des sociétés traditionelles et du monde antique*: 2139–2140. Paris: Flammarion.

BLACKBURN, Stuart 1988. *Singing of Birth and Death. Texts in Performance.* Philadelphia: University of Pennsylvania Press.

——— 1996. The Brahmin and the mongoose: the narrative context of a well-travelled tale. *Bulletin of the School of Oriental and African Studies* 59: 494–507.

BROCKINGTON, John 1998. *The Sanskrit Epics.* Leiden: E.J. Brill.

CHAUBEY, B. B. (ed.) 1987. *Viśvāmitra in Vedic and Post Vedic Literature.* Hoshiarpur: Panjab University.

DONIGER, Wendy 1991. Fluid and fixed texts in India. In: Joyce Burkhalter Flueckiger & Laurie J. Sears (eds.), *Boundaries of the Text.*

Epic Performances in South and Southeast Asia: 31–41. Ann Arbor: Center for South and Southeast Asian Studies.

FELLER, Danielle 2004. *The Sanskrit Epics' Representation of Vedic Myths*. Delhi: Motilal Banarsidass.

FITZGERALD, James L. 1983. The Great Epic of India as religious rhetoric: a fresh look at the *Mahābhārata*. *Journal of the American Academy of Religion* 51: 611–630.

——— 2002. The Rāma Jāmadagnya 'thread' of the *Mahābhārata*: a new survey of Rāma Jāmadagnya in the Pune text. In: Mary Brockington (ed.), *Stages and Transitions. Proceedings of the Second Dubrovnik International Conference on the Sanskrit Epics and Purāṇas*: 89–132. Zagreb: Croatian Academy of Sciences and Arts.

——— 2003. The many voices of the *Mahābhārata*. *Journal of the American Oriental Society* 123: 808–818.

——— 2006. Negotiating the shape of 'scripture': new perspectives on the development and growth of the *Mahābhārata* between the empires. In: Patrick Olivelle (ed.), *Between the Empires. Society in India 300 BCE to 400 CE*: 257–286. New York: Oxford University Press.

FLUECKIGER, Joyce Burkhalter 1996. *Gender and Genre in the Folklore of Middle India*. Delhi: Oxford University Press.

FOLEY, John Miles 1988. *The Theory of Oral Composition. History and Methodology*. Bloomington: Indiana University Press.

GOLDMAN, Robert 1972. Akṛtavraṇa and Śrīkṛṣṇa as narrators of the legend of the Bhārgava Rāma: à propos some observations of Dr. V. S. Sukthankar. *Annals of the Bhandarkar Oriental Research Institute* 53: 161–173.

——— 1977. *Gods, Priests, and Warriors. The Bhṛgus of the Mahābhārata*. New York: Columbia University Press.

——— 1978. Fathers, sons, and gurus: Oedipal conflict in the Sanskrit epics. *Journal of Indian Philosophy* 6: 325–392.

——— (trans.) 1984. *The Rāmāyaṇa of Vālmīki. An Epic of Ancient India*, vol. 1: *Bālakāṇḍa*. Princeton: Princeton University Press.

GOMBACH, Barbara 2000. Ancillary Stories in the Sanskrit Mahābhārata. Ph.D. Dissertation, Columbia University.

GRINTSER, Pavel 1974. *Drevneindijskij epos. Genezis i tipologiya*. Moscow: Nauka.

HARIYAPPA, H. L. 1953. *Ṛgvedic Legends through the Ages*. Pune: Deccan College.

HILTEBEITEL, Alf 1999. Reconsidering Bhṛguization. In: Mary Brockington & Peter Schreiner (eds.), *Composing a Tradition. Concepts, Techniques and Relationships. Proceedings of the First Dubrovnik International Conference on the Sanskrit Epics and Purāṇas*: 155–168. Zagreb: Croatian Academy of Sciences and Arts.

—— 2001. *Rethinking the Mahābhārata. A Reader's Guide to the Education of the Dharma King.* Chicago: University of Chicago Press.

—— 2005. Not without subtales: telling laws and truths in the Sanskrit Epics. *Journal of Indian Philosophy* 33: 455–511.

HONKO, Lauri 1998. *The Siri Epic as performed by Gopala Naika*, 2 vols. (Folklore Fellows Communications, 265–266.) Helsinki: Academia Scientarium Fennica.

HOPKINS, E. Washburn 1901. *The Great Epic of India. Its Character and Origin.* New York: Scribner (reprint: Calcutta: Punthi Pustak, 1969).

JONG, J. W. de 1975. Recent Russian publications on the Indian epic. *Adyar Library Bulletin* 39: 1–47.

KAPADIA, B. H. 1971. *Viśvāmitra in the Vedic, Epic, and the Puranic Literature.* Vallabh Vidyanagar: Sardar Patel University.

LAINE, James W. 1989. *Visions of God. Narratives of Theophany in the Mahābhārata.* Vienna: Institut für Indologie der Universität Wien.

LOMMEL, Hermann 1965–66. Vasiṣṭha und Viśvāmitra. *Oriens* 18–19: 200–227.

LORD, Albert B. 1960. *The Singer of Tales.* Cambridge, Mass.: Harvard University Press.

MAGNONE, Paolo 1993. The development of 'tejas' from the Vedas to the Purāṇas. *Wiener Zeitschrift für die Kunde Südasiens* 36 (Suppl.): 137–147.

Mahābhārata, Critical Edition. Ed. by V. S. Sukthankar, S. K. Belvalkar, P. L. Vaidya *et al.* 24 vols. Poona. Bhandarkar Oriental Research Institute, 1933–71.

MINKOWSKI, Christopher Z. 1989. Janamejaya's *sattra* and ritual structure. *Journal of the American Oriental Society* 109: 401–420.

—— 1991. Snakes, *sattras*, and the *Mahābhārata*. In: Arvind Sharma (ed.), *Essays on the Mahābhārata*: 384–400. Leiden: E.J. Brill.

MUIR, John 1868. *Original Sanskrit Texts on the Origin and History of the People of India, their Religion and Institutions*. 2nd edition. London: Trübner & Co.

MÜLLER, Friedrich Max 1860. *A History of Ancient Sanskrit Literature, so far as it Illustrates the Primitive Religion of the Brahmans*. London: Williams and Norgate.

NARAYAN, Kirin 1989. *Storytellers, Saints, and Scoundrels*. Philadelphia: University of Pennsylvania Press.

NARAYANA RAO, Velcheru 1993. Purāṇa as Brahminic ideology. In: Wendy Doniger (ed.), *Purāṇa Perennis. Reciprocity and Transformation in Hindu and Jaina Texts*: 85–100. Albany, NY: SUNY Press.

OLDENBERG, Hermann 1922. *Das Mahābhārata. Seine Entstehung, sein Inhalt, seine Form*. Göttingen: Vandenhoek und Ruprecht.

PARGITER, Frederick Eden 1913. Viśvāmitra and Vasiṣṭha. *Journal of the Royal Asiatic Society*: 885–904.

——— 1917. Viśvāmitra, Vasiṣṭha, Hariścandra and Śunaḥśepa. *Journal of the Royal Asiatic Society*: 37–67.

——— 1922. *Ancient Indian Historical Tradition*. London: Oxford University Press.

PROPP, Vladimir 1968. *Morphology of the Folktale*. 2nd edition. Austin: University of Texas Press.

PROUDFOOT, Ian 1979. Interpreting *Mahābhārata* episodes as sources for the history of ideas. *Annals of the Bhandarkar Oriental Research Institute* 60: 41–63.

RAHURKAR, V. G. 1964. *Seers of the Ṛgveda*. Pune: University of Pune.

RAMANUJAN A. K. 1999. Who Needs Foklore? In: Vinay Dharwadker (ed.), *The Collected Essays of A. K. Ramanujan*: 532–552. New Delhi: Oxford University Press.

Rāmāyaṇa. The Vālmīki-Rāmāyaṇa. Ed by Govindlal Hargovind Bhatt *et al.* 7 vols. Baroda: Oriental Institute, 1960–75.

REICH, Tamar 2001. Sacrificial violence and textual battles: inner textual interpretation in the Sanskrit *Mahābhārata*. *History of Religions* 41: 142–169.

ROTH, Rudolph 1846. *Zur Litteratur und Geschichte des Weda*. Stuttgart: A. Liesching & Co.

SHARMA, Umesh Chandra 1975. *The Viśvāmitras and the Vasiṣṭhas. An Exhaustive Historical Study, Vedic and Post-Vedic*. Aligarh: Viveka Publications.

SUKTHANKAR, V. S. 1936. Epic Studies, 6: The Bhṛgus and the Bhārata. A text-historical study. *Annals of the Bhandarkar Oriental Research Institute* 18: 1–76.

SUTTON, Nick 1997. Aśoka and Yudhiṣṭhira: a historical setting for the ideological tensions of the *Mahābhārata*. *Religion* 27: 333–341.

VASSILKOV, Yaroslav 1995. The *Mahābhārata*'s typological definition reconsidered. *Indo-Iranian Journal* 38: 249–256.

WEBER, Albrecht 1854. *Indische Studien. Beiträge für die Kunde des Indischen Alterthums*, vol. 2. Berlin: Dümmler.

WHITAKER, Jarrod 2000. Divine weapons and *tejas* in the two Indian Epics. *Indo-Iranian Journal* 43: 87–113.

——— 2002. How the gods kill: the *Nārāyaṇa astra* episode, the death of Rāvaṇa, and the principles of *tejas* in the Indian Epics. *Journal of Indian Philosophy* 30: 403–430.

WHITE, David Gordon 1993, You are what you eat: the anomalous status of dog-cookers in Hindu mythology. In: R. S. Khare (ed.), *The Eternal Food. Gastronomic Ideas and Experiences of Hindus and Buddhists*: 53–93. Albany: SUNY Press.

Unrooted Trees:
A Way around the Dilemma of Recension

WENDY J. PHILLIPS-RODRIGUEZ

One of the dark truths that has always haunted the textual critic when trying to find out the relationships between a series of manuscripts (MSS) is the dilemma of recension. That is:

> One cannot create a stemma until one knows what readings are archetypal, but one cannot determine what readings are archetypal until one has a stemma (Robinson & O'Hara 1992).

The core of this paradox resides in the fact that to establish the stemma, or the genealogical tree of a series of MSS we should know exactly which readings are original and which are not. However, if we already know that information, why should we need to build a stemma at all? (Donaldson 1970)

The reality is that at the beginning of the analysis the textual critic very rarely knows what the original text was like. He may have a fair idea of certain matters of style, language, metrics, etc. but only in very few cases would he be able to tell in the first instance an archetypal reading from a non-archetypal one. He has to go through the long and meticulous procedure of collating and analysing all the available copies of his text. Only later, he may venture an opinion as to what the archetype may have been like.

Throughout this discussion we will refer to the example of the *Mahābhārata*, behind which there has been a long and complicated process of transmission. Firstly, as an oral document it can be traced back to several centuries BC. Secondly, it kept growing and shaping itself freely in every recitation for several centuries till the time it was committed to writing and began a different phase of transmission in the first centuries AD. All this means that the text has gone through several centuries of undocumented oral transmission followed by at least ten more centuries of lost written evidence, and it has kept continually changing ever since.

When the editors of the different sections of the Critical Edition (CE) tried to reconstruct the text of the *Mahābhārata*, they necessarily had to build a stemma, or at least a general pedigree, to help them validate their editorial decisions. This pedigree was a guiding tool to determine the importance that each manuscript would have at the moment of choosing readings to reconstitute the text (Sukthankar 1933: xxx; see Figure 1).

As can be seen, at the end of the branches of the diagram we find the groups of extant MSS collated for the CE. The MSS have been classified according to the scripts they have been written in, except for the group called K. The MSS of the K group are Devanāgarī MSS that preserve a text distinctively related to what is considered to be one of the oldest texts, the Kashmirian version. If we continue looking upwards at the tree we find Greek characters representing lost archetypes. These intermediary archetypes subdivide the script groups into three main branches: roughly, Kashmirian (ν), North-eastern (γ and ε) and Southern (σ). The most significant level of branching is the big division between Northern and Southern recensions. At the top of the tree we have the archetype from which all copies are supposed to derive.

One thing we have to be aware of is the direct relationship between branching events and time line. Evidently, the more times a text is copied by hand the more chances are that it will suffer modifications, either intentionally or unintentionally. This means that the more we advance along the time line, the more

FIGURE 1

Pedigree of Manuscripts Proposed by V. Sukthankar in his Prolegomena

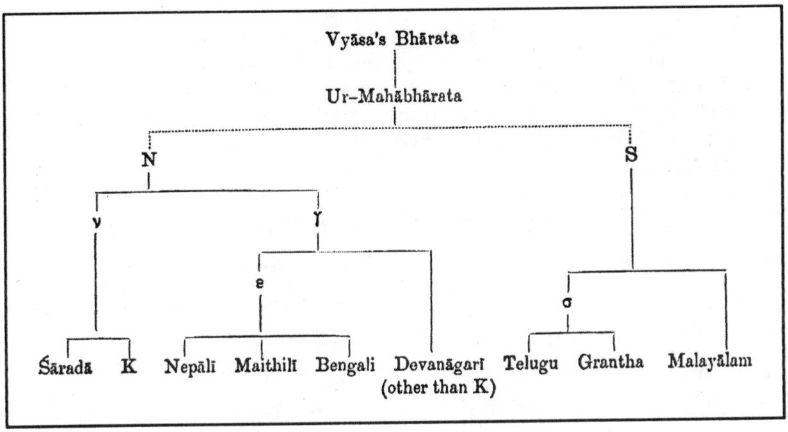

suffer modifications, either intentionally or unintentionally. This means that the more we advance along the time line, the more divisions we will find at the end of the diagram. That is exactly why the tree structure has been used so often to depict the historical relationships between MSS.

Now, going back to our central point: How was this pedigree made? How did the editors infer which readings were archetypal and which were later modifications? Close examination and collation of MSS can render very important information. However, to sort out the relationships between a series of MSS that contain thousands of variant readings is indeed a daunting task. Numerically speaking there are millions of possible combinations of trees.[1] Focusing on certain variants and relying on his own

---

[1] For a demonstration on how the number of possible stemmata increases exponentially whenever a new MS is added, see Flight 1990.

knowledge of the tradition, the only thing the scholar can do is to try to produce the most coherent stemma by trial and error. Let me quote here the words of Edgerton after presenting his hypothesis about the relationships between the MSS of the Sabhāparvan:

> The conclusions I have arrived at are based on very long and painstaking study of the variants of the whole work. They are radically different from such preconceptions as I had at the outset. They have changed repeatedly in the course of the investigation. They may not be right, but at least they have not been hastily adopted. (Edgerton 1944: xxxix)

During the last decades many advances have taken place in the practice of textual criticism. Besides electronic collation, the most important of them is the use of certain computational algorithms that can help us infer probable genealogical relationships between MSS (Barbrook *et al.* 1998; Howe *et al.* 2001). These algorithms, termed phylogenetic, were originally created in the field of evolutionary biology with the purpose of mapping the relationships between organisms and inferring the evolutionary history of species from their DNA sequences. Most probably, Lucy L. Rosenstein was the first to apply this kind of analyses to texts in Indian languages when editing her critical study and edition of *The Devotional Poetry of Svāmī Haridās*. In her chapter about textual criticism she gives a brief but illuminating explanation of the two methods she used to try to determine the historical relationships between MSS: phenetics and cladistics. Both analytic methods are borrowed from the field of evolutionary biology. However, they have an essential difference in the way they classify the material. As Rosenstein explains:

> The phenetic method assesses overall similarities and differences between objects (here MSS), with no regard given to how these similarities or differences may have arisen (Rosenstein 1997: 63).

This means that the algorithm does the clustering and ordination without making any 'historical' assumption. It just classifies elements according to how many equal or unequal traits exist between them. The common error ascribed to phenetics is that, on the basis of the overall degree of similarity, it tends to group together elements that may be genealogically unrelated. For example, a phenetic approach may group together bats, birds and flies on the basis that all of them have wings and can fly. However, these three organisms belong to completely different families, do not have any recent ancestor in common and developed the trait of wings through totally independent evolutionary paths. Therefore an analysis based in phenetics would show effectively how similar certain MSS are but would be completely unable to differentiate between true genealogical relationships (*homologies*) and mere parallelisms (*homoplasies*).

Cladistics, on the other hand, assumes evolution-like progressive relationships (phylogenies) between texts. This means that the algorithm does not work only by grouping MSS according to their agreements or their disagreements. Instead, it proceeds by looking at patterns of variation that are most likely to reveal genealogical relationships between them. The classifying principle behind this type of analysis is not overall similarity but common ancestry. Therefore, just as in traditional stemmatics, it considers that only variant readings can tell us something about the evolution of the text. This is because, once a MS has introduced a variant reading, there is a large probability that every copy from that exemplar will inherit that same variant. Hence, when several MSS have a similar pattern of variant readings it means that they are most probably genealogically related and therefore should be grouped in the same branch.

The results that these programs throw at us are in the familiar shapes of trees. These trees are called *phylograms* and are used as visual representations of the hypothetical kinship relationships between a series of elements, in our case of MSS. Phylograms are very much like the pedigree we have just seen except that they provide more information. This is because the analysis method is able to account for absolutely every single informative variant in each MS and can try the millions of possible combinations before hypothesising a number of optimal stemmata. For example, a phylogram of some episodes[2] of the Sabhāparvan looks like the diagram on the next page (Figure 2).

We made this phylogram by feeding the program *PAUP** (*Phylogenetic Analysis using Parsimony and other methods*) an accurate record of all the agreements and disagreements between the 31 MSS collated for the CE. This tree illustrates several of the relationships previously hypothesised by the editors except that it describes them with further detail and infers a larger number of intermediary archetypes. It has been particularly successful at identifying the three main subdivisions. All the MSS corresponding to the Kashmirian version are very closely located, except for K4 which is known to contain a contaminated text, mixing the Kashmirian version with that of Nīlakaṇṭha[3]. At this particular episode K4 is following closely Nīlakaṇṭha's version and therefore has been grouped with the Dn MSS. The North-eastern version consisting of MSS in Bengali, Devanāgarī, Newari and Maithili scripts appears to be very well differentiated. The same can be said about the Southern recension, which contains MSS in Telugu, Grantha and Malayālam scripts.

As can be observed, this phylogram has an orientation. To give a direction to this tree we had to tell the program what MS we believe to preserve the largest amount of archetypal readings.

---

[2] The material consists of (approx.) 1000 lines of text and corresponds to *sargas*: 43–47, 51, 59, 60, 64 and 65 of the Dyūtaparvan.

[3] Sukthankar's description of K4 in his Prolegomena says that it was 'copied from different exemplars; some parvans have the commentary of Nīlakaṇṭha, while others contain some old text tradition'. In the second volume of the edition, Edgerton added that the agreements of K4 with other MSS of the tradition 'seem to me to make it quite certain that the "old text tradition" noted by Sukthankar as blended in the MS. with a vulgate source, was no other than the Kashmirian' (Edgerton 1944: xii).

FIGURE 2

Phylogram of the 31 Manuscripts Used for the Critical Edition

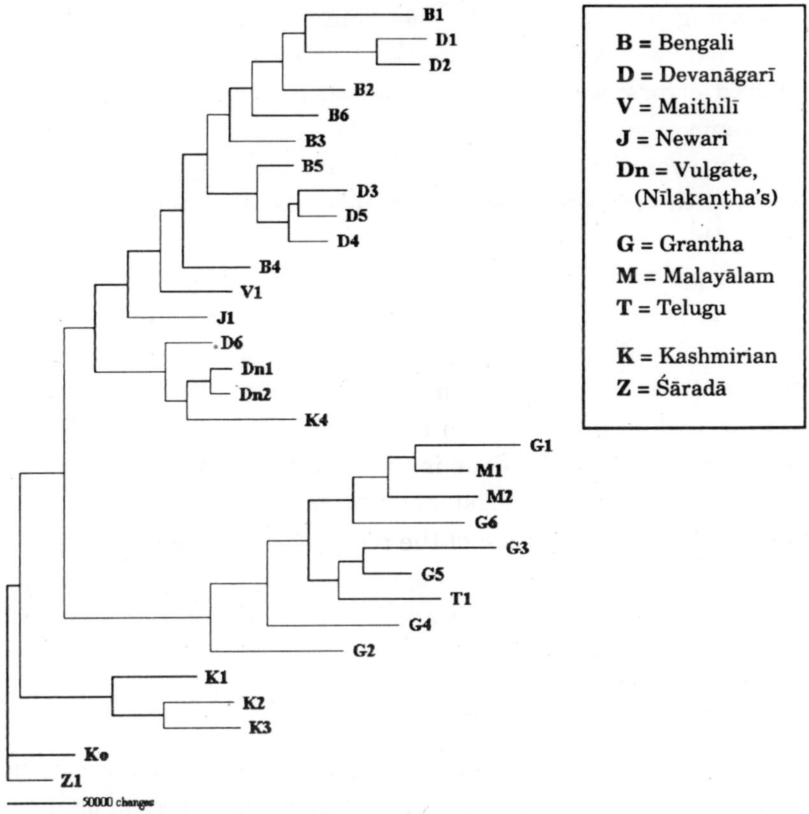

Following the hypothesis of the editors of the CE we instructed the program to take the Śāradā codex (Z1) as our best MS. However, we are left with the same problem as before. There still remains the element of doubt that the Śāradā codex in fact preserves the largest amount of archetypal readings. The truth is that we are basing all our analysis on a hypothesis that is highly subjective.

The necessity to give the stemma a direction was the explicit reason that discouraged Rosenstein from using cladistics when analysing her material. As she rightly points out, in a textual tradition 'we can mostly speak of shared variants but not of shared errors: we are not sure which is the original and which the innovation' (Rosenstein 1997: 64).

The evolutionary biologists, who are the people that normally work with cladistic programs, had found the same problem all along the way. In the history of evolution of life, which genes are archetypal and which are later modifications? What species gave place to what other species? For decades they have pondered such questions, arriving at the following conclusions: Since we do not have enough information about the events that took place in the past, it is impossible to solve this dilemma. Therefore, the only thing that can be done is to attempt a new kind of analysis that will require the least amount of guess-work from us. In particular, as the position of the root is the least certain aspect of the entire analysis, it makes sense to concentrate upon the part of the problem that can be addressed most accurately, namely inferring the interrelations between the extant MSS. In other words, if we do not know where to put the root of the tree, let us try to build *unrooted trees*.

An unrooted tree specifies only the nodes' interrelations but says nothing about the direction in which the evolution occurred. Therefore, it has the advantage that it does not require from us any subjective judgment and works only with the available data. Instead of taking one MS as the base for every combination of relationships, it takes all the MSS at the same time and organizes them according to their evolutionary patterns. An unrooted tree of the same section of the Sabhāparvan looks like the diagram opposite (Figure 3).

FIGURE 3

Unrooted Phylogram of the 31 Manuscripts Used for the Critical Edition

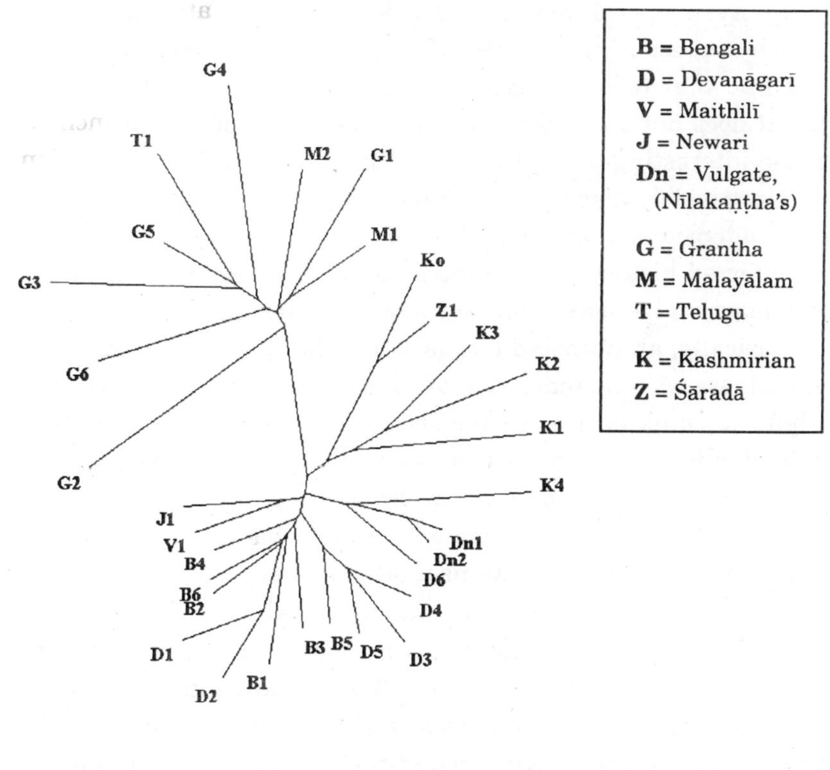

Besides reproducing most of the relationships that the editors had previously hypothesised, this unrooted phylogram has been able to suggest relationships that never before were observed so clearly. For instance, let us look at the way that MS B5

is attached to the branch of Devanāgarī MSS the same way that D1 and D2 appear in a branch mostly composed of Bengali exemplars. This, together with the fact that these two branches merge into one, gives us strong reasons to suspect that the difference in script is only superficial and they actually belong to a single group. Even our contaminated MS K4 is sensibly attached to the branch representing Nīlakhaṇṭha's version but separated by a long line that tells us that it contains several variant readings that it does not share with the other members of this branch. It is also interesting to see that the MSS of the Southern recension are more widely dispersed, a probable sign that they evolved in a more independent way than they did in the North. In the North, the inferred history is more intricate, with several intermediary exemplars and a more complex system of interrelations.

Basically, an unrooted tree is a tree that presents patterns of relatedness with no indication of time. In visual terms, it may be helpful to think of it as a tree seen from the aerial perspective. Instead of looking at it from the side as we did with the previous trees, we look at it from above where the top of the tree and the arrangements of the branches can be seen but the root lies underneath and therefore remains unseen by us.

Strictly speaking, however, an unrooted tree is more of a rhizome, a non-hierarchical structure whose centre can be anywhere, than a tree proper (Stolz 2006). The advantage of such a non-hierarchical structure is that it avoids stratification of data. It therefore provides a way around the dilemma of recension. As we mentioned before, the reason why Rosenstein decided to use phenetics over cladistics to analyze her data was precisely because she wanted to avoid making any subjective judgement about the direction of change:

> We can note changes in the text, but when we do not know which change is the 'right' reading, i.e. the original, or at least closer to the original, we cannot deduce the direction of change. This in turn

> means that we cannot construct a stemma in the traditional way. What, then, might we usefully do instead? (Rosenstein 1997: 62)

For some reason, however, she did not explore the possibilities that unrooted trees could provide for her:

> Since our aim as textual editors is to find the earlier forms, the 'unrooted' cladogram is not of much use to us (Rosenstein 1997: 64).

Instead, when she found herself forced to give an orientation to her trees she gave up her initial purpose and commented with disappointment that:

> Should we fail in our attempt to sort out MSS into a tree and build up a stemma, the phenetic analysis may at least enable us to see which MSS most closely resemble each other (Rosenstein 1997: 64).

I believe that had she reconsidered unrooted trees at the time when she had already abandoned the idea of building a 'stemma in the traditional way', she might have had a different view on the applicability of cladistics to textual traditions.

It is very important to notice that the construction of these trees does not make redundant the work or the knowledge of the scholar in any way. On the contrary, once the program is able to provide us with a series of hypotheses, it is the task of the scholar to verify them carefully. The great advantage of these tools is that they suggest where to look and empower the scholar by giving him control over large amounts of material that would otherwise remain unmanageable.

Working with unrooted trees, the manipulation of the data in different ways can render extremely interesting results. For example, it is possible to build trees of different sections of the text and compare them in order to see how stable the relationships between MSS are along the text. It is also possible to concentrate only in certain branches or groups of MSS and study their connections in further detail. We can try to give an answer to a whole new set of questions such as: How precise is the classification of the MSS by scripts? How many different streams of textual transmission can we identify? What is the distribution of variants along different passages and what can it tell us about MS evolution? In this way it will be possible to reach a much finer-grained understanding of the relationships between MSS, based rigorously on the available data.

Of course, at the end of the day unrooted trees will still leave us with the same old question: What was the archetypal *Mahābhārata* like? Since these new sharper tools will certainly tell us more about the relationships between our extant MSS we could use that knowledge to try to root our stemma. However, once again we would be stepping into the realm of uncertainty. Rather than to help us reconstruct an ideal text, the great advantage of these new methodologies is that they offer the possibility to study the *Mahābhārata* as a cultural phenomenon whose variations and ramifications have acquired an independent cultural value. Why privilege one version over the others? Are they not all valuable as separate interpretations, each with its particular history of evolution?

After facing the dilemma of recension for a long time, unrooted trees finally appear to provide us with a way to proceed. There is much work to be done in this line – several issues have to be discussed and refined before these tools can reach their full potential. For the moment, it is enough to know that there is a way past the old dilemma that will help us cope better with our uncertainties about the text. This, however, does not mean that the problems of editing works preserved in multiple versions are solved. Phylogenetic tools may be pushing the limits further just to let us see that the complexities involved in the transmission of texts may be even more intricate than we thought.

* * *

I would like to thank the National Council for Science and Technology (CONACyT, Mexico) for their generous support of my research.

REFERENCES

BARBROOK, A. C., C. J. HOWE, N. BLAKE & P. ROBINSON 1998. The phylogeny of *The Canterbury Tales*. *Nature* 394: 839.

DONALDSON, E. T. 1970. The psychology of editors of Middle English texts. In: E. T. Donaldson, *Speaking of Chaucer*: 102–118. London: Athlone.

EDGERTON, F. D. (ed.) 1944. *Sabhāparvan*. (Vishnu S. Sukthankar *et al.* [eds.], *The Mahābhārata*, vol. 2.) Poona: Bhandarkar Oriental Research Institute.

FLIGHT, C. C. 1990. How many stemmata? *Manuscripta* 34: 122–128.

HOWE, C. J., A. C. BARBROOK, M SPENCER, P. ROBINSON, B. BORDALEJO & L. R. MOONEY. Manuscript evolution. *Endeavour* 25: 121–126.

ROBINSON, P. & R. O'HARA 1992. Report on the Textual Criticism Challenge 1991. *Bryn Mawr Classical Review* 3.4: 331–337.

ROSENSTEIN, L. L. 1997. *The Devotional Poetry of Svāmī Haridās*. Groningen: Egbert Forsten.

STOLZ, M. 2006. Linking the variance: unrooted trees and networks. In: Caroline Macé *et al.* (eds.), *The Evolution of Texts. Confronting Stemmatological and Genetical Methods.* (Linguistica Computazionale, 24–25): 193–213. Pisa: Istituti Editoriali e Poligrafici Internazionali.

SUKTHANKAR, V. S. (ed.) 1933. *The Mahābhārata*, vol. 2: *Ādiparvan*. Poona: Bhandarkar Oriental Research Institute.

References for methods

ROBINSON P. 2005. *Collate. Interactive Collation of Large Textual Traditions*. Leicester: Scholarly Digital Editions.

SWOFFORD D. L. 2003, *PAUP*b.10, Phylogenetic Analysis Using Parsimony (*and Other Methods)* Version 4. Massachusetts: Sinauer Associates.

Upamās Occurring in Speeches: 'Abusive' Similes in the Sabhāparvan and Karṇaparvan

ANTONELLA COSI

INTRODUCTION

The analysis of *upamās* and the function they fulfil within the Epics reveal a variegated picture. Sanskrit Epics feature a wide range of similes, employed with different purposes within the narrative. The most common type of comparison occurring within the *Mahābhārata* is the 'decorative' instance. As shown in my previous paper (Cosi 2005), the study of the context in which *upamās* occur shows two important facts: similes can be employed with more than one purpose at a time and they can occur in sequences. Work carried out during my research on similes and themes within the Epics reveals that there are groups of comparisons which occur in relation to specific themes within the main narrative.

In her paper 'Remarques sur la function des figures de style dans le Rāmāyaṇa' (1996), Marie-Claude Porcher analyses the functions of figures of speech, particularly similes, in that Epic. In her research she points out a basic difference between the function of *upamās* occurring in the main narrative and within discourses. Similes are employed in order to 'reinforce antitheses, condemnations, denunciations etc.' in the former, whereas in the latter, comparisons are employed in order to 'underline the different speech-acts'. Research on similes within the *Mahābhārata* shows a far more differentiated use of similes within the different speech-acts of the various characters. It employs a wide range of discourses and dialogues: within the Sabhāparvan, for instance, the function of similes in speeches depends greatly on how each discourse is employed within the main narrative (whether to introduce different types of narrative digressions or didactic material). The *Mahābhārata* shows a basic difference between dialogues and discourses: while the latter display comparisons performing several functions, depending on the subject matter of the speech, the former display repetitive wording. *Upamās* in particular often employ a range of *upamānas*: animals, but also gurus, old men etc. Several passages are taken into account for the analysis of this type of comparison: dialogues in the Dyūtaparvan and the harsh exchange between Karṇa and Śalya in the final stages of the battle, which will lead to the death of Karṇa.

The aim of this research is to show how within the Sabhāparvan and the Karṇaparvan of the *Mahābhārata* a specific type of comparison repeatedly occurs within dialogues (as opposed to the more diversified field of similes occurring in speeches); the analysis reveals a set group of 'abusive'[1] *upamās* within the traditional stock-in-trade comparisons. The analysis of *upamās* within the Dyūtaparvan will first be taken into account, to show how similes are employed within discourses. Then I will focus on the 'abusive' similes, in order to highlight the different types featured elsewhere in the *Mahābhārata*. The small section containing the harsh exchange between Karṇa and Śalya will then be taken into account in order to show how the Karṇaparvan

---

[1] By the word 'abusive' I imply similes employed in order to insult somebody; therefore, regardless of the *upamānas* displayed by the simile, the word classifies comparison employed with a specific function.

employs more sophisticated 'abusive' similes, to highlight similarities and differences between the two types.

THE DYŪTAPARVAN AND ITS COMPOSITION

The Sabhāparvan is considered the key book of the *Mahābhārata*, for in it the events in the main narration lead to the loss of the kingdom and the consequent fratricidal war between the Kauravas and the Pāṇḍavas. Within the second book sub-sections, the Dyūtaparvan features the intense chain of events that will lead to the exile of the five protagonists of the *Mahābhārata*.

According to Edgerton,[2] the number of inconsistencies shown by the Dyūtaparvan's plot is consistent with the hypothesis of two parallel versions having been amalgamated into one. As later shown by Renate Söhnen-Thieme (1999), careful analysis of the sub-section reveals that the *triṣṭubh* verses only duplicate the contents of the *anuṣṭubh* verses. This inclines the author to consider the former later than the latter. The aspect concerning the evolution of the Dyūtaparvan will be considered as a background to the analysis of 'abusive' similes.

The Dyūtaparvan features a number of discourses and conversations among different characters, mainly between Vidura and Duryodhana, displaying a considerable number of similes.

Prior to the analysis of comparisons occurring within the section, it will be worth mentioning the sequence of events of the section taken into account.

SIMILES IN THE DYŪTAPARVAN

Similes in discourses feature a range of comparisons, employed with different purposes within the narrative. A number of *upamās* within the section occur in descriptive passages. Duryodhana, during his conversation with his uncle, profusely describes

---

2 Preface to the Sabhāparvan (Edgerton 1944: xxxiii).

the Pāṇḍavas and their riches. Talking to Śakuni, he tells him that the sacrifice held by the Pāṇḍavas was similar to the one Indra held among the Gods (*yathā śakrasya deveṣu tathābhūtaṃ mahādyute*, 2.43.20). He also says that his enviousness is burning him day and night, drying him like a small pond (*śuṣye toyam ivālpakam*, 2.43.21). He also describes the kings waiting at the gates of the Hall like Vaiśyas paying taxes (*vaiśyā iva karapradāḥ*, 2.43.25). Then Duryodhana talks about the previous attempts to kill the Pāṇḍavas, which all miserably failed, while Yudhiṣṭhira prospers like a lotus on the water (*vṛddho 'psv iva paṅkajam*, 2.43.33).

Similes appear again in *adhyāya* 46, after two chapters in which no comparison is employed. The first instances found are employed in the conversation between King Dhṛtarāṣṭra and his son Duryodhana. The king tries to convince his son to give up his envy and the decision to hold a dicing game to play against the Pāṇḍavas. The king lists all Duryodhana's possessions that make him shine like the lord of the gods in Heaven (*divi deveśvaro yathā*, 2.46.16). But the prince replies that his fortunes are not even comparable to Yudhiṣṭhira's: the Kauravas resemble servants if compared to him (*āvarjitāṃ ivābhānti*, 2.46.21).

In *adhyāya* 47 Duryodhana tells of the riches amassed by the Pāṇḍavas at the consecration, describing the horses they received as homage as having the swiftness of the wind (*anilaraṃhasaḥ*, 2.47.13). He also relates how some of the ambassadors of other countries have been refused at the gates of the Hall, although bringing many riches as gifts (2.47.16–18):

ekapādāṃś ca tatrāham apaśyaṃ dvāri vāritān ।
balyarthaṃ dadatas tasmai hiraṇyaṃ rajataṃ bahu ॥ 16 ॥
indragopakavarṇābhāñ śukavarṇān manojavān ।
tathaivendrāyuddhanibhān saṃdhyābhrasadṛśān api ॥ 17 ॥
anekavarṇān āraṇyān gṛhītvāśvān manojavān ।
jātarupam anarghyaṃ ca dadus tasyaikapādakāḥ ॥ 18 ॥

> I myself saw the One-footers excluded at the gate, after they arrived with large tribute in gold and silver. They brought horses, some of the colour of rain mites, others of parrots, fast as thought, some resembling the rainbow and others the clouds at twilight. They seized many-coloured wild horses as fast as thought, and the One-footers presented him with priceless gold.

The description continues in *adhyāya* 48, with more details about the tribute. In 2.48.5 Duryodhana tells of the honey and the yak-plumes, glittering like the moon (*śaśiprabhān*). In 2.48.19 he talks about the elephants, as big as mountains (*śailābhān*). In the next *adhyāya* the description concludes with the assembled kings and seers that resemble the seven seers in heaven who came to great Indra, the king of gods (*mahendram iva devendraṃ divi saptarṣayo yathā*, 2.49.12). The last instance in this chapter is in the lament of Duryodhana, who tells his father that, like a yoke tied by a blind man, all has come loose: the junior branch prosper and the senior one declines (*andheneva yugaṃ naddhaṃ*, 2.49.24). This kind of proverbial simile occurs several times across the section, particularly in the speeches made by the men in the assembly during the dicing game.

In *adhyāya* 50 comparisons are non-decorative and quite short. Here the similes add more emphasis to Duryodhana's speech, in order to convince his father to hold the dicing game. All similes seem to be proverbial, such as for instance in 2.50.21, where Duryodhana says that the king who does not contend is eaten up by the earth, like a snake eats up mice (*sarpo bilaśayān iva*). In 2.50.23, he expresses his concern over the flourishing of the Pāṇḍavas, by saying how they will, one day, cut the Kauravas' roots, like a swelling disease (*vyādhir āpyāyita iva*). A similar comparison is employed in 2.18.13 by Yudhiṣṭhira, before the departure of his brother for the conquest of Jarāsaṃdha. But while the former is talking about the cousins and the danger they represent for himself, the latter expresses his concerns about the expedition: Yudhiṣṭhira without his brothers feels miserable like a disease. The third instance is in 2.50.24, where a more articulated comparison is employed to describe the danger the Pāṇḍavas represent:

> *alpo 'pi hy arir atyantaṃ vardhamānaparākramaḥ* |
> *valmīko mūlaja iva grasate vṛkṣam antikāt* || 24 ||
>
> Though small, an enemy whose strength grows in an excessive way swallows one up, like an anthill swallows up the tree near whose roots it was born.

This simile, a longer one, is neither formulaic nor decorative. It is rather similar to a short story set up in order to explain Duryodhana's point of view.

The only simile in *adhyāya* 51 seems to be a rather proverbial one. Trying to convince his father, Duryodhana tells him that the indulgent person who avoids risks and protects himself perishes as if he were standing like straw that rots in the rainy season (*varṣāsu klinnakaṭavat*, 2.51.8).

In *adhyāya* 52 Vidura is sent to invite the Pāṇḍavas. At his meeting with Yudhiṣṭhira, Yudhiṣṭhira asks Vidura about his uncle, King Dhṛtarāṣṭra. Vidura replies by telling him that the old king is healthy and he sits in the midst of his kinsmen similar to Indra (*indrakalpaiḥ*, 2.52.6). After inviting him to the dicing game, Vidura tries to discourage Yudhiṣṭhira from accepting the challenge, but the king of Indraprastha replies that he is obliged to accept because he has to follow his destiny (2.52.18):

> *daivaṃ prajñāṃ tu muṣṇāti tejaś cakṣur ivāpatat* |
> *dhātuś ca vaśam anveti pāśair iva naraḥ sitaḥ* || 18 ||
>
> Fate steals one's reason like glare blinds the eyes. A man follows the will of the Disposer as if tied with fetters.

In this sentence there is all Yudhiṣṭhira's awareness of the events to come. Both simile and *utprekṣā* are proverbial rather than decorative. The last simile is in 2.52.27: the queen Gāndhārī surrounded by all her sisters-in-law resembles the constellation Rohiṇī surrounded by the stars (*tārābhir iva rohiṇīm*). The constellation of Rohiṇī does not shine at all: the simile probably refers to the voluntary blindness of Gāndhārī.

Two similes employing the gods and the sun are in *adhyāya* 53:

> *śuśubhe sā sabhā rājan rājabhis taiḥ samāgataiḥ* I
> *devair iva mahābhāgaiḥ samavetais triviṣṭapam* II 20 II
> *sarve vedavidaḥ śūrāḥ sarve bhāsvaramūrtayaḥ* I
> *prāvartata mahārāja suhṛddyūtam anantaram* II 21 II

> The Hall, o king, shone with the assembled kings as heaven with the lordly assembled gods. With all these Veda-wise champions, all like the sun incarnate, then, great King, the family game began at once.

The close symbolism between the sun and the gods in general has previously been analysed.[3] It is a very common parallel within the *Mahābhārata* (Sharma 1964: 33).

Two interesting similes are in *adhyāya* 54, when the game has already started and Yudhiṣṭhira is staking all his possessions. Both relate to the stake. In the first case he bets his chariot that is victorious and holy, resounding like the clouds or the sea (*meghasāgaranisvanaḥ*, 2.54.5). In the second instance he stakes his elephants, resembling clouds or mountains (*naga-meghanibhā*, 2.54.10). Both employ clouds as *upamāna*s.

In *adhyāya* 55 we find another kind of simile. The mood in the narration has changed: Vidura asks Dhṛtarāṣṭra to stop Yudhiṣṭhira from betting his riches. He tries to do so, although he knows that his words will not please Yudhiṣṭhira, any more than medicine for a moribund man (*mumūrṣor auṣadham iva*, 2.55.1). It is at this stage that the insults towards Duryodhana start. Vidura, trying to make him change his mind, tells Dhṛta-rāṣṭra that the evil-minded Duryodhana was destined to be the killer of the lineage of the Bhāratas since his birth, when he appeared shrieking like a jackal (2.55.2):

---

[3] In my paper at the 4th Dubrovnik International Conference on the Sanskrit Epics and Purāṇas (Sept. 2005).

yad vai purā jātamātro rurāva
gomāyuvad visvaraṃ pāpacetāḥ I
duryodhano bhāratānāṃ kulaghnaḥ
so 'yaṃ yukto bhavitā kālahetuḥ II 2 II

As soon as he was merely born he was shrieking like a jackal, the evil-minded Duryodhana, the slayer of the lineage of the Bhāratas, he will be the future cause of (our) Death.

This short simile is an anticipation of the list of insults with which Vidura will address Duryodhana.

Other instances of very short similes appear in this chapter: some of them are small comparisons added to a short story, such as in 2.55.4–5, a short tale about a mead-drinker, who, drunk, does not care about his kinsmen and loses everything; the comparison follows the story in 2.55.5 (*madhuvat*). A sequence of short similes is in 2.55.14–16. The first of them recalls a short story told in 2.55.13, where a man strangled his own birds, later repenting. Vidura tells Yudhiṣṭhira not to betray the Pāṇḍavas, as the man did with the birds (*pakṣihā puruṣo yathā*, 2.55.14). He also encourages the eldest Pāṇḍava to care about his brothers and not to pluck them, like a garland maker (2.55.15, *mālākāra iva*, 2.55.15). He then exhorts him not to burn them, like the charcoal-burner with the tree (*vṛkṣāṅgakarīva*, 2.55.16).

In *adhyāya* 56 there is Vidura's final exhortation, warning of Duryodhana's plot to take away his possessions from Yudhiṣṭhira (2.56.3):

duryodhano madenaiva kṣemaṃ rāṣṭrād apohati I
viṣāṇaṃ gaur iva madāt svayam ārujate balāt II 3 II

Duryodhana, in his folly, robs the kingdom of its safety, like a bull, because of its sexual excitement, breaks his own horn with strength.

The insult addressed to Duryodhana again displays an animal as an *upamāna*. The insults continue in the next *adhyāya*, where Duryodhana replies to Vidura's abuse (2.57.3):

utsaṅgena vyāla ivāhṛto 'si
mārjāravat poṣakaṃ copahaṃsi I

bhartṛghnatvān na hi pāpīya āhus
tasmāt kṣattaḥ kiṃ na bibheṣi pāpāt || 3 ||

Like a snake sitting on one's lap, like a cat who hurts the one who feeds it, they say that fratricide is the worst thing, so why Steward, are you not afraid of (such an) evil act?

Verse 2.57.15 features Vidura's reply:

na śreyase nīyate mandabuddhiḥ
strī śrotriyasyeva gṛhe praduṣṭā |
dhruvaṃ na roced bharatarṣabhasya
patiḥ kumāryā iva ṣaṣṭivarṣaḥ || 15 ||

A stupid man leads to no good, like a corrupt woman in the house of a scholar. He would not suit the bull of the Bharatas like a sexagenarian does not suit a girl.

Adhyāya 58 contains mostly very short similes, several of them in *triṣṭubh* verses. The similes are not decorative: in 2.58.14, Yudhiṣṭhira staking Sahadeva tells Śakuni to play against him like an enemy (*apriyavat*). In 2.58.17, sure that he is the victim of *adharma* deeds, Yudhiṣṭhira argues with his opponent, who wants to pluck the Pāṇḍavas like flowers (*sumanasāṃ*).

Then Śakuni tells Yudhiṣṭhira that he prattles like mad-men (*utkaṭā iva*, 2.58.19). When the Pāṇḍavas' elder brother stakes Arjuna, he compares him to a boat that carried everyone in battle (*naur iva*, 2.58.20).

The first longer comparison of *adhyāya* 58 occurs in 2.58.23, when Yudhiṣṭhira stakes Bhīmasena, comparing him to the thunderbolt-wielder, Dānava's foe (*yathā vajrī dānavaśatrur ekaḥ*).

A sequence of short comparisons refers to Draupadī when Yudhiṣṭhira bets her: in 2.58.33 she is compared to the goddess Śrī (*śrīsamānayā*), in 2.58.36 she is described as having a lotus-like face (*padmavat*) that shines like the jasmine (*mallikeva*). When Yudhiṣṭhira stakes Draupadī, the assembly raises its voice, and in the midst of the general confusion, Vidura, fearing bad consequences, seizes his face in despair, annihilated, wheezing like a snake (*niḥśvasan pannago yathā*, 2.58.40).

After the final stake, in which Draupadī has been lost, Vidura warns Duryodhana of the dangerousness of his misdeeds. He tells him that he is like a bamboo that comes into flower only to kill itself (*veṇur ivātmaghātī*, 2.59.5). He also adds that the ones who speak neither good nor bad of the householder, speak evil of the wise ascetic, barking like dogs (*śvanarāḥ sadaiva*, 2.59.9).

Similes in *adhyāyas* 60–63 are merely descriptive and decorative. Therefore the analysis on 'abusive' similes will not take these into account.

ANALYSIS OF 'ABUSIVE' SIMILES IN THE DYŪTAPARVAN

Within the Dyūtaparvan, two types of 'abusive' similes can be identifed. The first type employs animals as *upamānas*; this type also features among the Arghābhiharaṇaparvan similes. The insults perpetrated towards or by Duryodhana very often involve animal comparisons, such as cats, snakes (2.57.3ab) and bulls (2.56.3). Animals as *upamānas* are a common feature of similes within the Epics, although they can be employed in different ways. The bull, for instance, within the epic literature is generally a symbol of strength and power, but the image of a 'bull that because of its sexual excitement, breaks his own horn with strength' (2.56.3) obviously conveys a completely different meaning. In the Arghābhiharaṇaparvan, Śiśupāla insults Bhīṣma calling him a dog (2.34.19) and so does Bhīṣma when talking about the kings who back Śiśupāla up. The second type, curiously occurring in longer verses, employs a range of *upamānas*, including a sexagenarian, a boat and other images. These similes are employed in order to portray 'improbable' situations, such as a sexagenarian marrying a young girl. As will be shown later, this particular type of simile is often employed in order to insult.

THE EVOLUTION OF 'ABUSIVE' SIMILES IN THE KARṆAPARVAN

The Karṇaparvan features one of the most famous (and also interesting) word-exchanges of the Epics: the verbal fight between Karṇa and Śalya. The king of the Madras, who deeply despises the *sūta*'s son, begins his attempt to destroy Karṇa's energy (cf. Hiltebeitel 1976: 242). Similes occurring in the passage employ similar *upamānas*, such as animals, but in far more elaborate images. The first instance employed displays a mythological image (8.27.19):

> *bālyād iva tvaṃ tyajasi vasu vaiśravaṇo yathā* ।
> *ayatnenaiva rādheya draṣṭāsy adya dhanaṃjayam* ॥ 19 ॥
>
> From foolishness you are giving away wealth like Vaiśravaṇa, without any effort, o son of Rādha, you will behold Dhanaṃjaya (the winner of riches) today.

The absurdity of the situation is emphasized with two more images, one of which occurs in a longer verse (8.27.25, 33):

> *samudrataraṇaṃ dorbhyāṃ kaṇṭhe baddhvā yathā śilām* ।
> *giryagrād vā nipatanaṃ tādṛk tava cikīrṣitam* ॥ 25 ॥
>
> Your purpose is like someone wishing to cross the ocean with his arms after attaching a stone to his neck, or to someone leaping from a mountain summit.
>
> *bālaś candraṃ mātur aṅke śayāno*
> *yathā kaś cit prārthayate 'pahartum* ।
> *tadvan mohād yatamāno rathasthas*
> *tvaṃ prārthayasy arjunam adya jetum* ॥ 33 ॥
>
> Similar to a child trying to seize the moon while sitting on his mother's lap, in the same way, out of folly, standing on your chariot, you seek to vanquish the resplendent Arjuna today.

The 'animal' *upamāna* reappears a few verses later, but, unlike most similes occurring in the Dyūtaparvan, displays a double *upamāna* in order to emphasize differences between two characters (8.27.35–36):

siddhaṃ siṃhaṃ kesariṇaṃ bṛhantaṃ
bālo mūḍhaḥ kṣudramṛgas tarasvī I
samāhvayet tadvad etat tavādya
samāhvānaṃ sūtaputrārjunasya II 35 II
mā sūtaputrāhvaya rājaputraṃ
mahāvīryaṃ kesariṇaṃ yathaiva I
vane sṛgālaḥ piśitasya tṛpto
mā pārtham āsādya vinaṅkṣyasi tvam II 36 II

Your challenge to Arjuna today, o sūta's son, is like a young little deer would challenge a huge maned lion excited with wrath. Do not challenge that great hero of a prince, o sūta's son, like the jackal gratified with meat challenging the lion in the forest. Do not be destroyed encountering Pārtha.

As shown in the first section of this article, the Dyūtaparvan employs images of the lion (2.62.37), of lions and deer (2.64.12), and of the jackal (2.55.2). Although the instances occurring in the Dyūtaparvan fulfil a decorative function rather than being employed as abuse in speech-acts, it is interesting to point out how these types of *upamāna*s gather around passages whose main theme is developed through conversations or dialogues.[4]

Again, the lion and the jackal appear, along with the snake and Garuḍa (8.27.39–40):

siṃhaṃ kesariṇaṃ kruddham atikramyābhinardasi I
sṛgāla iva mūḍhatvān nṛsiṃhaṃ karṇa pāṇḍavam II 39 II
suparṇaṃ patagaśreṣṭhaṃ vainateyaṃ tarasvinam I
laṭvevāhvayase pāte karṇa pārthaṃ dhanaṃjayam II 40 II

Karṇa, going too far, you yell at that lion-man the Pāṇḍava like a jackal, out of foolishness, yells at the angry lion. Karṇa, like a snake for its own destruction challenges that best of birds, Vinatā's son, possessed of beautiful plumage and great activity, so you do with Dhanaṃjaya Pārtha.

Another short sequence shows a similar wider range of animal *upamāna*s (8.27.42–44):

---

[4] Similes in the Arghābhiharaṇa- and Śiśupālavadhaparvan also feature a similar range of *upamāna*s.

ṛṣabhaṃ dundubhigrīvaṃ tīkṣṇaśṛṅgaṃ prahāriṇam ׀
vatsa āhvayase yuddhe karṇa pārthaṃ dhanaṃjayam ॥ 42 ॥
mahāghoṣaṃ mahāmeghaṃ darduraḥ pratinardasi ׀
kāmatoyapradaṃ loke naraparjanyam arjunam ॥ 43 ॥
yathā ca svagṛhasthaḥ śvā vyāghraṃ vanagataṃ bhāṣet ׀
tathā tvaṃ bhāṣase karṇa naravyāghraṃ dhanaṃjayam ॥ 44 ॥

Karṇa, you challenge Dhanaṃjaya Pārtha (like) a calf challenging a smiting bull of sharp horns and neck thick like a drum. (Like) a frog (croaking) to a terrible and massive cloud yielding abundant showers of rain, you croak against Arjuna, who is Parjanya among men. Like a dog standing inside the house would bark at a tiger in the forest, so you bark at that tiger among men, Dhanaṃjaya.

A simile describing Arjuna and Kṛṣṇa precedes another short sequence highlighting the differences between the characters, in order to reinforce Śalya's statements (8.27.47, 51–52):

vyāghraṃ tvaṃ manyase 'tmānaṃ yāvat kṛṣṇau na paśyasi ׀
samāsthitāv ekarathe sūryācandramasāv iva ॥ 47 ॥

You consider yourself a tiger, till you see the two Kṛṣṇas standing on the same chariot like the sun and the moon.

yathākhuḥ syād biḍālaś ca śvā vyāghraś ca balābale ׀
yathā sṛgālaḥ siṃhaś ca yathā ca śaśakuñjarau ॥ 51 ॥
yathānṛtaṃ ca satyaṃ ca yathā cāpi viṣāmṛte ׀
tathā tvam api pārthaś ca prakhyātāv ātmakarmabhiḥ ॥ 52 ॥

Like a mole would be to a cat, like a dog to a tiger in relative strength, like a jackal to a lion and like a hare and an elephant, like falsehood and truth, like poison and nectar, so you and Pārtha are known to all for your personal deeds.

This short sequence confirms one of the main functions of 'abusive' similes: to describe the difference between the 'evil' and the 'good'. In one of the most famous passages of the *Rāmāyaṇa*, the abduction of Sītā, a very similar sequence of comparisons is employed in order to underline the distinctions between Rāvaṇa and Rāma (*Rāmāyaṇa* 3.45.40–42).[5]

---

[5] *yad antaraṃ siṃhaśṛgālayor vane, yad antaraṃ syandanikāsamudrayoḥ* ׀ *surāgryasauvīrakayor yad antaraṃ, tad antaraṃ dāśarathes tavaiva ca* ॥ 40 ॥ *yad antaraṃ kāñcanasīsalohayor, yad antaraṃ candanavāripaṅkayoḥ* ׀ *yad antaraṃ hastibiḍālayor vane, tad antaraṃ daśarathes*

After Śalya's long discourse aiming to discourage Karṇa, the son of Radhā replies describing his plan of challenging Arjuna with his weapons. A sequence of insults addressed to Śalya features no similes employed with the purpose of verbally abusing the king of the Madras. Despite the insults and the threats,[6] Śalya stresses again Arjuna's superiority over Karṇa with a simile employing again the jackal and the lion as *upamāna*s (8.28.57):

> *yatra vyastāḥ samastāś ca nirjitāḥ stha kirīṭinā* |
> *sṛgālā iva siṃhena kva te vīryam abhūt tadā* || 57 ||
>
> Where you were all defeated by the diadem-decked Arjuna, like jackals by a lion, what became of your prowess?

The final simile employed by Śalya finally closes the similes of 'difference', stating the apparent superiority of Arjuna (8.28.63):

> *kiyantaṃ tatra vakṣyāmi yena yena dhanaṃjayaḥ* |
> *tvatto 'tiriktaḥ sarvebhyo bhūtebhyo brāhmaṇo yathā* || 63 ||
>
> Here I will tell you, by which qualities Dhanaṃjaya is superior to you, like a Brāhmaṇa is superior to all the creatures.

Comparisons within the passage feature two types of simile also employed within the Sabhāparvan: similes displaying animals as *upamāna*s and others, featuring improbable situations. Whereas the former show features very close to the ones previously analysed, the latter show some important new characteris-

---

tavaiva ca || 41 || *yad antaraṃ vāyasavainateyayor, yad antaraṃ madgumayūrayor api* | *nyad antaraṃ sārasagṛdhrayor vane, tad antaraṃ dāśarathes tavaiva ca* ||
The same difference between a lion and a jackal in the forest, the difference between the ocean and a small pond, the difference between good wine and vinegar, such is the difference between the son of Dāśaratha and you. The same difference between the gold and the lead, the difference between sandal paste and mud, the difference between an elephant and a cat in the forest, such is the difference between the son of Dāśaratha and you. The same difference between a crow and Garuḍa, the difference between a seagull and a peacock, the difference between a crane and a vulture, such is the difference between the son of Dāśaratha and you.

[6] In 7.27.103 Karṇa threatens Śalya to crush his head with his club.

tics. The main aim of Śalya is to make Karṇa lose his temper, hoping that this will affect his ability to fight, and consequently advantage Arjuna in battle. In order to achieve that, he speaks using a sequence of similes whose aim is to insult Karṇa while glorifying Arjuna at the same time. The result is a double simile of 'difference' that will provoke Karṇa's anger. The similarities with the verses in the Araṇyakāṇḍa clearly show that this was a well known principle for epic composers.

Interestingly enough, some of the similes employing animals as *upamānas* carry the features of the 'improbable' type as well: the little deer challenging a lion (8.27.35), or the jackal fighting against the lion in the forest (8.27.36) clearly show situations that are impossible in real life.

CONCLUSION ON THE 'ABUSIVE' SIMILES

The analysis of 'abusive' similes employed in the passages analysed reveals some important facts. There are a few principles followed by the author in order to achieve an insult through comparisons. Animals are often employed, 'impure' animals in particular. The second type generally employs human beings or 'improbable' situations in general. But within the Karṇaparvan, the two types of comparisons are combined in some of the instances employing animals as *upamānas*. Similes in the latter passage appear to be an 'evolution' of *upamās* occurring within the Sabhāparvan. One of the most striking types of simile occurring within the Karṇaparvan is the *upamā* of difference. The instances within other passages occurring within the Sabhāparvan and similes within the Araṇyakāṇḍa show that the composers of the Epics were well aware of the principles behind the 'abusive' similes.

REFERENCES

COSI, Antonella 2005. The importance of contextual analysis in studies of similes: the case of the Śiśupāla episode in the Mahābhārata. Paper presented at the 4th Dubrovnik International Conference on the Sanskrit Epics and Purāṇas (2005).

EDGERTON, F. D. (ed.) 1944. *Sabhāparvan.* (Vishnu S. Sukthankar *et al.* [eds.], *The Mahābhārata*, vol. 2.) Poona: Bhandarkar Oriental Research Institute.

HILTEBEITEL, Alf 1976. *The Ritual of Battle. Krishna in the Mahābhārata.* Ithaca: Cornell University Press.

PORCHER, Marie-Claude 1996. Remarques sur la fonction des figures de style dans le Rāmāyaṇa. In: N. Balbir & G.-J. Pinault (eds.), *Langue, style et structure dans le monde indien. Centenaire de Louis Renou. Actes du Colloque international.* Paris: Libraire Honoré Champion.

SHARMA, R. K 1964. *Elements of Poetry in the Mahābhārata.* Berkeley: University of California Press.

SÖHNEN-THIEME, Renate 1999. On the composition of the Dyūtaparvan in the Mahābhārata. In: Mary Brockington & Peter Schreiner (eds.), *Composing a Tradition. Concepts, Techniques and their Relationships. Proceedings of the First Dubrovnik International Conference on the Sanskrit Epics and Purāṇas*: 139–154. Zagreb: Croatian Academy of Science and Arts.

The Heart in the *Mahābhārata*

SVEN SELLMER

EPIC PSYCHOLOGY

The following investigations about the heart in the *Mahābhārata* form part of a larger project of mine: the reconstruction of an 'epic psychology'. By that term I mean the implicit presuppositions that underlie the way mental events – taken in a wide sense: emotions, feelings, thinking, volition, etc. – are described in the epics. It is my working hypothesis that this implicit psychology makes up a coherent system and is, at least partially, independent of the explicit 'scientific' psychological theories found in other types of literature (and also in some passages of the epics themselves, most notably in the *Mokṣadharma*).[1] If this hypothesis is warranted, it can emerge only at the end of much more comprehensive and detailed analyses than I am able to present here.[2]

DIMENSIONS OF THE HEART

Of the nouns used to describe mental events in the *Mahābhārata* both words for 'heart' – *hṛd* and *hṛdaya* – belong to the most im-

---

1 'Scientific' is used as a broad term here, including pieces of introspective psychology, theories that are found in the medical literature, and also mystico-theological conceptions.

2 I am planning to present further material in future publications. A first draft of my project can be found in Sellmer 2009.

portant ones. This state of affairs is far from being exceptional, because the heart also plays a major role in many other psychological languages of different cultures.[3] One of the reasons for this seems to lie in the multi-layered richness and versatility of the heart language. In this paper I am going to use a three level model based on some phenomenological conceptions of the German philosopher Hermann Schmitz because it provides, in my conviction, a very useful tool for organizing and analyzing the Mahābhāratan evidence.[4] The following levels are distinguished:

(1) *The level of the physical body.* Here the heart is an organ consisting of muscles, blood, etc. (just like liver or kidneys) that can be seen and touched, and also literally be pierced, wounded, and the like.

(2) *The level of the felt body.* The concept of the felt body according to Schmitz can only be presented very roughly here.[5] The felt body consists of everything we feel 'in' or 'around' our body. So the body and the felt body are in a way connected, but they are not one and the same thing. If you close your eyes and try to feel your body, you will quite quickly notice that not all parts of your body are individually represented in your felt body (e.g., your single toes). On the other hand, when somebody feels a phantom limb, there is a part of the felt body where there is no body.[6] The

---

[3] To name only the closest 'relative' of the Indian epics, in the Homeric language we have even three words for 'heart' that all appear in psychological contexts (κῆρ, κραδίη and ἦτορ); cf. Sellmer 2009.

[4] For an application of this approach to the evidence in the *Ṛksaṃhitā* and for further bibliographical references see Sellmer 2004.

[5] The felt body (German: Leib, French: chair) was not invented by Schmitz, rather it is a common conception of modern phenomenology (dating back to Scheler, in fact). But he draws the most radical distinction between body (Körper) and felt body (Leib), which makes his theory especially apt for our purposes. The most comprehensive presentation of his investigations is Schmitz 1982. He applies his conceptions to questions of heart imagery in Schmitz 1969: 209–213, and, above all, in Schmitz 1996.

[6] It is important to make a clear distinction between (e.g.) the swollen thumb we see before our eyes (a part of the body) and what we feel when we have just hit it with a hammer (a part of the felt body). In everyday language

heart in this sense is a whole region with unclear borders in the upper middle part of our felt body where (mostly uncomfortable or painful) sensations may occur.

(3) Thirdly, there are several types of *non-concrete usages* where the heart often appears as a kind of 'inner space' containing thoughts, wishes and the like. (In modern English the mind plays a similar role.)[7]

THE *MAHĀBHĀRATA* EVIDENCE

The two terms I am concerned with in this paper are *hṛd* and *hṛdaya*, both probably etymologically connected to the English word 'heart' (though in a less straightforward manner than it may seem at first glance; cf. Szemerényi 1970). Being originally two independent words, in the *Mahābhārata* they virtually form one system, in which *hṛdaya* appears in the nominative, accusative, instrumental, ablative, genitive and locative case, *hṛd* almost exclusively in the locative (neither of them occurs in the vocative or dative). The figures (on the basis of the electronic text of the Critical Edition prepared by Tokunaga *et al.*, revised by Smith) are presented in the table on the next page.

Only in the instrumental and locative is there something like an opposition between the two words. Still, no difference in meaning can be discerned, the parallel occurrences seem to be due to metrical reasons, so it is safe to treat *hṛd* and *hṛdaya* as instances of one and the same lexeme.[8]

---

both levels are mixed up, and that is a common source of misunderstandings.

[7] I prefer to use the expression 'non-concrete' rather than 'metaphorical', because metaphorical language is not characteristic for the third level. There are also metaphors that involve the heart of the felt body level, but these refer, after all, to the felt body, which is more concrete than the space of thoughts.

[8] This is also the way they appear in Pāṇini's grammar (6.1.63 and 6.3.50–51).

	***hṛd* (only sg.)**	***hṛdaya***
N	–	70 (4 times pl.)
V	–	–
A	–	44 (once pl.)
I	6	29
D	–	–
Abl	–	14
G	–	22
L	125	35

A full and detailed picture of the specific functions of the heart words in the *Mahābhārata* can only be drawn by comparing other words belonging to the same semantic field, most importantly *manas*. But this is beyond the scope of this paper, in which I will confine myself to showing the *internal* structure of the heart language. In a first step, the main usages (excluding 'scientific' ones in the sense noted above, footnote 1) will be presented and exemplified by typical verses; the arrangement is primarily according to the phenomenological three level model, but also secondarily according to the occurring grammatical cases.[9]

First level

As my concern is mainly with psychology, the first, bodily level can be treated rather briefly. It appears exclusively in contexts of fighting, most often as the place in which a warrior is pierced or

[9] In the following, compounds are taken into account as well (which adds about 80 occurrences to the numbers given in the table above, apart from the extremely common noun *suhṛd*); cases in compounds are understood functionally. – I leave out the instances where 'heart' is used in the sense of 'essence', which is neither physiological nor psychological (e.g. *akṣa-hṛdaya*).

hit by arrows and other missiles. Both *hṛd* and *hṛdaya* are used promiscuously (about 20 occurrences in the locative each), they clearly function as metrical alternatives.[10] In some instances the whole region around the heart – i.e., the chest, not only the heart proper – is meant.[11]

Second level

Coming to the level of the felt body, there are several expressions that have to be taken into consideration. Firstly, the heart shakes (√*kamp*)[12] and trembles (*ud / pra-*√*vep*).[13] Insofar as the beating of the heart is outwardly observable, these occurrences also belong to the first level, but it is clear that the main point here is *felt* agitation, which is a question of the felt body. I also take it to be a re-reference to (negative) changes in the felt body when the heart is getting tired (√*glai*) or is sinking ([*pra-*]√*pat*), or is being destroyed (√*kṣi*).[14] The heart is further said to be torn asunder ([*vi-*]√*dṝ*)[15] and to be burnt, tormented and the like ([*pra-*/ *nir-*]√*dah*, √*du*, √*tap* [caus.], √*śuṣ*)[16] – sometimes with the addition of *iva*, sometimes without: quite stereotyped expressions, no doubt, but at least originally referring to the painful, quasi-

---

[10] At the beginning of the c *pāda*, for example, we have 4 times each: *hṛdi vivyādha* and *vivyādha hṛdaye*.

[11] Cf. 5.149.81cd: ... *kavacādīni hṛdy abhūvan nṛṇāṃ tadā* || 'cuirasses and the like decked the men's chests' (trans. van Buitenen).

[12] √*kamp* appears most often in the *bahuvrīhi* compound *hṛdayakampana* 'heart-shaking' (6.18.1, 22.18; 1.43.25), apart from one of the few occurrences where *hṛdaya* is found in the plural (*hṛdayāni cakampire*, 3.39.6).

[13] E.g. 5.22.30cd: *pravepate me hṛdayaṃ bhayena śrutvā kṛṣṇāv ekarathe sametau* || 'And with fright does my heart shudder when I hear the two Kṛṣṇas have joined on a single chariot' (trans. van Buitenen); other verses: 1.43.31; 3.58.24; 5.73.18.

[14] 12.54.16; 8.67.15; 6.114.83; 7.47.25.

[15] E.g. 5.80.41ab: *vidīryate me hṛdayaṃ bhīmavākśalyapīḍitam* | 'tormented by the arrow of Bhīma's words my heart is torn asunder' (+ 29 further passages).

[16] 1.5.21; 5.36.7; 6.61.3; 10.4.23, 27, 32; 5.26.7, 30.46; 6.41.13; 12.1.38, 335.31; 14.67.11; 5.45.21; 1.112.24; 5.58.17.

burning sensation in the heart region that may accompany the thinking about sad topics or the learning of bad news. When a person's heart, in spite of such a situation, is not 'torn asunder', this is attributed to the 'hardness'[17] of her heart; this feature of the heart thus represents something like a character trait. In the cases mentioned so far the heart appears in the accusative or nominative case. A cognate group of passages features the combination *hṛdayena* (once *hṛdā*) + med./pass. present (once past) participle, thus adverbially adding the information that the main action is done in a certain emotional state.[18]

Third level

The heart appears as the target of an action when (in comparatively few passages) it is said to be gladdened (*pra-√hlād* [caus.], *√prī*, *√hṛṣ* [caus.], *√nand*) and pleased (*√tuṣ* [caus.]).[19] Such emotions may often be accompanied by changes in the felt body, but these are not explicitly referred to, so the occurrences of this kind belong rather to the third, non-concrete level.

The vast majority of non-concrete usages show a different construction, however: Here, a mental event or entity is located

---

[17] The expressions used are: *aśmasāramaya*, *adrisāramaya*, *āyasa*, *vajrād dṛdhatara*, *vajrasāramaya*, and *durbhida*. Several times we find them applied in rhetorical self-blaming: the fact that my heart does not break and I continue to live in the face of, for example, the tragic death of a close person shows my lack of affection. Please note that in these instances the breaking of the heart is treated literally (on the body level), which does not contradict my hypothesis, however, that the *origin* of this expression is on the felt body level.

[18] E.g. 3.107.2–3: 'The strong-armed prince [Bhagīratha] heard that the Great-spirited Kapila had brought about that terrible end of his forebears, and that they had failed to go to heaven. The king of men entrusted the kingdom to his minister and with burning heart [*hṛdayena vidūyatā*] went to the slope of the Himālaya to do austerities there, wishing to propitiate the Ganges' (trans. van Buitenen). Similar verbs as the ones mentioned before appear in this construction: *pra-√vep* (1.43.31), *√du* (14.67.11), *vi-√du* (3.107.3, as quoted above, 1.53.5; 5.142.20), *√math* (1.145.11). The past participle mentioned is *kṛpāsaṃgṛhīta* (3.13.91).

[19] 1.185.17; 3.79.17; 5.88.7; 13.101.31; 3.79.17.

'in the heart' (in most cases *hṛdi*, but also sometimes *hṛdaye*). The range of things so located is quite wide: propositions, plans, wishes, wrath, but also persons and gods, and – in a few instances that touch upon the felt body level – suffering and trembling. In many cases it is not obvious if this way of expression is chosen for other than rhetorical or metrical reasons. But sometimes the context suggests that the location in the heart implies either that a wish, or the like, is particularly intense or that it is hidden to others.[20]

If something is 'in' the heart, it is no wonder that we also find a couple of passages where the possibility is envisaged (though denied) that something could 'leave' (mostly *apa-√sṛp*) the heart (ablative).[21]

A few words are connected with the heart by way of a genitive construction, most often 'peace' (*śānti* – 13 occurences).

Where we find the heart (*hṛdaya*, only in a few cases *hṛd*) used in the instrumental case, one has to distinguish between thc usages where the heart is further qualified by a participle that were mentioned under the former heading, and others where the pure instrumental occurs. In the latter instances, to do something 'with the heart' seems to have more or less the same implications as the location 'in the heart', i.e. intensity and/or privateness.

---

20 The nuance of intensity is more difficult to demonstrate, but the privateness of the things located in the heart emerges quite clearly in situations when there is a contrast between hiding and revealing them (in speech), e.g. 5.15.10 [Indrāṇī telling Nahuṣa]: *kāryaṃ ca hṛdi me yat tad devarājāvadhāraya* | *vakṣyāmi yadi me rājan priyam etat kariṣyasi* || 'Think about what I want, in my heart, to be done, Lord of the Gods: I will tell you if you will do me this favour.'

21 See 2.72.9 [The gods destroy a man by taking away his wits]: *buddhau kaluṣabhūtāyāṃ vināśe pratyupasthite* | *anayo nayasaṃkāśo hṛdayān nāpasarpati* || 'When destruction is immanent and his mind is beclouded, the wrong course appears as the right one and cannot be dislodged from his heart.' (trans. van Buitenen) (Also 5.34.79, 58.21, 141.44; 7.50.5; 12.283.20; 15.1.25, and [with *vi-√i*] 1.154.25).

In only three instances the heart functions as the subject of a transitive verb: it knows (√*vid*) (1.68.25; 2.18.7; 12.240.1).

Lastly, there are a few places where the heart is characterized by an adjective that describes an emotional-intellectual state.[22]

Overview

After that short survey of the main features of the heart language in the *Mahābhārata* it is now possible, with the help of the scheme below, to show how the rather wide-ranging usages are logically and semantically interconnected.

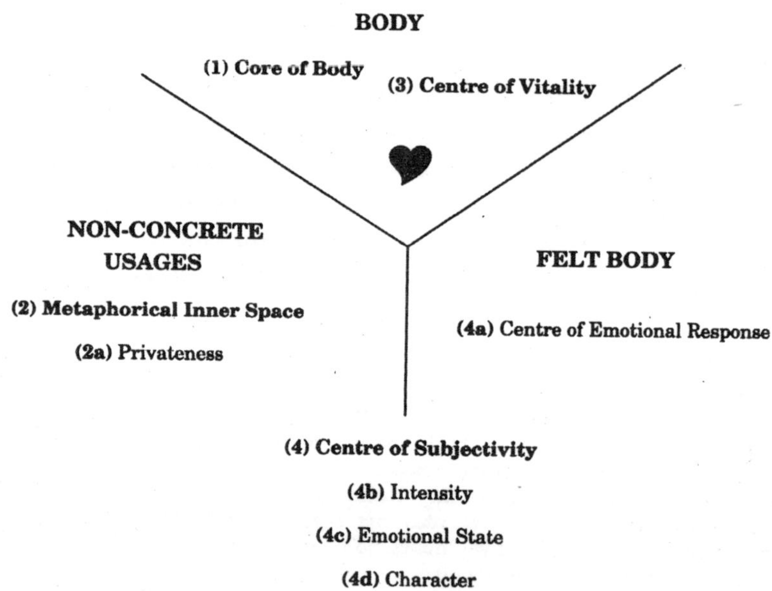

[22] *dvidhā* (3.59.23; 7.26.12) and *bhīta* (6.41.24).

The three levels (capital letters, bold type) are represented by the three segments of the scheme, words in bold type mark the main areas of meaning, more specific areas are given in standard type.

(1) The heart is an organ deeply hidden in the chest, in that sense it is the very core of the whole body, the one of its parts that is best hidden from the outside world. (2) And this is the link to the key non-concrete usage of the heart as 'inner space' whose main characteristic is the (2a) privateness of the things going on there. (Therefore it is the exact opposite to the public space of spoken words, to which we will return in a moment.) (3) Coming back to the body level, the heart is also the centre of vitality, because one immediately dies when it is wounded. (4) Therefore it is of utmost importance for every person; in this sense it is the centre of existential subjectivity, as well. This is a largely non-concrete notion, but it also has a phenomenological aspect because, again on the felt body level, (4a) the heart is one of the most important places where an emotional response manifests itself. The last three points in the scheme (4b–d) touch both upon the non-concrete and the felt body level. Intensity is a feature of an emotional response, but also naturally linked to the 'inner space' metaphor. Finally, emotional states and character traits are connected with the way a person's felt body responds (actually or habitually); at the same time, for outward observers, they belong to the inner world of the other person.

The heart-wounding word

All three levels of the heart language are touched upon simultaneously in a certain metaphor that therefore deserves to be briefly discussed, though it is neither particularly original, nor striking. I have in mind the word hitting and tormenting the heart like a dart or arrow. Sometimes this picture is touched upon only very briefly,[23] but the following passage expands it a little more (5.34.76–77):

karṇinālīkanārācā nirharanti śarīrataḥ ।
vākśalyas tu na nirhartuṃ śakyo hṛdiśayo hi saḥ ॥ 76 ॥
vāksāyakā vadanān niṣpatanti yair āhataḥ śocati rātryahāni ।
parasya nāmarmasu te patanti tān paṇḍito nāvasṛjet pareṣu ॥ 77 ॥

Arrows of various kinds may be pulled from a body,
But the dart of speech cannot be extracted, for it sticks in the heart.
From the mouth fly forth the arrows of speech,
Hit by which one suffers by day and by night;
They hit another in his weak spots –
The wise should not shoot them at other people.
(Trans. van Buitenen, slightly modified)

A short analysis of this simple metaphor may serve as a kind of summary of this paper:

(1) On the literal side, a real arrow pierces the body *organ* of the heart.

(2) On the felt body level there is the *pain* caused by the word (i.e., the metaphorical arrow).[24]

(3) Finally, the word (e.g., an insult), belonging to the public sphere, manages to enter and to wound, as it were, the private *sphere of consciousness*, just like an arrow penetrates the core of the physical body.

---

[23] It may even be condensed, as it were, to one of the compounds that both appear in the quoted verses: *vākśalya* or *vāksāyaka*.

[24] As noted above, I presume that originally there is a reference to *a real pain* as it is felt quite often on hearing very bad and shocking news.

REFERENCES

SCHMITZ, Hermann 1969. *Der Gefühlsraum.* (System der Philosophie, 3.1.) Bonn: Bouvier.

——— 1982. *Der Leib.* (System der Philosophie 2.1). 2nd edition. Bonn: Bouvier.

——— 1996. Leibliche Quellen der Herzmetaphorik. In: Georg Berkemer & Guido Rappe (eds.), *Das Herz im Kulturvergleich* (Lynkeus, 3): 23–31. Berlin: Akademie-Verlag.

SELLMER, Sven 2004. The heart in the Ṛg-veda. In: Piotr Balcerowicz & Marek Mejor (eds.), *Essays in Indian Philosophy, Religion and Literature*: 71–86. Delhi: Motilal Banarsidass.

——— 2009. Towards a semantics of the mental in the Indian Epics: comparative and methodological remarks. In: Petteri Koskikallio (ed.), *Parallels and Comparisons. Proceedings of the Fourth Dubrovnik International Conference on the Sanskrit Epics and Purāṇas, September 2005*: 181–191. Zagreb: Croatian Academy of Sciences and Arts.

SZEMERÉNYI, Oswald 1970. The Indo-European name of the 'heart'. In: Velta Rūķe-Draviņa (ed.), *Donum Balticum* (Festschrift Christian S. Stang): 515–533. Stockholm: Almqvist & Wiksell.

The Sāṃkhya-Yoga 'Manifesto' at Mahābhārata 12.289–290

JAMES L. FITZGERALD

THE TEXT-PAIR 12.289–290 IN THE MOKṢADHARMAPARVAN

The two chapters 289 and 290 of the Mokṣadharmaparvan (MDh) of the Śāntiparvan[1] inhabit a recurrent topos of the MDh, the juxtaposition of 'Yoga' and 'Sāṃkhya'.[2] The two chapters offer

---

[1] A very generous Faculty Research Award from the College of Arts and Sciences of the University of Tennessee made it possible for me to engage in a series of searching conversations on these two chapters with Professors Greg Bailey, Johannes Bronkhorst, Angelika Malinar, Peter Schreiner, and David White in 2005 and 2006. These conversations significantly deepened my apprehension of various themes in these texts and improved the translation of it in numerous ways large and small. I alone remain responsible for the final formulation of the translation and any errors or misunderstandings that remain are mine alone.

[2] There are five major instances of Sāṃkhya-Yoga text-pairs in the Mokṣadharmaparvan of the *Mahābhārata.* Each of these five pairs juxtaposes a teaching self-consciously labeled as '*yoga*' to a similarly self-conscious '*sāṃkhya*' teaching (or, in the case of 12.187, an '*adhyātma*' teaching closely resembling *some* of the later '*sāṃkhya*' teachings). The five instances are 12.187 and 188 (the 10th and 11th texts of the MDh), 12.231–232 (in the 28th text of the MDh, 'Śuka's Questioning of Vyāsa'), 12.289–290 (the 52nd and 53rd), 12.294.6–25 and 26–49 (within the 54th text of the MDh, 'The Dialogue of King Karāla Janaka and the Seer Vasiṣṭha'), and finally 12.303–304 (in the 56th text of the MDh, 'The Dialogue of King Janaka and the Seer Yājñavalkya'). The thematic juxtaposition of Yoga and Sāṃkhya is also brought up occasionally a few other times, e.g., in Vyāsa's instruction of Śuka at 12.228.27–38. The two concepts are mentioned as a pair, or virtually so, 50 times in the MDh, though all but six such oc-

clear and coherent presentations of doctrines explicitly labeled 'Yoga'[3] and 'Sāṃkhya',[4] presentations that are quite striking for reasons of form and content both. Two deliberately constructed elements bind the chapters together into an apparent text-pair, though, on the evidence available in the apparatus of the Pune edition, the manuscript tradition of the Śāntiparvan seems never to have perceived them as a unit. First the Bhīṣma–Yudhiṣṭhira frame provides an introduction to both texts, an introduction that is relatively detailed and doxographically interesting. This frame is resumed at the end of the *anuṣṭubh* Yoga exposition (12.289.57) and then at the beginning of the second chapter, 12.290. It is then closed reasonably tightly with the eight *śloka*s 12.290.94–101, following the *anuṣṭubh* exposition on Sāṃkhya. Besides the framing dialog that brackets the *anuṣṭubh* portions of these chapters, four obviously constructed questions are put in the mouth of Yudhiṣṭhira in the course of the two chapters,[5] a device which gives the longish text-pair a relatively high degree of unity and tautness and shows the care that went into the construction of the pair. Second, the end of each chapter is graced

---

currences are found in 12.289–290 (5 times), the Vasiṣṭha–Janaka dialog (8 times), the Yājñavalkya–Janaka dialog (13 times), and the *Nārāyaṇīya* (18 times). (The other six occurrences are at 12.47.34–35, 50.32, 189.4, 228.28, 231.3 [both in 'Śuka's Questioning of Vyāsa'], and 308.25 [in Sulabhā–Janaka]. The pairing of the terms occurs fourteen other times in the Pune text of the *Mahābhārata*, 6 of them in the *Gītā* (2.39, 3.3, 5.4, 5 (bis), 13.24). The other 8 are: 3.2.14, 211.21; 13.14.159, 16.25, 18.53, 74.24, 135.139; and 18.5.33).

3 My general practice in translating and writing about the texts collected in the Mokṣadharmaparvan is to use and translate the name most commonly found in the colophons of the manuscripts used for S. K. Belvalkar's Critical Edition of the Śāntiparvan (*Mahābhārata* 1933–66, vols. 13–16). The colophons of 25 of the 26 manuscripts of the critical edition that give a name to *adhyāya* 12.289 label it *yogavidhiḥ*. The other label is *yogakathanam*.

4 Ten critical edition manuscripts label *adhyāya* 12.290 simply *sāṃkhyam*. Four call it *sāṃkhyādhyāyaḥ*, three *sāṃkhyapraśaṃsā*, two *sāṃkhyayogaḥ*, two *sāṃkhyajñānam*, two *sāṃkhyajñānavidhiḥ*, one *sāṃkhyapraśnaḥ*, one *sāṃkhyavarṇanam*, and one *sāṃkhyarītikathanam*.

5 At 12.289.10, 289.42, 290.52, and 290.76–79.

with a carefully composed passage in classical *upajāti triṣṭubhs*, passages which each assimilate the preceding teaching to an anthropomorphically represented 'Nārāyaṇa'. These concluding *triṣṭubh* passages give the text-pair another level of formal and substantial unity beyond that provided by the Bhīṣma–Yudhiṣṭhira framing. But the *triṣṭubh* passages alter the substance of the prior *anuṣṭubh* expositions significantly. While the text-pair, read without the two *triṣṭubh* codas, seems to have been constructed as a kind of extended 'manifesto' of the '*darśanas*'[6] of 'Yoga' and 'Sāṃkhya', the *triṣṭubh* passages dramatically subordinate Yoga and Sāṃkhya to a theology of the divine Lord Nārāyaṇa,[7] bringing the text-pair into the ambit of the other Nārāyaṇa-materials of the *Mahābhārata* in some fashion or other.[8] A careful look at the contents of this text-pair makes interesting reading in the history of Indian philosophy and may also give us some small but definite insight into the history of the Mokṣadharmaparvan, some of the themes, ideas, and movements behind it, and even a bit about the history of the being called 'Nārāyaṇa'.

Apart from what has already been noted, this text-pair also occupies an intriguing position at the beginning of a striking series of texts that constitutes the end of the Mokṣadharma an-

6 This term is clearly sometimes used in these texts and in passages of the subsequent Vasiṣṭha and Yājñavalkya instructions in a sense very close to 'theory' or 'theoretical system', a comprehensive, ordered vision of a complex and murky set of realities. See for example 12.294.44–47. So too Deussen & Strauss 1906: 592 and Hauer 1958: 208.

7 It is the case that all major relevant passages of the *anuṣṭubh* Sāṃkhya exposition do assign an important place to 'Nārāyaṇa'. The *triṣṭubh* codas, however, offer elegantly constructed statements that use Yoga and Sāṃkhya to glorify Nārāyaṇa. It is also quite conceivable that the *triṣṭubh* codas were composed by someone other than the redactor of the *anuṣṭubh* text pair.

8 These materials have been surveyed by Reinhold Grünendahl in the volume of *Nārāyaṇīya-Studien* (Schreiner *et al.* 1997). See in particular Grünendahl's chapter 7, 'Zur Stellung des Nārāyaṇīya im Mahābhārata', pp. 197–240. See also Couture 2005.

thology. Going into this fact in detail involves too long and highly involved a discussion at this time, so I give here only a brief sketch of the basic facts.[9] Table 1 gives an overview of the last fifteen of the MDh's sixty-three texts. The five texts that make up the 54th through the 58th texts of the collection all center upon the ethical commitment of kings named Janaka. Four of these texts make these Janakas the recipients of either long and detailed expositions of philosophical doctrines that lead to the ethical stance of *nivṛtti*, withdrawal from actions, or of short exhortations (by an unnamed Bhārgava seer, at 12.297, and by the Seer Pañcaśikha, at 12.307) to the ethical stance of *pravṛtti*: adherence to performance of one's rightful duties as a king, predominantly making donations and performing Vedic rituals. (These short sermons are delivered against the background of issues arising from the soteriological perspectives: preferring *dharma* to *kāma*, in 12.297, and readying oneself for death, in 12.307.) A fifth text presents a Janaka king arguing that Pañcaśikha had taught him how to perform his duties and enjoy the soteriological benefits of *nivṛtti* at the same time; this argument is refuted by the female renouncer and practitioner of *yoga*, Sulabhā (12.308). The next three texts of the collection argue a similar refutation of the *pravṛtti* ethic by showing the superiority of Śuka to Vyāsa. Vyāsa, who is portrayed here as allied to Janaka's *pravṛtti* perspectives, is Śuka's biological father through the intervention of Śiva. Śuka rejects his father's and Janaka's exhortations to become a householder and produce offspring, and then, after a period of sentimental hesitation, turns his back on his father and escapes the world. The father is portrayed as a virtuous scholar of the Vedas, but one who, like the Janakas, lacks the inner

[9] A less comprehensive, but slightly more detailed account of part of this phenomenon was reported in my annotated translation of the Sulabhā–Janaka dialog from the MDh (Fitzgerald 2002: 647–649). J. W. Hauer (1958) perceived a bifurcation in the MDh at this point: see p. 195 and his discussion of the importance of Janaka's court in the history of Sāṃkhya and Yoga at pp. 205–208.

strength and detachment to gain Absolute Freedom from the world. This remarkable set of eight texts near the end of the MDh is bracketed at its beginning and end by two text-complexes that subsume the ethical debate of these eight texts into different arguments regarding the supreme God Nārāyaṇa. The long *Nārāyaṇīya*, the 62nd and penultimate text of the MDh, advocates exclusive loving devotion to Nārāyaṇa within a general *pravṛtti* ethic of the performance of Vedic rituals to the Gods and ancestors. The text-pair 12.289–290 which concerns us here, the 52nd and 53rd texts of the MDh, assimilates the expositions of Yoga and Sāṃkhya in these two texts to Nārāyaṇa, with no real mention of the *pravṛtti* concerns of the *Nārāyaṇīya*.[10] The text preceding the 289–290 pair (12.288, 'The Song of the Golden Goose (Prajāpati) [sung to the Sādhya Gods]') and the final text of the MDh, 12.340–353 'The Narrative Praising Gleaning as a Way of Life') are both texts more typical of the overall ethical and cultural orientation of the MDh as a whole: that is, both are grounded in brahminic values and perspectives and, while offering soteriological advice and raising the theme of *mokṣa*, both remain grounded in the *pravṛtti* side of the ethical divide; neither advocates renunciation.[11] Was the series of five Janaka texts deliberately composed as a unit culminating in Sulabhā's refutation of Janaka at 12.308? Were 12.297 and 12.307 composed for the express purpose of creating such a dialectical series? Was Vyāsa allied with the Janakas in the mind of the putative creator of the Janaka series? Is the humiliation of Vyāsa by Śuka intended as a polemical parallel to Sulabhā's humiliation of Janaka in 12.308? Is a single redactor responsible for the insertion of

[10] Though there is an interesting and potentially relevant mention of *pravṛttidharma* in Yudhiṣṭhira's question at 290.79. See below in the discussion of the Sāṃkhya exposition.

[11] The *uñchavṛtti* is a heroic form of pious *gārhasthya* or *vānaprastha*. The point of it as an ethical form is to adhere to one's proper *svadharma* as a brahmin, rather than take up the work of another *varṇa*, even if one must starve.

TABLE 1

The Final Fifteen Texts of the MDh

Adhy	# in MDh	Text Identification	Purport of Text
278	49	The birth of the Seer Śukra.	Etiological story explaining why Uśanas Kāvya is called Śukra ('shining; semen').
279–287	50	(86m) The song of the Seer Parāśara.	Exhortation to pursue *mokṣa* through *dharma karman*; with long exposition of aspects of *dharma*.
288	51	The song of the golden goose (Prajāpati) [to Sādhya gods].	Exhortation to *dama* and the cultivation of 'The Real.'
289	52	The rule of Linking	Yoga and *aiśvarya*. The yogin becomes **Nārāyaṇa**.
290	53	Sāṃkhya (total knowledge)	Sāṃkhya leads to **Nārāyaṇa**, Sāṃkhya knowledge is a manifestation of **Nārāyaṇa**.
291–296	54	(86n) The dialogue of King Karāla Janaka and the Seer Vasiṣṭha.	Long, systematic set of instructions centering around Sāṃkhya-Yoga. At the end Bhīṣma characterized Karāla Janaka as still one step from absolute perfection.
297	55	King Janaka's instruction by a Bhārgava seer.	Exhortation to 'Janaka' to pursue *dharma* instead of *kāma*.
298–306	56	(86o) The dialogue of King Janaka and the Seer Yājñavalkya.	Long set of instructions similar to Vasiṣṭha's. At the end of them Bhīṣma related that Janaka Daivarāti renounced his kingdom and lived as an ascetic.

307	57	The dialogue between King Janaka and the Seer Pañcaśikha.	Pañcaśikha lectured on the inevitability of death and then recommended adherence to Vedic sacrifices and generosity.
308	58	The dialogue between King Janaka and the woman renouncer Sulabhā.	The self-styled '*mukta*' King 'Janaka' is discredited as a '*mukta*' king by a wandering female *bhikṣuṇī*.
309	59	The Seer Vyāsa's barley-gruel lesson to his son Śuka.	Lecture given by Vyāsa to Śuka; interpreted by the frame as Vyāsa spurring Śuka to serious 'soteriological' effort.
310–315	60	(86p) The origin of Śuka	Suka's birth, education, interview with 'Janaka', and his eventual journey of yogic penetration to *brahman* by smashing through a Himalayan peak.
316–320	61	(86q) Śuka's flying against the mountain peaks.	
321–339	62	(86r) The book of the Nārāyaṇa cult.	Long, complex text supporting rituals directed to the gods and ancestors within the framework of exclusive devotion to **Nārāyaṇa**.
340–353	63	(86s) The story about gleaning as a way of life.	Narrative praising gleaning as a way of life.

both the 'Nārāyaṇized' pair 12.289–290 before this Janaka–Vyāsa series and the *Nārāyaṇīya* after it? Did the same person who inserted 12.289–290 in the MDh compose the *triṣṭubh* codas assimilating the Yoga and Sāṃkhya teachings of those chapters to Nārāyaṇa? These are intriguing questions, but they shall have to wait for the occasion of a thorough and comprehensive study of this very large set of texts. The focus for now is the first element of this putative series.

THE FRAMING, THE EXPOSITIONS, AND THE CODAS OF THE TEXT-PAIR

I will survey each of the elements of the text-pair distinguished just below in Table 2 and try to separate and illuminate the main threads of continuity that constitute the text-pair. I will pay particular attention to the framing passages and the *triṣṭubh* codas, highlighting the most important features of them and their bearing upon the overall nature and structure of the text-pair. I will account for the compiled expositions of Yoga and Sāṃkhya with summary analyses and descriptions. Table 2 provides a schematic overview of the two chapters.

TABLE 2

The General Structures of Mbh 12.289 and 12.290

289	MDh #52	THE REGIMEN OF LINKING (*yogavidhi*)	
A	1–10	Outer frame	Sāṃkhya and 'Sāṃkhyas' vis-à-vis Yoga and 'Yogas'
B	11–56	Exposition of Yoga	Emphasis upon *bala* for breaking *bandhanas*. Descriptions of *samādhāna* and *dhāraṇā* meditation and emphasis upon *bala* needed to carry those out and to stay on path.
C	57	*Anuṣṭubh* outer frame transition	Yoga originated in many non-brahmin traditions, but all of the highest *yoga* was definitively fixed among brahmins.
D	58–62	*Upajāti triṣṭubh coda*	The yogin fuses with beings up and down all levels of the cosmos; he becomes a marvelously powerful being who acts while his (higher) Self is **Nārāyaṇa** and his (lower) self, his body, is coterminous with the entire cosmos.

290	MDh #53	**Total knowledge (*sāṃkhya*, COMPREHENSIVE ENUMERATIVE UNDERSTANDING)**	
A	1–4	**Outer frame resumes**	Continuation of outer (Bhīṣma–Yudhiṣṭhira) frame
B	5–75	Exposition of Sāṃkhya	Prescribes knowledge of all life-realms and consequent understandings of their good and bad points; knowledge of the most general ontological layers of the universe and understanding of general processes of birth and rebirth.
C	76–94	The final stages of Sāṃkhya comprehension	Sāṃkhyas have no awareness in *mokṣa*, and their penultimate knowledge is a disembodied and supra-sensual knowledge, unlike that of *yogeśvaras*.
D	94–102	**Closing of the entire *anuṣṭubh* outer frame**	The greatness of the Sāṃkhya body of knowledge. The texts of the ancient Vedic seers, the Yogas, and the Sāṃkhyas are all 'speaking of' the transcendent *brahman*. The body of Sāṃkhya knowledge is the manifest form (*mūrti*) of that which has no form (*amūrti*).
E	103–110	*Upajāti triṣṭubh coda*	All important forms of knowledge have entered into Sāṃkhya, which can be effective and is always salubrious. This body of knowledge is issued by **Nārāyaṇa** at creation, then is supported by him, and finally is swallowed by him at the end of world.

THE INITIAL FRAMING PASSAGE

One of the most interesting elements of the text-pair is the first framing passage, the first ten *śloka*s of 12.289.

PASSAGE 1

12.289.1–10: Initial frame of text-pair
(289.11 provided for context)

yudhiṣṭhira uvāca
sāṃkhye yoge ca me tāta viśeṣaṃ vaktum arhasi
tava sarvajña sarvaṃ hi viditaṃ kurusattama || 1 ||
bhīṣma uvāca
sāṃkhyāḥ sāṃkhyaṃ praśaṃsanti yogā yogaṃ dvijātayaḥ
vadanti kāraṇaiḥ śraiṣṭhyaṃ svapakṣodbhāvanāya vai || 2 ||
anīśvaraḥ kathaṃ mucyed ity evaṃ śatrukarśana
vadanti kāraṇaiḥ śraiṣṭhyaṃ yogāḥ samyaṅmanīṣiṇaḥ || 3 ||
vadanti kāraṇaṃ cedaṃ sāṃkhyāḥ samyagdvijātayaḥ
vijñāyeha gatīḥ sarvā virakto viṣayeṣu yaḥ || 4 ||
ūrdhvaṃ sa dehāt suvyaktaṃ vimucyed iti nānyathā
etad āhur mahāprājñāḥ sāṃkhyaṃ vai mokṣadarśanam || 5 ||
svapakṣe kāraṇaṃ grāhyaṃ samarthaṃ vacanaṃ hitam
śiṣṭānāṃ hi mataṃ grāhyaṃ tvadvidhaiḥ śiṣṭasaṃmataiḥ || 6 ||
pratyakṣahetavo yogāḥ sāṃkhyāḥ śastraviniścayāḥ
ubhe caite mate tattve mama tāta yudhiṣṭhira || 7 ||
ubhe caite mate jñāne nṛpate śiṣṭasaṃmate
anuṣṭhite yathāśāstraṃ nayetāṃ paramāṃ gatim || 8 ||
tulyaṃ śaucaṃ tayor yuktaṃ dayā bhūteṣu cānagha
vratānāṃ dhāraṇaṃ tulyaṃ darśanaṃ na samaṃ tayoḥ || 9 ||
yudhiṣṭhira uvāca
yadi tulyaṃ vrataṃ śaucaṃ dayā cātra pitāmaha
tulyaṃ na darśanaṃ kasmāt tan me brūhi pitāmaha || 10 ||

bhīṣma uvāca
rāgaṃ mohaṃ tathā snehaṃ kāmaṃ krodhaṃ ca kevalam
yogāc chittvādito doṣān pañcaitān prāpnuvanti tat || 11 ||

This introduction is uncharacteristically lengthy and informative for the MDh. It introduces the subject of the juxtaposition of Sāṃkhya and Yoga, and asks what is the difference (*viśeṣa*) between them. Important for the charting of ancient Indian intellectual history, 'Sāṃkhya' and 'Yoga' are clearly used here as labels for particular ideas, doctrines, or procedures, and these labels are extended to people who identify with those ideas, doctrines, or procedures, people identified as 'Sāṃkhyas'[12] and others identified as 'Yogas'.[13] As we shall see, the text-pair as a whole does well address Yudhiṣṭhira's initial request to learn the *viśeṣa* of Sāṃkhya and Yoga. And Bhīṣma's initial remarks introduce the ensuing expositions quite well, offering two differentiating characterizations of each school that adhere closely to what will follow. At the same time, Bhīṣma presents these differences in terms of behaviors common to contending schools of abstract thought. These two schools of thought are said here to be made up of brahmins and to be contentious rivals of each other. Members of each school glorify their own particular kind of thought or practice, their own 'side' (that is, *pakṣa*), as a unique way to gain *mokṣa*. The partisans of each *pakṣa* proclaim their *pakṣa*'s superiority on the basis of reasons or arguments (*kāraṇa*s) which Bhīṣma characterizes as self-serving (in 289.6), and they seek to advance or elevate their own side (*svapakṣodbhāvana*, 2). Bhīṣma tells us in verses 6 and 8 that both are systems of thought (*mata*s) with regard to fundamental reality (*tattve*) and with regard to knowledge (*jñāne*). Switching from his observations of their forms of behavior to a judgment of his regarding their ultimate value (and echoing an argument made explicitly in

---

[12] I count 32 instances in the *Mahābhārata* of the word *sāṃkhya* used as a plural noun referring to 'adherents of Sāṃkhya'. 26 of these occur in the Nārāyaṇa-bounded Janaka–Vyāsa section of the MDh noticed above. The passage under examination here provides clear confirmation that *sāṃkhya* sometimes refers to a 'school of thought' with adherents called Sāṃkhyas.

[13] I count 39 instances of *yoga* being used as a plural noun in the *Mahābhārata*, where I believe it refers to persons, but since persons 'doing *yoga*' are not necessarily the same as persons adhering to a 'school of "Yoga"' (i.e., could be simply the same as *yogins*, a term with a couple dozen independent occurrences in the *Mahābhārata*) this usage is inherently more ambiguous. It is clear, however, that we do have some school of 'Yoga' with adherents in the text-pair here and in the same Nārāyaṇa-bounded Janaka–Vyāsa section described earlier. The word *yoga* in the plural occurs 31 times in this stretch of MDh texts, almost identical to the count for *sāṃkhya* in the plural (see the previous footnote).

other parts of the epic), Bhīṣma states that Sāṃkhya and Yoga both lead their respective followers along the highest soteriological trajectory when followed as taught (8cd). They are also alike in respect of the general way of life of their adherents (*śauca*, *dayā*, *vrata*s). 'But their theories are not the same' (9d).

Bhīṣma points out a fundamental difference in basic epistemology in 7ab: Those affiliated with Yoga are grounded upon direct observations, while Sāṃkhyas are based on the settled determinations of normative bodies of teaching, *śāstra*. But the most telling theoretical difference between Sāṃkhya and Yoga that will form the subject of this text pair is well indicated right at the beginning of this comparison, when Bhīṣma cites what he represents as the characteristic rationales of each school.

YOGA'S CHARACTERIZATION OF SĀṂKHYA IN THE INITIAL FRAMING PASSAGE

Bhīṣma exemplifies the characteristic arguments of the two *pakṣa*s, Yoga and Sāṃkhya, in *śloka*s 3–5, and I shall move into his first exposition, his presentation of Yoga in 12.289, by way of these. Bhīṣma cites the characteristic argument of Yoga first, but since this argument does, in fact, set the agenda for the whole of the *anuṣṭubh* text-pair, I will take it up after treating the argument he offers for Sāṃkhya in *śloka*s 4 and 5. That is a positive self-description that previews the long exposition of Sāṃkhya in 12.290 quite well: There is one and only one way to gain *mokṣa* upon leaving the body, and that is to become *virakta* after coming to know and comprehend all the courses or pathways (*gati*s) in the *viṣaya*s. (The word *viṣaya* means more in this context than just the five sensory realms – here it is whole systems of sense objects as experienced by whole orders of living beings, so I render *viṣaya*s here as 'experiential realms' or 'life-realms'.[14])

[14] The need for this understanding of *viṣaya* becomes clear if one looks at the beginning of the exposition of the Sāṃkhya *jñāna*, *śloka*s 5 and 6 of

289.5cd goes on to quote Sāṃkhyas (at least that is the most economical interpretation of '*mahāprajñās*' in 5c) to the effect that Sāṃkhya is a, or the, 'theory', *darśana*, of *mokṣa*. Prior to this epitome of Sāṃkhya, Yoga's characteristic argument was asserted aggressively and negatively in *śloka* 3a: '*anīśvaraḥ kathaṃ mucyet*', 'How can someone who is *anīśvara* gain *mokṣa*?' Regardless exactly how we understand the compound *anīśvara*, this Yoga argument is clearly presented as an attack on Sāṃkhya from the Yoga point of view.

This little *pāda* has had powerful effects on the Western understanding of Sāṃkhya over the years. The main line of thinking has taken the compound as a *bahuvrīhi* to be rendered as 'one who does not have a lord, or God', that is 'an atheist'.[15] The

12.290 (in Passage 4 below). The first such *viṣayas* listed there are those of *manuṣyas*, Piśācas, Rākṣasas, Yakṣas, *uragas*, and Gandharvas. Franklin Edgerton, however, persisted in the sense 'sense-objects' (Edgerton 1924: 10). Deussen & Strauss (1906: 598) understood it rightly as 'Bereiche'.

[15] So E. W. Hopkins (1901: 104–105) and Deussen & Strauss (1906: 592) 'Wie kann einer ohne Gott (*īśvara*) erlöst werden?' So too Hauer (1958: 200 and 208), and most recently John Brockington (2003: 15). Erich Frauwallner did not address our text-pair directly, as he discussed only those Sāṃkhya texts of the MDh he judged most important from the perspective of the development of the Sāṃkhya philosophical doctrines (Frauwallner 1953), based in part on his earlier article (Frauwallner 1925). We may infer Frauwallner's thoughts on 12.289.3 from his general remarks on the introduction of the idea of 'God' into the developing 'systems' of Indian philosophy (Frauwallner 1953, vol 1: 425–427). Like Richard Garbe before him, he sees the concept of God to be essentially alien to Sāṃkhya, and even with the classical school of Yoga where '... hat die Vorstellung von einem höchsten Gott ... eine Stelle gefunden' and where '... man kam dazu, in dieser Vorstellung ein besonderes Kennzeichen des klassischen Yoga-Systems zu sehen', the concept of God was essentially alien: 'weder die alte Überlieferung darauf Gewicht legt, noch die Stellung im System ein solches Urteil rechtfertigt' (Garbe 1917: 426–427). As will become clear below, I think Frauwallner's views on the subject are based on too circumscribed a reading of these texts. Johannes Bronkhorst's paper, 'God in Sāṃkhya' (1983), shows that the concept of God did play a role in later Sāṃkhya thought. 12.289–290 and other texts of the MDh (such as 12.228) testify to a more complicated history of the idea of *īśvara* than Frauwallner saw. See Angelika Malinar's consideration of the *bhūtātman* in her *Rājavidyā* (Malinar 1996: 196–203). Especially when read against the background of van Buitenen's classic studies '*Akṣara*', (1959) and 'The Large Ātman', (1964), Malinar's discussion here lays the

atheism of Sāṃkhya is one of the ideas at the core of the main line of interpretation of earlier scholars,[16] a view that Franklin Edgerton rightly thought to be a fundamental misprision of epic Sāṃkhya. But, as Edgerton argued in 'The Meaning of Sāṃkhya and Yoga', the issue of whether or not there is some ('sacred') transcendent reality at the heart of the universe is never a serious question in Brahmin-based thought prior to the (post-epic)[17] rise of Sāṃkhya's absolute ontological severance of *puruṣa* and *prakṛti* (*ātman* and *brahman*) and its postulation of a plurality of *puruṣa*s. Thus there was no possibility that Yoga was charging Sāṃkhya with atheism in 289.3a. Edgerton too took the word *īśvara* to mean 'lord' or 'God', and he construed the compound analogously to the usage of *uttama* ('highest') and the *bahuvrīhi* based on it, *anuttama* ('what has nothing higher'). Thus *anīśvara* is a *bahuvrīhi* meaning '[that principle of a person (*tattva*)] that has no lord, that is, that has no superior'. Thus, according to Edgerton, *anīśvara* here signifies the highest principle within a person, 'that which has no lord above it', 'the soul'.[18] According to

---

groundwork for a wider conceptualization of the idea of *īśvara* and the role it might play in some epic accounts of *yoga* and soteriology, one that finds important application in the interpretation of 12.289.3.

[16] For a brief partial discussion of turn-of-the-(nineteenth)-century views on Sāṃkhya's 'atheism', see Edgerton 1924: 7–14. On the purported atheism of Sāṃkhya, see also comments in the preceding footnote.

[17] According to Edgerton, that is. '... there is no passage, I believe, in the epic which attributes to it [Sāṃkhya, JLF]' a belief contrary to 'a belief in a [unitary, JLF] World-Soul (Brahman) or God' (Edgerton 1924: 26). Edgerton then goes on to discuss the interpretation of the question of the unity or plurality of *puruṣa*s as posed in 12.338, which, he concludes, teaches a form of monotheism (Edgerton 1924: 28–29). On the other hand, Edgerton fails to consider the possible interpretations of 12.308.176 (*jñasya ekatvapṛthaktvayoḥ*), in which, I believe, *jña* refers to the soul. In the context here Sulabhā is arguing to Janaka that any contact of their bodies has not involved any essential contact of them, if both are *mukta*. The claimed *ekatva* and *pṛthaktva* of the soul could imply radical distinctiveness and integrity on the part of numerous souls.

[18] Edgerton 1924: 11–12 and 1970: 291, n. 1. Construed as a possessive *bahuvrīhi* adjective, the word *anīśvara* signifies 'having no lord, supreme' (Edgerton 1924: 11).

him (and I think he is right with regard to the epic), the word *anīśvara* 'never means "an atheist", or "without God"' (1970: 291, n. 1).

I agree with Edgerton that the question here is not 'Is there some ultimate, all-encompassing sacred reality or not?' But I think Edgerton's interpretation of the meaning of *īśvara* in *an-īśvara* is far wide of the mark – the word does not mean 'the highest metaphysical principle in a person', 'soul'. The word *īśvara* and its common synonym *īśa* primarily signify the adjective 'powerful' and, substantively, 'a powerful being'. The word can mean 'lord', or 'God', because of this more fundamental sense. Years ago, based largely on the exposition of Yoga that follows in 12.289, I proposed to take the compound as a simple *tatpuruṣa*[19] (rather than a *bahuvrīhi*) meaning 'one who is not strong', that is 'one who is weak'.[20] If we turn now to Bhīṣma's *anuṣṭubh* exposition of Yoga, we shall see that it supports this more general understanding of the word *īśvara*.

THE YOGA EXPOSITION

The presentation of *yoga* here, commencing with *śloka* 11 and running up through *śloka* 56 just prior to the beginning of the *triṣṭubh* coda, is dominated by the idea of strength, *bala*, and fortitude of will, the soteriological importance of which it argues or exemplifies in a number of ways. This *bala* is the power to break out of all forms of bondage (*karman*, *lobhaja bandhanas*, the senses) (11–23); it is the ability to enter at will into all other beings (including Gods and *ṛsis*), the ability to withstand the power even of an angry Yamadeva, to make many thousands of clones of oneself and act in diverse ways in each and every one of them, to have complete control of all forms of bondage (24–28, to

[19] *Aṣṭādhyāyī* 2.2.6 '*nañ*'.

[20] I am referring to earlier, unpublished drafts of my translation of 12.289.

be *bandhana-īśa*, 28c); it is the fortitude to hold the mind fixed in difficult *dhāraṇā* meditations ('Holding-Meditations', 29–41, 54–55), it is the might to overcome various internal temptations and infirmities (such as lust and laziness) (47–49), and it is the perseverance to adhere to the path of *yoga* which is never easy (50–53). Besides the great powers mentioned above,[21] what one is said to gain through *yoga* is to 'get back to "That Place"[22] when those afflictions[23] are all gone' (12)[24] – 'spotless, [they] go the propitious way that goes farthest' (14) [25] – and, of course, *mokṣa.*[26]

The exposition falls into five distinct sections that are unified thematically (as described above) but not rhetorically or argumentatively. The first section (11–28) is a medley of *yoga* characterizations emphasizing the strength, *bala*, *yoga* requires and the *īśvaratva* ('being an *īśvara*') it generates. The second section (29–41) gives examples of intense concentration (*ātmanaḥ samādhānam*) and relates some of the soteriological effects of such concentration and the 'Holding-Meditations' (*dhāraṇās*). Yudhiṣṭhira then interrupts (42) to ask what foods the yogin eats to gain strength and Bhīṣma discusses foods and fasting for several *śloka*s (43–49). Fourthly, in *śloka*s 50–53, Bhīṣma offers an ex-

[21] These 'powers of attainment (through *yoga*)', *siddhi*s, play an important part in the description of *yoga* at 12.228.27–38. I am grateful to David White for many particular insights into *yoga* and *yogaiśvarya* during a series of conversations on this and other texts in the Spring of 2006 (see footnote 1). I am particularly grateful that David White insisted upon drawing out the many implications of Boris Oguibenine's paper (1984) for a general understanding of *yoga*. David White is currently working on all the stories of *yoga* in the *Mahābhārata* – not looking primarily at the soteriological uses of the idea – and this work of his will fill in, in some real way, that priority and predominance of *yoga* that Peter Schreiner wrote about a few years back (1999).

[22] An unspecified reference to Absolute Reality.

[23] The five afflictions of passion, error, affection, desire, and anger (12.289.11).

[24] *prāpnuvanti tathā yogās tatpadam vītakalmaṣāḥ* ‖ 12 ‖

[25] ... *yogāḥ paraṃ mārgaṃ gacchanti vimalāḥ śivam* ‖ 14 ‖

[26] 12.289.13c, 18d, 28, 31, 35, etc., etc.

tended statement regarding how terribly difficult and challenging the path of *yoga* is, 'if a brahmin gets onto the path of Yoga Linking[27] and finds it comfortable, then he should get off of that path, for tradition teaches that he must have many faults'.[28] Fifthly and lastly, in *ślokas* 54–56, Bhīṣma returns to the subject of *dhāraṇā* meditations and their difficulty and the reward of carrying them out as prescribed (*mokṣa*). *Śloka* 57 is part of the framing of the *adhyāya* and 58–62 make up the *triṣṭubh* coda.

An apt summation of the *anuṣṭubh* Yoga exposition is contained in *ślokas* 28 and 39–41 (which I present in slightly simplified translations):

> Certainly the powerful yogin, who has complete control over the bonds that hold him, has at his disposal the absolute power to escape them at will. (28)[29]
>
> A yogin who has concentrated in great observances focused on his navel, head, heart, etc., who perfectly links his self to the Subtle Self (39–40), will burn off his past deeds good and bad and gain spotless wisdom. If he wishes, he may take up the final Linkage and quickly get Absolute Freedom. (41)[30]

The term *īśvara* and its cognates and parallels in this exposition confirm directly that the word applies here to human beings

---

[27] At this time I am testing 'Linking' as a technical translation formula for soteriological uses of the word *yoga* in the epic. The term may prove too abstract, not as good as 'harnessing' or 'yoking', which seem too concrete. 'Meditation' narrows the scope of *yoga* too much; 'regimen' emphasizes the human behavioral side too much. As I have learned from David White, it is the *bandhus* that are most important in *yoga*, the lines of interconnection, transfer, and control between two different entities.

[28] *yogamārgaṃ tathāsādya yaḥ kaś cid bhajate dvijaḥ* | *kṣemeṇoparamen mārgād bahudoṣo hi sa smṛtaḥ* || 53 ||

[29] *balasthasya hi yogasya bandhaneśasya pārthiva* | *vimokṣaprabhaviṣṇutvam upapannam asaṃśayam* || 28 ||

[30] *nābhyāṃ kaṇṭhe ca śīrṣe ca hṛdi vakṣasi pārśvayoḥ* | *darśane sparśane cāpi ghrāṇe cāmitavikrama* || 39 || *sthāneṣv eteṣu yo yogī mahāvratasamāhitaḥ* | *ātmanā sūkṣmam ātmānaṃ yuṅkte samyag viśāṃ pate* || 40 || *sa śīghram amalaprajñaḥ karma dagdhvā śubhāśubham* | *uttamaṃ yogam āsthāya yadīcchati vimucyate* || 41 || Note the element of volition expressed here explicitly; the *yogeśvara* here seems to be some kind of allomorph of the Mahāyāna Buddhist idea of the *bodhisattva*.

who gain power and mastery through the practice of *yoga*. It is *īśvaras* who can enter when they will (*avaśāḥ*) into Prajāpatis, seers, Gods, and the prolific universal elements (*mahābhūtas*, 289.24).[31] The powerful yogin is a 'Lord, or Master' over bonds.[32] Yama, anger, and death do not 'rule over the yogin with his unlimited fiery energy' (*na ... īśate ... yogasyāmitejasaḥ*, 25). So while my earlier interpretation of *anīśvara* as 'weak' in 3a was headed in the right direction, it stopped short of the full truth. The word here signifies 'Lord', 'powerful Master', a very powerful agent, a being who has creative, elemental control of the *bhūtas* (the five elements) of which the whole material world is made, an *īśvara*.[33] In the *anuṣṭubh* teaching of 12.289 this word refers to an embodied human being who has, through the use of *yoga*, forcibly 'linked' himself upward (through lower elements within himself) to the higher reality that gives rise ultimately to the *bhūtas* (also called the *mahābhūtas*), the universal elements that are the matrix for the material world (see van Buitenen 1964: 186; Malinar 1996: 198–203), a *yogasiddha*.[34] Such an ascent requires great power (*bala*) to subject the lower structures of one's body and one's mind to the higher ones and thus restrain and alter their normal functioning. And, when one is successful (*siddha*), this ascent brings one to, or makes one, or amounts to one's becoming, some kind of powerful 'master' or 'lord' (*īśvara*)

---

[31] *prajāpatīn ṛṣīn devān mahābhūtāni ceśvarāḥ* || 24 ||

[32] *bandhaneśa*, 12.289.28c.

[33] Now as earlier I take the fundamental sense of *īśvara* here from *śloka* 24: *viśanti cāvaśāḥ pārtha yogā yogabalānvitāḥ* | *prajāpatīn ṛṣīn devān mahābhūtāni ceśvarāḥ* || 24 || I now render this *śloka*: 'Men practicing Yoga Linking who are mighty Lords filled with power from Linking enter at will into Progenitors, Seers, Gods, and the elements, son of Pṛthā.' It was, however, a close discussion of 12.289 (especially its *triṣṭubh* coda) in the Spring of 2005 with Angelika Malinar (see footnote 1), that pushed me all the way from my early interpretation of *īśvara* as simply 'strong, powerful' to the conviction that the word is indeed referring to *yogasiddhas* as *īśvaras*.

[34] As one indication that numerous human beings reach this status is the occurrence of the plural *īśvarāḥ* at 289.24d. See too 12.290.3.

at one of the highest levels of being, at a node that is a gateway to the highest *tattva* and *mokṣa*, on the one hand, and a locus for the use and control of the lower orders of being that derive from this one.[35] The exposition of Yoga in chapter 12.289 militates strongly against the overly 'spiritual' and exclusively meditational way of understanding the praxis of *yoga* that has tended to dominate the imaginations of all who have taken the condemnations of the *yogasiddhis* by the Buddha, other parts of the MDh,[36] and Patañjali seriously – myself very much included.[37]

---

[35] In this regard see Couture 2005, which discusses instances of the *yoga* of Nārāyaṇa found in a section of the *Harivaṃśa*. I might here offer an incidental observation that from the point of view of a human being striving upward toward *mokṣa* and the highest being (or, perhaps better imagined to be *inward*, toward the metaphysical center), the heart of the procedure as often depicted in the MDh is a series of upward, or inward, directed linkings of lower *ādhyātmika* elements to higher ones (I refer to the various ontologically telescoping *ādhyatmika tattvas*), and the consequent successive transformations of the lower structures by the linkage with the higher ones (restraint, quieting, purification, simplification), until the lower structures are neutralized, thus allowing the knowledge of the principal entity at the top or in the center. But from the point of view of an *īśvara*, whether God (eventually denominated *parameśvara*) or a *yogasiddha*, *yoga* is a downward directed linking in which the *īśvara* penetrates, inhabits, and controls lower entities – which are also *ādhyātmika* from his point of view – in order to use them for his purposes (or, later, for his purposeless *līlā* [*aprayojana karman*]). The word *yoga* signifies in both cases a binding, linking, harnessing, or yoking between two distinct realities, though in one case the motivation comes from 'below' in the hope of becoming like what is 'above', and in the other the initiative comes from 'above' and makes some use of structures that exist 'below'. Both orientations are clear in the classic chariot metaphor of the third *valli* of the *Kaṭha-Upaniṣad*: from the point of view of the conscious being 'above', the *ātman* riding in the chariot of the body is the 'consumer', the 'enjoyer', of the 'senses and mind when they are linked' (I read *Kaṭha* 3.4cd this way: *ātmā-indriyamano yuktaṃ bhoktā-iti-āhur manīṣiṇaḥ*). From the point of view of the 'upward striving' *mumukṣu*, he, the riding *ātman*, does or does not get where he is going (*tat padam*, *Kaṭha* 3.7, 8) as he is 'clean' (*śuci*) and has linked or applied (*yukta*) his mind to govern the senses or not. The same bidirectionality animates the 'cosmic gloss' of *yoga* at *Kaṭha* 6.11: *yogo hi prabhavāpyayau*.

[36] 12.190.7, as pointed out by Edgerton 1924: 46.

[37] The general translation formula I used for '*yoga*' in the first part of my translation of the Śāntiparvan was 'Regimen, or Discipline, of (*yoga*) Meditation'. I had spent a fair amount of time settling upon it and had a full battery of arguments justifying the fact that, and the ways that, it reflected

So the Yoga question challenging Sāṃkhya in *śloka* 3 is: Is it really possible, in the quest to gain the highest beatitude, to get to the critical ontological 'node' (*tattva*) that is the 'gateway' to *mokṣa* other than by *forcibly* (*balena*) 'ascending' *directly through* one's own substantial self (both 'physical' and 'psychic')? So, I construe 289.3a to say: 'How can one who is not a powerful Lord become Absolutely Free?' Yoga, as presented consistently in chapter 289, argues that the powers one acquires through the special linkings effected in this practice of Linking are a way to *mokṣa*. By imputation, the question in 3a implicitly argues it is the only way to *mokṣa*.

CONCLUDING FRAME TO THE *ANUṢṬUBH* EXPOSITION ON YOGA

Bhīṣma concludes the Yoga exposition with another very interesting historical observation.

PASSAGE 2 **12.289.57: Outer frame at end of Yoga exposition**
nānāśāstreṣu niṣpannaṃ yogeṣv idam udāhṛtam I *paraṃ yogaṃ tu yat kṛtsnaṃ niścitaṃ tad dvijātiṣu* II 57 II

a deliberate backward reading of Patañjali into epic usage. It reasoned that epic soteriological *yoga* always culminated in some kind of operations that occur within what we Westerners would call 'the mind' thus it was always a kind of 'meditation'. In conversations in the Spring of 2005 (see footnote 1), Johannes Bronkhorst challenged my reading 'meditation' into any instances of '*yoga*' in which the Sanskrit word *dhyāna* is not a part of the immediate discourse. Though I would still not systematically restrict the English 'meditation' only to '*dhyāna*', I have accepted the thrust of this objection and believe it is important to find a way to render the regimen of '*yoga*' in English in some way that does not systematically exclude the non-'mental' elements of a process that works on the whole *ātman* in all its extensions from its absolutely abstract center.

> What has been declared to you arose among those who subscribe to the Method of Linking relying on diverse Learned Traditions – but it was among brahmins that the whole of the supreme Method of Linking was definitively fixed. (57)

The reference to brahmins implies that the 'diverse Learned Traditions' include the various *nāstika* traditions.

The shift of subject from *yoga* and its results to an historical observation about the school or *śāstra* of *yoga* is a clear return to the framing of the Yoga exposition. Ignoring this shift, the *triṣṭubh* coda returns to the subject of the yogin's activity and his result.

THE *TRIṢṬUBH* CODA FOLLOWING THE EXPOSITION ON YOGA

Five classical *triṣṭubh* stanzas conclude the Yoga exposition. First I give the text and then the translation.

PASSAGE 3

12.289.58–62: *Triṣṭubh* coda ending the Yoga exposition

paraṃ hi tad brahma mahan mahātman |
brahmāṇam īśaṃ varadaṃ ca viṣṇum ||
bhavaṃ ca dharmaṃ ca ṣaḍānanaṃ ca |
ṣaḍ brahmaputrāṃś ca mahānubhāvān || 58 ||
tamaś ca kaṣṭaṃ sumahad rajaś ca |
sattvaṃ ca śuddhaṃ prakṛtiṃ parāṃ ca ||
siddhiṃ ca devīṃ varuṇasya patnīṃ |
tejaś ca kṛtsnaṃ sumahac ca dhairyam || 59 ||
tārādhipaṃ vai vimalaṃ satāraṃ |
viśvāṃś ca devān uragān pitṝṃś ca ||
śailāṃś ca kṛtsnān udadhīṃś ca ghorān |
nadīś ca sarvāḥ savanān ghanāṃś ca || 60 ||
nāgān nagān yakṣagaṇān diśaś ca |
gandharvasaṃghān puruṣān striyaś ca ||
parasparaṃ prāpya mahān mahātmā |
viśeta yogī nacirād vimuktaḥ || 61 ||

kathā ca yeyaṃ nṛpate prasaktā |
deve mahāvīryamatau śubheyam ||
yogān sa sarvān abhibhūya martyān |
nārāyaṇātmā kurute mahātmā || 62 ||

The fact is,[38] exalted one,[39] a man employing the Method of Linking[40] should enter into and then fuse with[41] (61cd) That, the Highest, The Mighty All,[42] *brahman*,[43] then the Lord Brahmā, the favor-

[38] This phrase registers the particle *hi* of 289.58a. I mention this for this sentence is drawing words in from several different *śloka*s.

[39] Addressing Yudhiṣṭhira. 'Exalted' is the formula I normally employ to render *mahātman* when it is applied to persons. In this passage, however, *mahat* is used repeatedly with its old Vedic overtones of greatness, prolixity, and power. As a glance at the Sanskrit shows, this address of Yudhiṣṭhira occurs immediately after the adjective *mahat* referring to *paraṃ tad brahma*, which must be displaced in the English to follow the verbal phrase (which occurs in the Sanskrit only at the end of the series of 'fusions' or 'linkages', down in 61cd; in the English it has been moved up to the top). The vast contrast between the all-comprehending and mighty *brahman* and Yudhiṣṭhira is intentional; even more salient is the contrast between *mahātman yudhiṣṭhira* and the yogin who is the subject here. In 61c the yogin is described as '*mahān mahātmā*', which refers to this yogin's being coterminous with the 'Tremendous All', who is now, himself, the 'subjectivity embodied in the Tremendous All' (*mahātman*). (In this case *mahātman* signifies both the *ātman*, 'the embodied subjectivity' of, in, the *mahat* and he whose self, particularly the lower, body-portion of his self, is the *mahat*. There is no difference in reference or meaning.) For some of the background history of the concept of the *mahān ātman* see van Buitenen 1964. Let me also note here gratefully my debt to Angelika Malinar. Prior to our discussions in the Spring of 2005, in cosmological contexts I tended to emphasize only the comprehensiveness, the 'universality', encoded in the word *mahat*. She impressed upon me the fundamental importance of seeing the tremendous power or energy that also forms part of the concept. I bear full responsibility for my English renditions of the word ('vast and tremendous' and 'The Mighty All'), which may impress some as awkward or cumbersome.

[40] The subject and verb governing this long, descending series of entities occurs after the list of objects, in 289.61cd.

[41] 'Should enter into and fuse with' renders *parasparaṃ prāpya ... viśeta* in 61cd.

[42] = *mahat*.

[43] I take this series to represent the (temporary) exercise of *aiśvarya* by the successful yogin (see descriptions of his doing so at 289.26–27 and see 289.41 above describing his option to take up the 'final linkage' and quickly gain *mokṣa*). This passage too describes his finally gaining *mokṣa* sometime

granting Viṣṇu, Bhava,[44] the God Dharma, the six-faced God,[45] the six majestic sons of Brahmā, (58) the Attribute noxious Darkness,[46] the Attribute tremendous Energy, the Attribute pure Lightness, the highest Generative Matrix,[47] Complete Attainment,[48] the Goddess who is the wife of Varuṇa,[49] all of Fiery Energy,[50] the vast and mighty firmness,[51] (59) the spotless moon amidst the stars, the All-Gods, the serpents, the ancestors, all the mountain peaks, the terrific oceans, all rivers, clouds, forests, (60) elephants, mountains, the throngs of Yakṣas, the quarters of the sky, the throngs of Gandharvas, and men and women. (61abc) Thus vast and tremendous, his body that Mighty All,[52] he gains Absolute Freedom shortly afterward.[53] (61d)

King, this account focused upon this god of tremendous power and thought[54] is splendid! Transcending all mortal practitioners of Yoga Linking, this one – his body that Mighty All, his soul Nārāyaṇa – acts.[55] (62)

Stanzas 58–61 describe a series of 'fusions' the yogin should (may?) effect once he has 'ascended' to a level of being one step below *mokṣa*. Having reached this level, he should (may?) (evi-

---

after he effects these fusions, sometime after he is done acting with the *mahat* as his body, Nārāyaṇa as his soul (289.62).

44 = Śiva.

45 = Skanda.

46 Attribute = one of the three *guṇas*: Darkness = *tamas*, Energy = *rajas*, Lightness = *sattva*.

47 'The highest Generative Matrix' = *parā prakṛtiḥ*.

48 The complete mastery (*siddhi*) over the elements gained by a person through yogic linking from the lower elements within himself.

49 = the element water (Varuṇa is the Lord of Waters).

50 = the element fire, which would either include the sun or have the sun as its entry-point.

51 A reference to the Earth, probably intended to refer to the element earth. While these last three suggest the series of the five great elements, ether and wind are missing. We have instead here, the triad of the three worlds: the *svarga* of the stars and sun, the *antarikṣa* of the waters, and the earth.

52 'Vast and tremendous' = *mahān*, 'his body that Mighty All' = *mahātmā*.

53 This sentence is the second half of 290.61d.

54 Referring to the yogin who has gone this path and become a god.

55 My thanks to Angelika Malinar for her insights into this description of the yogin's becoming one with Nārāyaṇa.

dently reversing the direction of his movement) go to (*prāpya*) and then enter into 'That' supreme *brahman*, which is called *mahat*, a term being appropriated from the old Upaniṣadic accounts of the *mahān ātman.*[56] From this reality referred to with the neuter gender, he should or may (evidently still moving 'downward' or 'backward') do the same for the 'Lord' Brahmā, then Viṣṇu, Śiva,[57] Dharma, Skanda (?), and the six sons of Brahmā (who are sometimes, in cosmogonies, secondary progenitors of different species of beings). Next he enters *tamas*, *rajas*, and *sattva*, and '*parā prakṛti*'. Next *siddhi*, water, fire, and earth. Then the moon and stars, the Viśvedevas, serpents, Ancestors, and several geological features and earthly beings, ending with men and women. The series is curious and needs to be analyzed in conjunction with all other such series in the MDh. What is clear, however, is that the yogin, located above and beyond *param brahman*, can move up and down and around within the orders of (manifested)[58] being, entering any entity at any level within it. And this yogin, who has become a 'God of tremendous power and thought' (*deva mahāvīryamati*, 61b), has become both *nārāyaṇātman* and *mahātman*, in which I see a play upon the regular bifurcation of '*ātman*' between a higher, purer form (*ātman* unqualified, *paramātman*, *sūkṣma ātman* here at 289.40, 49) and the lower form that is qualified and, in some sense, embodied[59] (*bhūtātman*, *dehin*, often). I take *mahātman* to refer to his 'lower' half, his being the 'self' in the entirety of the *mahat*

---

[56] See van Buitenen 1964: 114; his remarks on pp. 108–109 are also broadly relevant here.

[57] Do these three form a lateral triad or a vertical series of beings? Perhaps the whole group of anthropomorphic deities between the neuter *brahman* and *tamas* occupy a single level of ontology.

[58] Interestingly, this account makes no use the language of *vyakta* and *avyakta*, which *are*, however, found in 12.290.

[59] It is important we always bear in mind that the word *ātman* can refer to the 'soul' and to the 'body' and to every sense of 'self' embracing both these extremes simultaneously or any part of the continuum between them.

(i.e., *param brahman* and all that is contained within it either latently or actually), his being a self that has that *mahat* as his body. The description in 62d refers to the yogin who 'acts' after having accomplished some or all the fusions listed in 58–61 and before he becomes *vimukta*.

THE SECOND TEXT OF THE TEXT-PAIR, 12.290

After 289's complicated pastiche of pronouncements on *yoga* around various themes of strength and power, 12.290 turns to Sāṃkhya with this resumption of the outer *anuṣṭubh* frame that ended at 289.57.

PASSAGE 4

12.290.1–4: Continuation of frame begun in 12.289.1–11 (290.5–6 provided for context).

yudhiṣṭhira uvāca
samyak tvayāyaṃ nṛpate varṇitaḥ śiṣṭasammataḥ
yogamārgo yathānyāyaṃ śiṣyāyeha hitaiṣiṇā || 1 ||
sāṃkhye tv idānīṃ kārtsnyena vidhiṃ prabrūhi pṛcchate
triṣu lokeṣu yaj jñānaṃ sarvaṃ tad viditaṃ hi te || 2 ||
bhīṣma uvāca
śṛṇu me tvam idaṃ śuddhaṃ sāṃkhyānāṃ viditātmanām
vihitaṃ yatibhir buddhaiḥ kapilādibhir īśvaraiḥ || 3 ||
yasmin na vibhramāḥ ke cid dṛśyante manujarṣabha
guṇāś ca yasmin bahavo doṣahāniś ca kevalā || 4 ||

(*jñānena parisaṃkhyāya sadoṣān viṣayān nṛpa*
mānuṣān durjayān kṛtsnān paiśācān viṣayāṃs tathā || 5 ||
rākṣasān viṣayāñ jñātvā yakṣāṇāṃ viṣayāṃs tathā
viṣayān auragāñ jñātvā gāndharvaviṣayāṃs tathā || 6 ||)

This framing passage is lively and personal; that is, it seems to have been composed with some care to fit into the Bhīṣma–

Yudhiṣṭhira frame of the Śāntiparvan. On the other hand, it is more perfunctory than the previous two in 289. Yudhiṣṭhira requests an explanation of Sāṃkhya, since his general subject is their difference (*viśeṣa*), particularly the difference in their *darśanas*. The single most interesting fact in this passage is that Bhīṣma identifies the sages of Sāṃkhya, 'Kapila and so on', as *īśvaras*, thus seeming to assimilate them to the successful yogins described in 289. On its face, this description of the Sāṃkhya teachers would seem to be one response to the Yoga critique of Sāṃkhya quoted by Bhīṣma in 289.3a. If the originators of the Sāṃkhya teaching were *īśvaras*, that is *yogasiddhas*, then the *anīśvara* criticism is rebutted with an argument from authority: Sāṃkhya's teaching may not entail becoming an *īśvara*, but that teaching was put forward by men who were *īśvaras*.

THE SĀṂKHYA EXPOSITION

The basic formula of Bhīṣma's basic exposition here is well indicated by *ślokas* 290.5–6 cited in Passage 4 just above. 'Having known' (*jñātvā*) various 'spheres of life' (*viṣayas*) – those of humans, Piśācas, Rākṣasas, Yakṣas, serpents, Gandharvas, etc., including *īśvaras* of *yoga*, Prajāpatis, and Brahmā – and having thus 'understood' (*vijñāya*) various general features of the lives and transmigrations of embodied beings, including the good and bad features of the Vedas, Yoga, and Sāṃkhya; then having 'known' the ten general strata of worldly being[60] – Lightness, Energy, Darkness, Higher Mind, Ether, Lower Mind, Higher Mind, the Mighty Universal, Darkness, Energy, and Lightness – and having understood from that knowledge the spectacle of dissolution (*pralaye prekṣaṇam*), 'those fully endowed with knowledge and understanding, who have been steeped in good rea-

[60] This is my interpretation of the series enumerated in 290.14–16.

sons, attain radiant Absolute Freedom, just as tiny bits of matter go up to the highest sky' (290.17).[61]

This passage of 13 *śloka*s is followed by a second, much longer list of 'knowings' and 'understandings', one that goes over many of the things mentioned in the first passage, but adding further insights and realizations about them. This second passage runs from 290.18 to 69 with a brief interruption between 290.52 and 55. In 290.51 Bhīṣma mentioned 'the faults that are based in oneself' (51) and Yudhiṣṭhira demanded clarification. (As Bhīṣma's lecture at this point has consisted of an immensely long series of as yet unresolved gerund phrases, we may speculate that Yudhiṣṭhira's question here is simply a rhetorical cry for help!)

After the interruption Bhīṣma continues with the second listing. In 290.61–68 Bhīṣma lays out an extended metaphor in which the processes of life are mapped against terrifying aspects of an ocean that must be crossed. Then, successful sages (*munayaḥ siddhāḥ*) who have crossed this ocean by means of knowledge (*jñānayogena*) enter the clear sky (69). He then offers a more detailed description of the final course (*gati*) by which the enlightened person exits the world and gains *mokṣa* than given in the first passage compiled in this exposition. These sages ascend upward through different levels of the cosmos to 'Lord Nārāyaṇa' (*nārāyaṇaṃ prabhum*, 74b).[62] The Lord, 'pure Mind' or 'a purified soul' (*śuddhātman*, 74c), then carries them himself to the *paramātman* (*prabhur vahati śuddhātmā paramātmānam ātmanā*, 74cd), whence they never return: thus the *paramā gati* of *nirdvaṃdvānāṃ mahātmanām* (75ef). So the Sāṃkhya way of salvation does take the successful follower through the ontological node of lordship.

---

[61] *jñānavijñānasaṃpannāḥ kāraṇair bhāvitāḥ śubhaiḥ* | *prāpnuvanti śubhaṃ mokṣaṃ sūkṣmā iva nabhaḥ param* || 17 ||

[62] The word *prabhu* can be a synonym of *īśvara*.

At this point Yudhiṣṭhira asks an uncharacteristically long and truly interesting substantive question, one not provoked by anything Bhīṣma has just mentioned: (I paraphrase 290.76–79)

> Do these sages have any recollection or awareness once they have reached the highest place? If they did, then *mokṣa* would be faulty and *pravṛtti dharma* would be a superior way to live.[63]

The author's framing of Bhīṣma's response here is once again doxographically very interesting:

> You have properly posed a very difficult question, son; there is confusion about this question even among those who are enlightened, O bull of the Bharatas. There is here, a most important fundamental reality. Learn of it as I declare it exactly. (80) This is the supreme insight of the exalted followers of Kapila.[64]

In the theory that is laid out following this, not only is Yudhiṣṭhira's question of 76–79 answered; more significantly, the *an-īśvara* challenge of 289.3a is refuted to the satisfaction of the author of the text-pair. The real issue explained here is how someone who is not an *īśvara*, someone who lets the embodied soul and its senses 'go to sleep' rather than making continued use of them (as does the *yogeśvara*), has knowledge for a time after separating from the biological body and then gains *mokṣa*, where

---

[63] *yudhiṣṭhira uvāca* || *sthānam uttamam āsādya bhagavantaṃ sthira-vratāḥ* | *ājanmamaraṇaṃ vā te smaranty uta na vānagha* || 76 || *yad atra tathyaṃ tan me tvaṃ yathāvad vaktum arhasi* | *tvad ṛte mānavaṃ nānyaṃ praṣṭum arhāmi kaurava* || 77 || *mokṣadoṣo mahān eṣa prāpya siddhiṃ gatān ṛṣīn* | *yadi tatraiva vijñāne vartante yatayaḥ pare* || 78 || *pravṛtti-lakṣaṇaṃ dharmaṃ paśyāmi paramaṃ nṛpa* | *magnasya hi pare jñāne kiṃ nu duḥkhataraṃ bhavet* || 79 ||

[64] *bhīṣma uvāca* || *yathānyāyaṃ tvayā tāta praśnaḥ pṛṣṭaḥ susaṃ-kaṭaḥ* | *buddhānām api saṃmohaḥ praśne 'smin bharatarṣabha* | *atrāpi tattvaṃ paramaṃ śṛṇu samyaṅ mayeritam* || 80 || *buddhiś ca paramā yatra kāpilānāṃ mahātmanām* | I am not certain upon what basis Belvalkar took *śloka*s 80 and 81 as 6-*pāda śloka*s, but I think that segmentation is misleading. 80ef and 81ab form an injunction followed by a rationale for it and they should have been bound as a single *śloka*. 81cd and 81ef also then make a normal *śloka*. The word *buddhi* in 81a should *not* be construed here as referring to any mental organ. Bhīṣma is praising the theory he is about to lay out.

all consciousness is gone. Yudhiṣṭhira's question thus contains within it an implicit criticism of *yogaiśvarya*, and provides the occasion for the theoretical counter to the straw-man *anīśvara* claim of 289.3a. Here is the passage in translation.

> The sense-faculties of the embodied soul are aware of their own proper body,[65] king. These[66] are the instruments[67] of the Mind;[68] it is subtle, but it sees by means of them. (81) When these are separated from the Mind they vanish like wooden fences, no doubt of it, like foam upon the great ocean. (82)
>
> After the embodied soul and its faculties have gone to sleep, O scorcher of your enemies, then the subtle one moves everywhere, like the wind in the sky. (83) O Bhārata overlord, he sees fully, he touches fully, being completely aware as before. (84) All the sense faculties, like serpents that have had their venom removed, are dissolved each into its own place according to rule, because he is not a Lord.[69] (85) But the Mind,[70] steps into every one of the places of each of those sense faculties and ranges along their subtle courses, there is no doubt of it. (86) Pervading all the derivatives of Lightness, and then all those of Energy, and all those of Darkness, and the derivatives of the Higher Mind, Bhārata, (87) and likewise those of the Lower Mind, the derivatives of Ether and those of Wind, O you who are ever Virtuous, and those of Fire, (88) Water, and Earth – having pervaded them with its whole self and its own derivative attributes it is the Knower of the Field, Yudhiṣṭhira, son of Pṛthā. (89) Lord, as pupils go to their exalted teacher,[71] the mind goes to the Knower of

---

[65] If we see 290.59 (which describes Sāṃkhyas as leaving behind their 'biological bodies,' the *prajākṛtaṃ deham*, at some point in the process) to be pertinent here, then the reference here must be to some 'body-principle' more subtle than the biological body, perhaps some kind of *sūkṣma śarīra* '(subtle body') found in other texts. In which case the 'biological body' would be the *sthūla śarīra* ('gross body').

[66] = the sensory faculties.

[67] *kāraṇa* = *karaṇa*, as elsewhere in the MDh.

[68] = the *ātman*.

[69] *anīśatvāt*. If these men were 'Lords', powerful 'masters' of Yoga linkings (*īśvaras*), their senses would not be withdrawn in this way. They would continue to make use of them as they operated upon and acted within the material world.

[70] = *ātman*.

[71] Though tempted to read 90 (and 93) as some kind of interpolated asides, I believe it is possible to read them integrally. If we take *ātman* in 90a as the equivalent here of *dehin*, then we can imagine that when the

> the Field and so do actions good and bad; and the faculties go to this one too. (90) Moving out past the Generative Matrix[72] he[73] goes to the Never-Changing Self, the Supreme Self Nārāyaṇa, which is beyond all the pairs of oppositions, which is beyond the Generative Matrix. (91) Freed from good and bad *karma*, having entered into that wholesome Supreme Self that has no attributes, he does not return, Bhārata. (92) But the sense faculties return in the course of time to the Lower Mind that has remained behind, like pupils obedient to their teacher. (93)[74]

This form of mind, the 'Knower of the Field' (*kṣetrajña*), continues to operate in the places and along the courses of the senses within *prakṛti*, but separately from the senses and the embodied mind. It is capable of cognizing the qualities of entities derived from Lightness, Energy, and Darkness (both 'physical' – those derived from the elements ether, wind, fire, water, and

---

dehin and its *indriyas* awaken and find the higher *ātman* (= *kṣetrajña*) headed out past *prakṛti* to Nārāyaṇa, they run after him out of attachment, not wanting to be left behind; like pupils might run after their teacher. As will be pointed out in 93, to sustain the same reading, they will be sent back. Why the *manas* had remained behind (*śiṣṭam*, 93a) is not clear. It might not be far-fetched to see in these two *ślokas* a humorous dig at the yogins, whose *aiśvarya* involves their continued use of these lower faculties.

[72] = *prakṛti*.

[73] = the *kṣetrajña*.

[74] *indriyāṇy api budhyante svadehaṃ dehino nṛpa* | *kāraṇāny ātmanas tāni sūkṣmaḥ paśyati tais tu saḥ* || 81[c–f] || *ātmanā viprahīṇāṇi kāṣṭhakuḍyasamāni tu* | *vinaśyanti na saṃdehaḥ phenā iva mahārṇave* || 82 || *indriyaiḥ saha suptasya dehinaḥ śatrutāpana* | *sūkṣmaś carati sarvatra ṇabhasīva samīraṇaḥ* || 83 || *sa paśyati yathānyāyaṃ sparśān spṛśati cābhibho* | *budhyamāno yathāpūrvam akhileneha bhārata* || 84 || *indriyāṇīha sarvāṇi sve sve sthāne yathāvidhi* | *anīśatvāt pralīyante sarpā hataviṣā iva* || 85 || *indriyāṇāṃ tu sarveṣāṃ svasthāneṣv eva sarvaśaḥ* | *ākramya gatayaḥ sūkṣmāś caraty ātmā na saṃśayaḥ* || 86 || *sattvasya ca guṇān kṛtsnān rajasaś ca guṇān punaḥ* | *guṇāṃś ca tamasaḥ sarvān guṇān buddheś ca bhārata* || 87 || *guṇāṃś ca manasas tadvan nabhasaś ca guṇāṃs tathā* | *guṇān vāyoś ca dharmātmaṃs tejasaś ca guṇān punaḥ* || 88 || *apāṃ guṇāṃs tathā pārtha pārthivāṃś ca guṇān api* | *sarvātmanā guṇair vyāpya kṣetrajñaḥ sa yudhiṣṭhira* || 89 || *ātmā ca yāti kṣetrajñaṃ karmaṇī ca śubhāśubhe* | *śiṣyā iva mahātmānam indriyāṇi ca taṃ vibho* || 90 || *prakṛtiṃ cāpy atikramya gacchaty ātmānam avyayam* | *paraṃ nārāyaṇātmānaṃ nirdvaṃdvaṃ prakṛteḥ param* || 91 || *vimuktaḥ puṇyapāpebhyaḥ praviṣṭas tam anāmayam* | *paramātmānam aguṇaṃ na nivartati bhārata* || 92 || *śiṣṭaṃ tv atra manas tāta indriyāṇi ca bhārata* | *āgacchanti yathākālaṃ guroḥ saṃdeśakāriṇaḥ* || 93 ||

earth – and 'mental' – those of the Higher and Lower Minds). Ultimately this form of the soul goes out past the whole system of the Generative Matrix (*prakṛti*) to the Nārāyaṇātman, which is beyond *prakṛti* and all the pairs of oppositions (12.290.91). Here the Nārāyaṇātman is the *paramātman*, where the soul lodges permanently (12.290.91–92). Thus Yudhiṣṭhira's question at 76–79 is answered implicitly. The *kṣetrajña* is capable of the cognition of entities within *prakṛti*, but once beyond *prakṛti*, having entered into the one being beyond *prakṛti* and beyond all distinctions, there is no longer anything of which to be cognizant.

This theory of the *kṣetrajña*, articulated as a counter-theory to Yoga's postulated theory of *mokṣa*, is consistent with Sāṃkhya's thematization of *jñāna*. The main general point of the Sāṃkhya exposition is its argument for a soteriological method distinct from that of the Yoga school and its notions of *īśvaratva*. It proposes a method of freeing oneself from the painful processes of the world (both 'physical' and 'psychological') that pointedly does not involve the intense struggles involving force that characterize Yoga. The overall argument of the text-pair claims on behalf of Sāṃkhya that indeed *it is possible* to reach that soteriologically critical node without the forcible 'ascent' through the substantial structures of one's being. *Īśvaratva* is involved in this Sāṃkhya theory, but here one reaches the *īśvara* Nārāyaṇa by comprehensive, itemized knowledge (*jñāna*) and understanding (*vijñāna*) alone and one does not become an *īśvara* himself.

THE CLOSING FRAME OF THE *ANUṢṬUBH* TEXT

Bhīṣma follows his presentation on the *kṣetrajña* in response to Yudhiṣṭhira's question with a concluding series of observations that praises Sāṃkhya as an historical phenomenon in the world, in a way similar to his praise of *yoga* in the framing *śloka* at 289.57. Bhīṣma makes these points at the end of the *anuṣṭubhs* of 290: 1) Sāṃkhya is effective for the gaining of *mokṣa* (94); 2) Sāṃkhya adherents gain *mokṣa* with a kind of knowledge that knows no equal in the world (95–96ab); 3) Sāṃkhyas and Yogas both postulate the transcendent source of all, *brahman*, described by the highest seers, the Gods, and all men who know the textual traditions (96cd–100); and 4) Sāṃkhya is the manifest form of that which is formless (101). This passage closes the general exposition of Sāṃkhya, praises Sāṃkhya as an important body of knowledge (which connects to the basic theme of the exposition), and then affirms the ultimate 'referential unity' of the Vedic tradition and Sāṃkhya and Yoga (all are speaking of *brahman*). This last point connects to the point regarding the convergence of Sāṃkhya and Yoga in the same final result that was asserted in the opening frame (289.8cd). These three points, especially the last one, offer an apt conclusion to the two *adhyāya*s conceived as a single text-pair. The final point made in 290.101 – the praise of Sāṃkhya in somewhat mystical terms ('Sāṃkhya is the *mūrti* of the *amūrti*') – points ahead to the *triṣṭubh* coda that begins very shortly in 290.103. I present here the text and translation of the whole passage.

PASSAGE 5

12.290.94–102: Closing of outer Bhīṣma–Yudhiṣṭhira frame

(Bhīṣma is concluding:)

śakyaṃ cālpena kālena śāntiṃ prāptuṃ guṇārthinā
evaṃ yuktena kaunteya yuktajñānena mokṣiṇā || 94 ||
sāṃkhyā rājan mahāprājñā gacchanti paramāṃ gatim
jñānenānena kaunteya tulyaṃ jñānaṃ na vidyate || 95 ||
atra te saṃśayo mā bhūj jñānaṃ sāṃkhyaṃ paraṃ matam
akṣaraṃ dhruvam avyaktaṃ pūrvaṃ brahma sanātanam || 96 ||
anādimadhyanidhanaṃ nirdvaṃdvaṃ kartṛ śāśvatam
kūṭasthaṃ caiva nityaṃ ca yad vadanti śamātmakāḥ || 97 ||
yataḥ sarvāḥ pravartante sargapralayavikriyāḥ
yac ca śaṃsanti śāstreṣu vadanti paramarṣayaḥ || 98 ||

sarve viprāś ca devāś ca tathāgamavido janāḥ
brahmaṇyaṃ paramaṃ devam anantaṃ parato 'cyutam ॥ 99 ॥
prārthayantaś ca taṃ viprā vadanti guṇabuddhayaḥ
samyag yuktās tathā yogāḥ sāṃkhyāś cāmitadarśanāḥ ॥ 100 ॥
amūrtes tasya kaunteya sāṃkhyaṃ mūrtir iti śrutiḥ
abhijñānāni tasyāhur mataṃ hi bharatarṣabha ॥ 101 ॥
dvividhānīha bhūtāni pṛthivyāṃ pṛthivīpate
jaṅgamāgamasaṃjñāni jaṅgamaṃ tu viśiṣyate ॥ 102 ॥[75]

Son of Kuntī, he who is rich in virtues, has the appropriate knowledge, and desires Absolute Freedom is able to gain complete quiescence in a short time when he is engaged in this way. (94)

The very wise followers of the school of Sāṃkhya Discrimination, king, go the course that goes the farthest with this Philosophy; no Philosophy equal to it is known. (95) Have no doubt of this, Sāṃkhya Philosophy is the highest system of thought.

The undecaying, fixed, unchanging, primordial, and everlasting *brahman* (96) that has no beginning, middle, or end – the eternal cause beyond all the pairs of opposites; what those who are completely calm speak of as 'at the very peak'; (97) that out of which and into which all the transformations of the emissions of creation and the dissolutions of the end-time proceed; that which the highest seers speak of and praise in the traditions of learned Teaching; (98) that which all seers, and Gods, and men schooled in traditional learning speak of as the supreme, the God known to brahmins, infinite and never fallen from the highest (99) – seeking that, brahmins, thinking of its good qualities, speak of it; likewise adherents of the Method of Linking who have linked themselves completely and followers of the Total Knowledge,[76] who have unlimited vision. (100)

[75] I am at a loss to connect this final *anuṣṭubh* of the text-pair either to what precedes or to what follows in the *triṣṭubh* coda. And it is also hard to see any particular motivation for the interpolation of 102. My best suggestion is that it had somehow become attached to one or another set of the *śloka*s that were combined to form this text-pair, but the redactor did not come to fit it in anywhere in his exposition. I am averse to relying upon the explanation of 'agentless utterances', but such scraps do get pushed to the end of all deliberate compositions and occasionally are not edited out in the final redaction for one reason or another. It is also conceivable, if highly unlikely, that the composer of the *triṣṭubh* codas truncated the closing frame. The frame does not close as tightly as many in the MDh do, with a pat statement of Yudhiṣṭhira's satisfaction with what he has heard.

[76] As with 'Linking' for the word *yoga,* I am testing the viability of using 'Total Knowledge' as a technical translation formula for the name Sāṃkhya. As Edgerton argued earlier, Sāṃkhya is an intellectual method (*jñāna*)

Total Knowledge is said to be the solid image of that which has no solid form, son of Kuntī; this body of thought is that by which it[77] is to be recognized, O bull of the Bharatas. (101)

There are two kinds of beings on this earth, O lord of the earth, those called mobile and those called immobile. The mobile are superior to the others. (102)

THE *TRIṢṬUBH* CODA FOLLOWING 12.290

The *triṣṭubh* coda following the Sāṃkhya exposition tracks closely the most important theme of the closing *anuṣṭubh* frame: the importance of Sāṃkhya as a form of knowledge that is coincident with all historically important knowledge, or even the repository of it. It then goes on to restate elegantly the mystical theological point that Sāṃkhya knowledge itself is an embodied manifestation of the formless transcendent reality, here the Lord Nārāyaṇa.

PASSAGE 6 **12.290.103–110: *Triṣṭubh* coda ending the Sāṃkhya exposition**
jñānaṃ mahad yad dhi mahatsu rājan । *vedeṣu sāṃkhyeṣu tathaiva yoge* ॥ *yac cāpi dṛṣṭaṃ vividhaṃ purāṇaṃ* । *sāṃkhyāgataṃ tan nikhilaṃ narendra* ॥ 103 ॥ *yac cetihāseṣu mahatsu dṛṣṭaṃ* । *yac cārthaśāstre nṛpa śiṣṭajuṣṭe* ॥

rather than an active method (*karman*); Sāṃkhya, very obviously in this text-pair, but in other contexts too, emphasizes comprehensive knowledge of an extended plurality of entities. Intrinsically important to Sāṃkhya is *parisaṃkhyā*, 'enumeration of the whole set', and consequent *vijñāna*, '(analytic) understanding'. The formula 'Total Knowledge' is somewhat awkward, but insofar as Sāṃkhya is basically a proper name that surfaces in the epic, the awkwardness of it dissipates some. 'Total Knowledge' is certainly less cumbersome than the more explicit 'Comprehensive Enumerative Knowledge'. For a stimulating and thoughtful discussion of "counting" and "seeing" as modes of knowing in ancient Indian religions see Gómez 1999.

77 = that which is devoid of solid form, the infinite *brahman*-like God.

jñānaṃ ca loke yad ihāsti kiṃ cit ।
sāṃkhyāgataṃ tac ca mahan mahātman ॥ 104 ॥
śamaś ca dṛṣṭaḥ paramaṃ balaṃ ca ।
jñānaṃ ca sūkṣmaṃ ca yathāvad uktam ॥
tapāṃsi sūkṣmāṇi sukhāni caiva ।
sāṃkhye yathāvad vihitāni rājan ॥ 105 ॥
viparyaye tasya hi pārtha devān ।
gacchanti sāṃkhyāḥ satataṃ sukhena ॥
tāṃś cānusaṃcārya tataḥ kṛtārthāḥ ।
patanti vipreṣu yateṣu bhūyaḥ ॥ 106 ॥
hitvā ca dehaṃ praviśanti mokṣaṃ ।
divaukaso dyām iva pārtha sāṃkhyāḥ
tato 'dhikaṃ te 'bhiratā mahārhe ।
sāṃkhye dvijāḥ pārthiva śiṣṭajuṣṭe ॥ 107 ॥
teṣāṃ na tiryaggamanaṃ hi dṛṣṭaṃ ।
nāvāg gatiḥ pāpakṛtāṃ nivāsaḥ
na cābudhānām api te dvijātayo ।
ye jñānam etan nṛpate 'nuraktāḥ ॥ 108 ॥
sāṃkhyaṃ viśālaṃ paramaṃ purāṇaṃ ।
mahārṇavaṃ vimalam udārakāntam ॥
kṛtsnaṃ ca sāṃkhyaṃ nṛpate mahātmā ।
nārāyaṇo dhārayate 'prameyam ॥ 109 ॥
etan mayoktaṃ naradeva tattvaṃ ।
nārāyaṇo viśvam idaṃ purāṇam ॥
sa sargakāle ca karoti sargaṃ ।
saṃhārakāle ca tad atti bhūyaḥ ॥ 110 ॥

All the vast knowledge that occurs in those great bodies of learning, king – in the Vedas, among the adherents of the School of Total Knowledge, and in the Method of Yoga Linking – and which is known to us, diverse, and ancient, king, has come into[78] Total Knowledge. (103) And also what is observed in the vast Itihāsas, and in the teachings on wealth and power that the learned are fond of – whatsoever knowledge there is among the public – the whole vast

[78] I am grateful to Peter Schreiner for suggesting this reading of the participles *āgata* and *vihita* here and in verse 105 (see footnote 1). I had originally thought the stanzas were saying that these important bodies of knowledge had come from 'Sāṃkhya', as did Edgerton (see his translation of 290.104 in 1970: 300), though Deussen & Strauss, *ad loc.* (12.303.108, 1906: 607) and Hopkins (1901: 111, n. 2) are both non-committal as to any direction of movement.

extent of that has come into Total Knowledge, exalted one. (104) Inner calm is observed In Total Knowledge, king, so too the highest power; subtle knowledge has been accurately expressed there, and ascetic observances that are light and easy have been properly enjoined there. (105)

Truly, should there be any catastrophic upset of it,[79] son of Pṛthā, followers of the School of Total Knowledge always go easily to the Gods,[80] and having visited with them, their business completed, they then fall again among strict brahmins. (106)

Having abandoned the body the followers of Total Knowledge enter into Absolute Freedom, as Gods enter heaven. Therefore, prince, son of Pṛthā, brahmins delight in the highly worthy School of Total Knowledge of which the learned are fond. (107) Of those brahmins who are passionately devoted to this knowledge, no rebirth as an animal is ever observed, no downward course at all, nor any dwelling with the wicked or the stupid. (108)

The extensive Body of Total Knowledge is supreme and primordial, a great ocean that is clean and beloved by exalted men. The exalted Nārāyaṇa supports the whole, immeasurable Body of Total Knowledge, king. (109)

I have declared to you, O God among men, this fundamental reality. Nārāyaṇa issues this entire primordial knowledge at the time of creation and he swallows it again at the time he draws the world back in. (110)

THE CODAS CONSIDERED TOGETHER

These two codas closely resemble each other in style and language, and at the same time each is closely related thematically to the *adhyāya* which it concludes. But in the ways in which they are alike, they stand dramatically apart from their preceding chapters. The fundamental concern over *īśvaratva* in the *anuṣṭubh* pair is whether the *mumukṣu* becomes an *īśvara* or not. And while the coda of 289 illustrates the *yogasiddha* having become a *yogeśvara*, he is one having merged into the *īśvaratva* of Nārāyaṇa. The *īśvaratva* of Nārāyaṇa is also the theme of the coda of 290. More in 290 than in 289, but actually in both codas,

---

[79] = some circumstance disrupting their progress on the path before they have reached its end, *mokṣa*.

[80] Failing to reach *mokṣa*, they join the Gods in heaven.

the focus of attention has shifted from the *mumukṣu* to God, an anthropomorphic divine being who possesses and exercises powerful control over the whole world, material and intellectual. In both codas the ideal human being is depicted as participating in the substance of God (who is embodied as 'Sāṃkhya Knowledge' in 12.290) or the substance and activity of God, the powerful creative and controlling 'Lord', *īśvara*, in 12.289. These *triṣṭubh* codas are very elegantly[81] constructed passages glorifying *yogaiśvarya* and *sāṃkhyajñāna* in terms of a proto-theology of Nārāyaṇa.

In 289, the *triṣṭubh* coda depicts the concept of Yoga Linkages from 'above', that is from the position of a human who has reached the ontological level called Nārāyaṇa in 290, just one ontological step below *paramātman*, a very dramatic representation.[82] Having reached this level, the yogin has come to have all

---

[81] There is but one flaw in these two *triṣṭubh* passages comprising 52 otherwise perfectly realized *upajāti* verses. 12.190.109b reads the hypermetric *vimalam* in positions 5–8 of the verse. It is hard to allay the suspicion that the use of the word 'flawless' here – in such a way that it creates a glaring fault in the verse – is a deliberately ironic mistake. The mistake cannot be explained easily. One might think that *vimala* here simply was pronounced as '*vimlam*', which would heal the verse's meter. This suggestion loses all purchase, however, when we survey all other instances of the word in the epic's *triṣṭubh*s. The word is used 17 times in the epic's 17,928 *triṣṭubh-jagatī* lines, once in a *mālinī*, and 139 times in *anuṣṭubh ślokas*. 15 of the 17 uses in *triṣṭubh-jagatī* verses are *upajātis* (the other two occur in common sorts of *pāda*s in non-classically mixed stanzas). In no instance other than 12.290.109b does the word *vimala* perturb the meter in any way at all, demanding some exceptional explanation. Nor is there any help from the variant readings. There is a widely distributed northern variant, *vipulam*, 'extensive', which leaves the meter equally flawed. The variant *vipulam* does, however, remove all trace of irony from the line, a change of meaning which reinforces Belvalkar's correct choice of *vimalam* on the grounds of the usual Pune editorial principles. As to the motive for this supposed irony, I can only suggest that as the author here is praising the spotlessly clean ocean of 'Total Knowledge' (Sāṃkhya) that comes from the Lord God Nārāyaṇa and is supported by him, he may have put the 'flawless' flaw into his composition as a note of humility before God.

[82] It is justifiable to impute to this passage the notion that its Nārāyaṇa is the same as that of 290 since in 289 there is no theorizing of Nārāyaṇa otherwise. As I will say more fully shortly, I think both *triṣṭubh* codas were

being that is below this level for his 'lower self,' his 'body,' and Nārāyaṇa as his 'higher self,' his 'soul'. By reaching 'downward' within the cosmos, that is to entities that appeared subsequently from the substance of anterior, 'higher' entities, the yogin lord, the powerful *īśvara*, 'acts' (*kurute*), though his motives or ends are not mentioned. In the second of these two chapters, 290, the *triṣṭubh* coda characterizes Sāṃkhya as an infinitely extensive body of knowledge which Nārāyaṇa sends forth at the time of creation, swallows at the *saṃhāra*, and supports in the meantime (*śloka*s 109–110 in Passage 6). Both of these are highly developed theological ideas that seem more typical of the Purāṇas than the epic.

CONCLUDING OBSERVATIONS

The two *triṣṭubh* codas impart to this pair of texts quite an elegant and dramatic theological sheen. Each in its own way contributes powerfully to a sense that Yudhiṣṭhira's initial request to know the difference between Sāṃkhya and Yoga has been well answered. The Lord Nārāyaṇa stands at the end of each coda and crowns the argument of each *adhyāya*. First, the successful practitioner of Yoga becomes co-terminous with Nārāyaṇa and acts (*kurute*), his soul and the Lord's soul coincident (*nārāyaṇātman*), himself the self within the 'body' of the entire manifest world (*mahātman*). Second, the exalted and comprehensive body of all valuable knowledge that is Sāṃkhya, itself an outward manifestation of the formless transcendent one, is emitted by the Lord Nārāyaṇa at creation, maintained by the Lord Nārāyaṇa, and then drawn back into the Lord Nārāyaṇa at the end of time. The yogin acts within the mind and body of God; the Sāṃkhya wisdom is itself the body of God. The fundamental difference of

---

composed subsequent to the creation of the *anuṣṭubh* text-pair. Thus 'Nārāyaṇa' in the coda of 289, as in the coda of 290, likely derives from the *anuṣṭubh* passages of 290.

the two schools of thought has been made luminously clear, as has the fact each one brings its adherent to the Lord Nārāyaṇa and then *mokṣa*.

While these striking theological poems serve well as gems aptly shaped to reflect the teachings of the *anuṣṭubh* texts anterior to them, it is not likely they were composed by the same person who collated, edited, and framed those *anuṣṭubh* texts. Apart from their great distance from the literary qualities of the two *triṣṭubh* passages, the *anuṣṭubh* texts were collated and framed as a text-pair prior to the composition of the *triṣṭubh* codas. The *triṣṭubh* codas interrupt the flow of the frame between the end of the first *anuṣṭubh* exposition (at 289.57) and the frame's resumption at the beginning of 290. At the end of 290 the *triṣṭubh* coda repeats some of the matter of the *anuṣṭubh* closing frame. If the person responsible for the *anuṣṭubh* text-pair composed the *triṣṭubh* codas, he ignored his framing *ślokas*. I think it much the more likely possibility that the *triṣṭubh* codas were composed by a second hand at a later time. The *triṣṭubh* author was likely loath to delete the pre-existing framing *ślokas* and simply added his codas onto the *adhyāya*s as ornaments which would be entirely welcome.

If we read the two *adhyāya*s without the *triṣṭubh* codas, we find two amalgamations of Yoga and Sāṃkhya teachings that have been compiled into a unified presentation with introduction, transitions, and a conclusion. Someone wove this material into a unified text-pair for some purpose. And because the author's actions were purposeful and consistent, it is not too difficult to discern what that purpose was. The purpose behind the production of 289–290 as a text-pair was to glorify and justify 'Sāṃkhya' as an important and soteriologically effective body of 'philosophical knowledge' in a context in which some school called 'Yoga' held 'the high ground'. In particular, our author wanted to justify Sāṃkhya's 'knowledge' against the supposed claim that only the person who had effected the difficult Yoga linkings and had acquired its cosmic mastery could effect *mokṣa*.

Again ignoring the *triṣṭubh* codas (which, in each case, only augment the basic theme of each *anuṣṭubh* exposition), the framing *ślokas* form a reliable guide to the author's argument. Citing the arguments Yogas and Sāṃkhyas make to justify their theory and praxis, Bhīṣma cited for Yoga the attack on Sāṃkhya that turns upon the issue of becoming an *īśvara*. He then cites for Sāṃkhya the critical, and effective-of-*mokṣa*, connection between comprehensive knowledge and *vairāgya*. The *anuṣṭubh* compilations that follow present the case that *yoga* requires power and produces lordship and *mokṣa*,[83] and then the case that the comprehensive knowledge and understanding involved in Sāṃkhya result in one's coming to know ever higher cosmic realities, until one reaches the Lord Nārāyaṇa, who conveys one to the Supreme Embodied-Subjectivity (*ātman*) of the universe, the *paramātman*, and infinite *mokṣa*. Upon the conclusion of the second of the two long passages making this argument in 290, there is the extremely interesting *kṣetrajña* theory that has a close bearing on the issue of the difference between the direct and embodied knowledge involved in *yoga* and the disembodied, non-sensual knowledge of Sāṃkhya. Thus, against the putative argument of 289.3a for the necessity of *īśvaratva*, our author has fortified Sāṃkhya theory with three forms of *īśvaratva*: the original formulators of Sāṃkhya were *īśvaras* (290.3), the ultimate Sāṃkhya *gati* leads to and through the *prabhu* Nārāyaṇa (290.23, 74, 91), and the disembodied *kṣetrajña* is superior to the thoroughly embodied *yogeśvara* (290.81–93).

---

[83] But notice that 289 does not actually make any case that becoming an *īśvara* through *yoga* is the only way to gain *mokṣa*. That assertion, which is implicit in 289.3a, is rhetorical 'spin' put on the Yoga argument by our author. That assertion is the straw man that occasions the entire editorial and compositional effort of the author of the *anuṣṭubh* text-pair here.

Upon the completion of this discussion of perception and knowledge in the higher reaches of the soul's journey to *mokṣa* (290.93), the redactor closed his text-pair with direct praise of Sāṃkhya and then offered an elaborately asserted formulation that the speech of Sāṃkhya teachers and Yoga teachers and the supreme seers (of the Vedas) had in each case the transcendent *brahman* as its object. If Sāṃkhya was in need of justification, if Sāṃkhya needed defending against arguments from Yoga, then this author has provided that defense and justification in good measure.

REFERENCES

BROCKINGTON, John 2003. Yoga in the *Mahābhārata*. In: Ian Whicher & David Carpenter (eds.), *Yoga. The Indian Tradition*: 13–24. London: RoutledgeCurzon.

BRONKHORST, Johannes 1983. God in Sāṃkhya. *Wiener Zeitschrift für die Kunde Südasiens* 27: 149–162.

BUITENEN, J. A. B. van 1959. *Akṣara*. *Journal of the American Oriental Society* 79: 176–187.

——— 1964. The large ātman. *History of Religions* 4.1: 103–114.

COUTURE, André 2005. The meaning of the words *yoga* or *yogin* in the Puṣkaraprādurbhāva section of the *Harivaṃśa*. In: Petteri Koskikallio (ed.), *Epics, Khilas, and Purāṇas. Continuities and Ruptures*: 389–409. Zagreb: Croatian Academy of Sciences and Arts.

DEUSSEN, Paul & Otto STRAUSS 1906. *Vier philosophische Texte des Mahābhāratam*. Leipzig: F. A. Brockhaus.

EDGERTON, Franklin 1924. The meaning of Sāṃkhya and Yoga. *American Journal of Philology* 45.1: 1–46.

——— 1970. *The Beginnings of Indian Philosophy*. Cambridge, Mass.: Harvard University Press.

FITZGERALD, James L. 2002. Nun befuddles king, shows *karmayoga* does not work: Sulabhā's refutation of King Janaka at MBh 12.308. *Journal of Indian Philosophy* 30.6: 641–677.

FRAUWALLNER, Erich 1925. Untersuchungen zum Mokṣadharma, 2. *Wiener Zeitschrift für die Kunde des Morgenlandes* 32: 179–206.

FRAUWALLNER, Erich 1953. *Geschichte der indischen Philosophie*. 2 vols. Salzburg: Otto Müller Verlag.

GARBE, Richard 1917. *Die Sāṃkhya Philosophie. Eine Darstellung des indischen Rationalismus, nach den Quellen*. 2nd edition. Leipzig: H. Haessel Verlag.

GÓMEZ, Luis O. 1999. Seeing, touching, counting, accounting: Sāṃkhya as formal thought and intuition. *Asiatische Studien / Études Asiatiques* 53.3: 693–711.

HAUER, J. W. 1958. *Der Yoga: Ein indischer Weg zum Selbst. Kritisch-Positive Darstellung nach den indischen Quellen mit einer Übersetzung der massgeblichen Texte*. 2nd edition. Stuttgart: W. Kohlhammer Verlag.

HOPKINS, Edward Washburn 1901. *The Great Epic of India. Its Character and Origin*. New York: Charles Scribner's Sons.

Mahābhārata, Critical Edition. Ed. by V. S. Sukthankar, S. K. Belvalkar, P. L. Vaidya *et al.* 19 vols. Poona. Bhandarkar Oriental Research Institute, 1933–66.

MALINAR, Angelika 1996. *Rājavidyā: Das königliche Wissen um Herrschaft und Verzicht. Studien zur Bhagavadgītā*. Wiesbaden: Harrassowitz Verlag.

OGUIBENINE, Boris 1984. Sur le terme *yóga*, le verbe *yuj-* et quelques-uns de leurs dérivés dans les hymnes védiques. *Indo-Iranian Journal* 27: 85–101.

SCHREINER, Peter 1999. What comes first (in the *Mahābhārata*): Sāṃkhya or Yoga? *Asiatische Studien / Études Asiatiques* 53.3: 755–777.

SCHREINER, Peter, Reinhold GRÜNENDAHL, Angelika MALINAR & Thomas OBERLIES 1997. *Nārāyaṇīya-Studien*. (Purāṇa Research Publications Tübingen, 6.) Wiesbaden: Harrassowitz Verlag.

The Boar Shakes the Mud off: A Specific Motif in the *Varāhakathā* of the Great Epic and Purāṇas

YAROSLAV VASSILKOV

Years ago Professor Georgy A. Zograph, head of the South and South-East Asia Department at the Institute of Oriental Studies, St. Petersburg, advised me, then a postgraduate student with a keen interest in the *Mahābhārata*, to look into some records of Indian tribal, non-Aryan mythology contained in the books on Indian languages and folklore (such as *Linguistic Survey of India* etc.) in the Department's library. I then failed to see in what way these primal myths of tribal communities could be useful for my studies of the Sanskrit epic. Since then I have had plenty of opportunities to confirm the wisdom of my teacher's word of advice.[1] In the present paper I shall provide one more example to demonstrate that looking into the mythologies of the Indian tribal world can sometimes shed light on obscure motifs in the epic and Purāṇic myths.

The starting point for this study was provided by a version of the *Varāhakathā* in the *Viṣṇupurāṇa*, *adhyāya* 4 of its first book. It describes the awakening of Prajāpati Brahmā (here regarded as a form of Nārāyaṇa) at the end of the *pralaya*: the god sees

---

[1] In particular, the acquaintance with some tribal myths and rituals proved to be useful for the study of the *gandharvas* and *apsarases* as classes of mythical beings in Indian mythology; see Vassilkov 1989–90.

that the Earth has disappeared under the waters. He takes the form of an 'amphibian' animal, the boar,[2] and plunges into the ocean.

It is here that we meet with an unusual and strange motif. When the divine Boar, after his cosmogonic dive, came out onto the surface, with the Earth on his tusks, the waters, struck by the 'wind of his mouth' (*mukhānila*), reached heaven and washed/purified (*prakṣālayām āsa*) the great sages residing in the Janaloka (the sphere of the holy men). The *munis*, scattered by the wind of his breath (*śvāsānila*), fell from the Janaloka. When the Boar rose out of the ocean, with his belly wet with water (*jalārdrakukṣi*), and started shaking off (*vidhunvat*) his 'scriptural body' (*vedamayaṃ śarīram*), they found shelter among the bristles of the animal (ViP 1.4.27–29).[3]

There is something strange in the variable repetition of the same motif.[4] The reason for the repetition seems to be that the original sense of the episode was already obscure to the Purāṇic narrators. Folklorists know very well that when one deals with folklore texts, certain indistinction, illogicality and vagueness will usually arise due to the misunderstanding of a half-forgotten archaic motif which belonged to an ancient tradition that had been mostly overcome by the time of the text's composition. Seemingly, this is precisely what we have in our passage from ViP 1.4.

---

[2] Traditionally, the boar was regarded as an animal with two natures, belonging both to the earth and the water element. Some Purāṇas explain that in order to rescue the Earth from the bottom of the ocean the god had to assume the form of the boar as the animal that likes playing in water (*Kūrmapurāṇa* 1.6.7: *jalakṛḍāsu ruciram*; *Brahmāṇḍapurāṇa* 1.1.5.11: *jalakṛḍāsamucitam*); cf. Prasad 1989: 76.

[3] *Viṣṇupurāṇa* 2002. The key word *vidhunvat* 'shaking' in this translation is rendered as 'trembling' (p. 24: the Boar, as he rose up, stands 'trembling').

[4] Firstly, winds and waters fly in all directions at the moment of the Boar's rise out of the ocean, then, after his rise, the same happens again when the Boar starts shaking his wet body.

At this moment one begins to suspect whether the motive of the Boar shaking his body, expressed by the verb √*dhū*, had once had much more importance and had referred, in fact, to the main action of cosmogony?

This conjecture is partly confirmed by a story from book 12 (Śāntiparvan) of the *Mahābhārata*: its *adhyāya* 333 deals with the origin of the three *piṇḍa*s – rice-balls presented to the three generations of the dead (fathers, grandfathers and great-grandfathers) in the *śrāddha* rites. Having risen from the ocean and having established the Earth at her proper place, the Boar whose body was wet and covered with mud (*jalakardamaliptāṅga*) suddenly shook off (*vidhūya sahasā*) three balls of mud from his tusks, and performed, with them, as with the *piṇḍa*s, the archetypal *śrāddha* ritual (Mbh 12.333.11–14). The key verb √*dhū* reappears in 333.17 where the *piṇḍa*s are called *daṃṣṭrābhyāṃ pravinirdhūtā* ('shaken off from the tusks'). As we can see, in this chapter the action of the 'shaking off' (*vi*-√*dhū*) undoubtedly has a creative function: 'shaking off mud' is the way in which the central element of the *śrāddha* ritual was created.

This gives us grounds to believe that an ancient version of the Boar story must have existed earlier in India. In this version the divine diver created the Earth precisely by way of shaking mud off his wet body. Is it possible to find, in the historical development of the *Varāhakathā*, any traces of such a version? In order to answer this question, let us take a brief retrospective look at the history of the Boar myth in South Asia.

All versions of the *Varāhakathā* following the *Viṣṇupurāṇa* represent the *avatāra* doctrine in its classical form. In them, the action of Viṣṇu in the form of the Boar is not a cosmogony any more: the Earth had been created long before, but then it was stolen by an *asura* and hidden under the waters. Viṣṇu appears here not as the creator, but as the guardian of the good and the chastiser of the evil (which is his function in all classical *avatāra* stories). This is why the central event in the late versions of the

Varāha myth is not the Boar's dive into the ocean, but his fight with the demon who had stolen the Earth, the *asura* Hiraṇyākṣa.

The version of the *Viṣṇupurāṇa* is a transitional one: the Boar here is certainly reinterpreted already as an aspect of Nārāyaṇa (Viṣṇu) and fulfils his function as the guardian of the universe, but first Nārāyaṇa takes the form of the ancient creator god Prajāpati-Brahmā (ViP 1.4.1–5) and only after that does he dive into the ocean; the myth still retains its meaning of cosmogony.

In the epic and Purāṇic versions preceding the *Viṣṇupurāṇa* (*Mārkaṇḍeyapurāṇa*, *Rāmāyaṇa*) the Boar is always identified with Brahmā; in the late Vedic Saṃhitās and Brāhmaṇas he is identified with Prajāpati. There are also some grounds to believe that in the late Vedic period Prajāpati in his 'rise to higher rank' (Gonda 1986) appropriated the function of an earlier divinity, already known in the Early Vedic period – Varāha, the Boar, who brought the Earth onto the surface of the waters. It should be noted that in one of the earliest late Vedic texts, *Maitrāyaṇisaṃhitā* of the Yajurveda tradition, this Boar (called simply *ādivarāha*, the Primeval Boar) is not yet identified with Prajāpati. In the *Śatapathabrāhmaṇa* Varāha (who seems to be in the context identical to Prajāpati) bears the name Emūṣa, which provides us with a link to the mythology of the Boar in the Early Vedic period.

In the *Ṛgveda* (1.61.7; 7.69.14, 77.6, 10) Emuṣa is the name of the demonic Boar, the guardian of the *asuras*' goods, with whom Indra and Viṣṇu fight (the 'goods' of the *asuras* consist of a buffalo herd and *odana*; – rice cooked with milk). Eventually, Indra kills the Boar (*varāha*), piercing the mountain, on the other side of which the enemy was hiding. Obviously, the story reproduces the pattern and the motif well-known from the myths of Indra who killed his other enemies, Vṛtra and Vala. In another version of the story (from the *Taittirīyasaṃhitā*), the Boar guards the treasure of the *asuras* on the other side of seven hills. Indra pierces the hills, as well as the Boar's body, with the sharp blades of *kuśa* grass, and then Viṣṇu, the god of the sacrifice, carries off the body 'as a sacrifice' for the gods. Most Vedic scholars agree that the name of the Boar (Emuṣa) used in this story is non-Aryan and, most likely, Austroasiatic (see e.g. Kuiper 1950: 17; Gonda 1954: 137 n.; Mayrhofer 1956–76, vol. 1: 128–129).

However, in the same Early Vedic tradition the Boar appears not only as the antagonist of the gods, but also as the protagonist (the hero) of the cosmogonic myth. It can be suggested that it is the meaning of the Boar's image in the *Bhūmisūkta* of the *Atharvaveda* (12.1.48): 'The Earth, found by the Boar (*varāheṇa*), opens itself to (or: becomes expanded for, *vijihīte*) the wild pig/hog (*sūkara ... mṛga*).' We can find extended versions of the same myth in some Late Vedic texts (*Kāṭhakasaṃhitā*, *Kapiṣṭhalakathasaṃhitā*, *Taittirīyasaṃhitā*): here Prajāpati, in the form of a boar, dives and brings from the ocean's bottom a piece of clay (*mṛd*), which turns into the Earth (Prasad 1989: 6–7). At the beginning, the Earth is small, like a pig's snout, but then it opens itself, or becomes larger / expands 'for the Boar'.[5] It is in the context of this version that the *Śatapathabrāhmaṇa* gives the Boar, in his role of the Earth-diver, the name 'Emūṣa', and this connects him historically with the demonic Boar Emuṣa of the *Ṛgveda*.

---

[5] The verb *vijihīte* (√*hā* with prefix *vi*) is usually translated as 'opens itself', 'expands', but its precise meaning remains far from clear ('What does the verb *ví jihīte* mean precisely?' – Gonda 1954: 137). The context (the Earth opens/expands itself for or before the Boar) enables us, as it seems, to suggest in particular the meaning of readiness for copulation. Side by side with the motif of the originally small Earth's expansion or growth in front of the Boar, we find in the Late Vedic texts (Saṃhitās and Brāhmaṇas) another motif: impregnation of the Earth by the Boar. According to the *Taittirīyasaṃhitā* (7.1.5), after the Earth 'extended' in front of the Boar, he begat on her the gods, Vasus, Rudras and Ādityas (Prasad 1989: 7). The motif reappears sometimes in much later Purāṇic versions (see their review in Gonda 1954: 141–143), but there the offspring of the Boar's union with the Earth usually is Naraka – the demonic antagonist of Kṛṣṇa.

F. B. J. Kuiper was convinced that the Vedic Aryans had applied by mistake the name of the non-Aryan demonic Boar to their own cosmogonic Boar (= Prajāpati); in fact, both mythological animals were, according to him, entirely different (Kuiper 1950: 18). This view was criticized by Jan Gonda who believed that the use of the same name in both cases makes the two myths at least somehow connected (Gonda 1954: 139). According to a more recent author, 'the dramatis personae of both the myths are the same. But no definite relation between the functions of the two myths has yet been established' (Prasad 1989: 5–6). However, the relation between the two 'Boar myths' may show up more distinctly if we take into account that the Boar's name (or the common noun for 'boar') was known to the Vedic Aryans in many variants (*emūṣa*, *emuśa*, *āmukha*, *vāmamuṣa*, *vāmamoṣa*; Prasad 1989: 5), and that in *Maitrāyaṇisaṃhitā* the Primeval Boar has no name at all and it is not identified with any Vedic god. Does not all this mean that the name (or noun) and the myth itself were borrowed by the Aryans from an alien ethnic and linguistic tradition? In this case the myth of the Boar Emuṣa as the enemy of the Vedic gods and the myth of the Boar as the form of the Brahmanic creator god Prajāpati may be contrasted as two different attempts at the alien myth's adaptation, or two subsequent stages in the process: first the alien god is borrowed in a negative context, as a demonic antagonist of the Vedic gods, but later the Boar (Emūṣa) appears as the cosmogonic diver reinterpreted as the incarnation of the Vedic Prajāpati.

It must be noted that the boar (Skt. *varāha*; Av. *varāza*)[6] might have played a certain role in the mythology of Indo-Iranians before the Aryan migration to India. Both Iranians and

[6] Both words cannot be etymologised on the basis of Indo-Iranian or Indo-European and are regarded by the majority of specialists as the words of unknown origin (see Mayrhofer 1956–76, vol. 3: 150).

Vedic Indians worshipped the boar as the embodiment of fierce energy, impetuosity and reckless bravery. We can refer in this connection to the representations of the boar in the art of the Bactria-Margiana archaeological complex, such as the famous bronze mace in the shape of a furious boar (Amiet 1986, fig.171). Both in Ancient Iranian and Vedic mythology some gods take the form of a boar in order to fight with the forces of evil. But nowhere in the available records of Iranian mythology can we find any traces of the 'Boar as the cosmogonic diver' motif. This gives us some grounds to suggest that the Vedic Aryans made their first acquaintance with this motif in India where they received it from a local cultural substratum.

Can we say with any precision which particular local substratum (one of so many) it might have been? The IVC materials are of no help to us: a boar is never found among the animal figures on the seals; archaeologists agree that the pig was not yet domesticated and played no role in the IVC economy (Kenoyer 1998: 165). According to M. Witzel (1999, fn. 4), the domestication of the pig 'seems to be an eastern phenomenon (thus Munda?); cf. Munda and Sino-Tib. [words for] "pig" and cf. the ancient boar cult on the Nicobar Islands'.[7] At the same time it must be noted that ethnologists have recorded mythological narratives about the Boar as the Earth-diver among Indian tribal people speaking both Austroasiatic and Dravidian languages. A brief review of such myths given below is based mostly on the Catalogue of myth and folklore motifs in America and Eurasia by Yu. E. Berezkin which contains at the moment over 40,000 summaries of folk narratives (see http://www.ruthenia.ru/folklore/berezkin – in Russian).

---

[7] We should probably add to it that the Munda-speaking Kharia of Eastern India worshipped Ghorea, the boar-god (Gonda 1954: 135). It is also noteworthy that the origin of pig-breeding and rice agriculture was traced by special studies to the South-East Asian centre of producing economy, i.e. to the Austroasiatic homeland (Shnirelman 1989).

Apart from India the myth of the animal Earth-diver (Thompson motif A 812; Thompson 1955–58) has been registered in two other regions of the world: Siberia and North America. In the myths of American Indians the creator god usually sends, one by one, various animals down to the bottom of the water to find some earth. Finally one of them (often a beaver, or a musk-rat) comes back with a piece of mud; the creator uses it, and then it expands to become the Earth. Much more widespread is the variant in which the Earth-diver is an aquatic bird (in some American Indian traditions, among the palaeoasiatic peoples of North Siberia, the Finno-Ugrians, some South Siberian peoples etc.).

In India, among non-Aryan-speaking Indian tribes, we sometimes find versions with an aquatic bird (the Dravidian-speaking Gonds, the Aryan-speaking Baiga). But specifically Indian and totally unknown outside India is the version in which the Earth-diver appears in the image of the Boar (Austroasiatic Bondo, Dravidian Koya, Bison-Horn Maria [Māḍiyā] Gonds etc.). In the myth of the Didani tribe from Orissa, the creator god is unable to find any earth needed for the creation of a new world, but finally finds a little piece of earth stuck to the tail of a boar. There is no mention of diving, but we may well suggest that it is a reduced version of the myth and that there had been an earlier, complete version, in which the boar dived and returned with a piece of earth on its tail.

Verrier Elwin recorded among the Mundas in Orissa the story of a boar who brought some earth from the depths of the ocean *on his tusks* – as in the classical Hindu myth. But it seems that the most popular in the Indian tribal world is the version in which the Boar performs his cosmogonic action in a very specific way. It can give us a clue to the strange motifs in the *Viṣṇu-purāṇa*: the Boar dives to the bottom, rolls itself in mud, then re-appears on the surface of the ocean and shakes itself violently.

The mud from its bristles flies in all directions. That was how the Earth was created.

Here is the myth of the Dravidian-speaking Koya: there was the flood, the Earth sank away in the waters, and in order to save it, Bhima's daughter jumped from heaven into the water, turned herself into a boar who then rolled itself in the mud at the bottom of the ocean, came out of the water and shook the mud off. In the myth of Dravidian-speaking Bison-horn Māḍiyās, the motif of the Boar as the diver combines with a typically Austroasiatic motif of the first human pair: brother and sister, who float over the waters of the flood in a hollow pumpkin. They cannot find a piece of dry land to live on. This is why Mahaprabhu, the god, sends the Boar to find some earth under the waters. The Boar rolls itself in mud on the back of the great underwater tortoise, then comes out of the water and shakes the mud off. The mud flies in all directions and then turns into the Earth.

In a story known among the Bondos – the Munda ethnic group from Orissa – the initial part of the Earth-diver myth contains the same Austroasiatic cosmogonic motif: 'Brother and sister floating in a pumpkin'. In order to provide them with a place to live and means of existence, Mahaprabhu sent the Boar to the underworld (below the ocean), where the animal stole some seeds of the seven fruit-bearing trees belonging to the local god, Kermo Deota, then rolled itself in mud, came out onto the surface and violently shook the mud off. The drops of mud fell upon the waters, dried and eventually turned into the Earth.

What particular ancient ethnic tradition the 'Earth-diving boar' motif originally belonged to is at the moment difficult, and probably impossible, to say, because numerous mythological motifs in South Asia had acquired the *areal* distribution and *areal* characteristics in times immemorial. Elements of many traditions, including the archaic Indo-Aryan, Dravidian, Austroasiatic and more ancient substrata, had long ago melted together and formed a kind of alloy.

We would be able to make a reasonable suggestion about an ethnic tradition from which the motif of the 'diving boar' originally came if there were some traces of the motif somewhere else, outside India. For example, if there were some instances of the motif's appearance in South East Asia and South China, i.e. in the Austroasiatic homeland, we would have enough grounds to suggest that the motif is of Austroasiatic origin. Indo-European parallels would permit us to give priority to the Indo-Aryan tradition, and a discovery of the 'diving boar', for example in the written sources of ancient Elam would lead us to the suggestion of a Dravidian origin, in the light of the well-known Elamo-Dravidian hypothesis. But, unfortunately, the motif of the 'diving boar' is unique; it has never been registered anywhere outside India.

The same must be said about another motif: the creation of the Earth by way of shaking off mud, clay or dirt from one's body. Again, we cannot find it anywhere outside India. According to Yury Berezkin's impressive database of folk narrative motifs there is only one exception. In the myth of the Mansi (the Ugrian-speaking tribal people of Western Siberia), an aquatic bird – the loon – creates the Earth after diving into the primeval ocean and after bringing, from the bottom, a piece of mud in its beak. But in a single version out of many, the loon does it by way of shaking the mud off its feathers. Is it enough to suggest that the proto-Indo-Aryans had borrowed the motif from their Ugrian neighbours in the Southern Urals or in the Volga region, and had brought it with them to their new Indian home? Or should we suggest that the motif was brought to India by the Dravidians, since specialists still continue considering a possibility of very early Uralic-Dravidian linguistic and cultural contacts? In my opinion, the argument can hardly be convincing in either case. The loon is a notorious diver, and every diving animal or bird often shakes itself immediately after getting out of water. Consequently, the motif of the cosmogonic diver shaking itself after diving can have appeared among the Mansi independently. It was simply based on a naturalistic observation. Most probably, the coincidence between the single Mansi variant of the Earth-diver myth and the Indian tribal myth of the diving boar is due to pure chance.

If we were to choose between the Dravidian, the Austroasiatic and the Indo-Aryan traditions in our search for a more plausible source of the Boar story in the Vedic and the Hindu Sanskrit literatures, it seems to me that a more likely suggestion would be the priority of the Austroasiatic tradition. Indeed, does not the name of Emūṣa have an Austroasiatic etymology? Does not the myth of the diving boar often appear in association with other cosmogonic motifs (such as the 'brother and sister in a pumpkin') which are undoubtedly Austroasiatic? These facts make it highly probable that the Vedic Aryans had received the myth of the diving boar from the lore of the Indian Austroasiatic (Munda-speaking) tribes.[8]

This does not mean, however, that the boar myth itself is necessarily of Austroasiatic origin. Its absence in the mythology of the Austroasiatic homeland – SE Asia – is a very strong argument against it. In the present state of our knowledge it seems most probable that the Munda (or Para-Munda) tribes (and perhaps some Dravidian tribes too) had themselves inherited this myth from a more ancient local substratum.

Now, if we return to our epic and Purāṇic versions of the *Varāhakathā*, we must pose and answer one more question. How was it possible that the authors of such Hindu texts as the *Mahābhārata* and the *Viṣṇupurāṇa* – texts written in the language of the cultural elite, Sanskrit – from time to time reproduced (no doubt, unconsciously) in the story of the Boar, the

---

[8] Influence in the opposite direction is much less probable. As I have tried to show in the present paper the data of the tribal myths enable us to elucidate the obscure motifs in the Vedic and Hindu versions of the Boar myth, but not the other way round.

unique motif, which is not represented explicitly in any Sanskrit text known to us, but was obviously represented in the source from which the Vedic tradition had borrowed the Boar story – the mythology of some Munda, or Para-Munda ethnic groups, or even of an earlier, pre-Austroasiatic Indian autochthonous substratum? How can we explain the fact that a motif of a foreign origin, almost completely forgotten, persistently reappears in the Sanskrit texts, without the will of the authors and to their obvious bewilderment?

Previously, similar cases were sometimes explained through the concept suggested by the Russian philologist and philosopher Mikhail Bakhtin: the 'memory of the genre'. But this concept has never been clearly defined; the mechanics of its functioning have never been described; and in our case, unfortunately, the 'memory of the genre' concept does not work.

Quite recently, a research project has started, which promises, as it seems, to lead us one day to the solution of this problem. A group of scholars in Russia (Yu. E. Berezkin from St. Peterburg, S. A. Borinskaya and A. V. Korotayev from Moscow, V. V. Napol'skikh from Izhevsk) have begun a new study of this particular myth (the 'Earth-diver myth'). Unfortunately for us they have used only the versions in which the diver is an aquatic bird. The special aim of their study was to make clear in what way the geographical distribution of the 'Earth-diver' myth correlates with the spread of 1. language families and 2. genetic characteristics. This is the conclusion to which they have come, according to the brief preliminary report which this group published in 2005:

> The correlation between myths and genetic characteristics is much stronger than the correlation between myths and languages. In other words, it is easier, in the process of historical development, to exchange an old language for a new one than to part with old cosmogonic myths. (Berezkin *et al.* 2005: 205)

This means, if I understand it correctly, that sometimes heirs of certain genetic lines are able to reproduce some motifs from the mythology of their distant physical ancestors even in spite of the fact that the society has already adopted a new language and a new culture.

In the case of the Sanskrit epic and Purāṇic narrators, we may suggest that the forgotten ancient motifs could be periodically revived in their minds as a result of occasional contacts with the traditions in which these motifs continued to exist. But can we imagine any real cultural contacts between Brahmanic authors of such texts as the Mokṣadharmaparvan and *Viṣṇupurāna*, and the members of the Munda or Dravidian tribal societies?

It seems to me that there was not a single chance for a possibility of such contacts. It would therefore perhaps be more reasonable to assume the mediation of the archaic (pre-classical) Indo-Aryan traditions (village traditions, common folk traditions) which continued to exist for centuries side by side with the highly developed Sanskrit civilization. True, these archaic folk traditions had occasional contacts with the tribal people, could have borrowed some myths from them, and later, in their own turn, might have influenced the Great tradition of the Sanskrit literature. The recurrent waves of 'archaization' constitute a well-known phenomenon in the cultural history of the Vedic and Hindu India.

REFERENCES

AMIET, P. 1986. *L'Age des échanges inter-iraniens, 3500–1700 avant J.-C.* Paris: Musée du Louvre.

BEREZKIN, Yu. E., S. A. BORINSKAYA, A. V. KOROTAYEV & V. V. NAPOL'SKIKH 2005. *Sravnitel'nyj analiz areal'nogo raspredeleniya mifologicheskikh motivov i geneticheskikh linij. VI kongress ehtnografov i antropologov Rossii. Tezisy*. St.-Peterburg: MAE RAN.

GONDA, Jan 1954. *Aspects of Early Viṣṇuism*. Utrecht: N. V. V. Oosthoek's Uitgevers Mij. [reprint: Delhi: Motilal Banarsidass, 1993].

—— 1986. *Prajāpati's Rise to Higher Rank*. Amsterdam: North-Holland Publishing Company.

KENOYER, J. M. 1998. *Ancient Cities of the Indus Valley Civilization*. Oxford: Oxford University Press – American Institute of Pakistan Studies.

KUIPER, F. B. J. 1950. *An Austro-Asiatic Myth in the Rigveda* (Mededelingen der Koninklijke Nederlandse Akademie van Wetenschappen, N.S., Deel 13, No. 7). Amsterdam: N. V. Noord-Hollandsche Uitgevers Maatschappij.

MAYRHOFER, M. 1956–76. *Kurzgefasstes etymologisches Wörterbuch des Altindischen*. 3 vols. Heidelberg: Carl Winter-Universitätsverlag.

PRASAD, Maheshwari 1989. *Some Aspects of the Varāha-Kathā in Epics and Purāṇas*. Delhi: Pratibha Prakashan.

SHNIRELMAN, V. A. 1989. *Proiskhozhdeniye proizvodyashchego khozyajstva*. Moskva: Nauka.

THOMPSON, Stith 1955–58. *Motif-Index of Folk-Literature*. 6 vols. Copenhagen: Rosenkilde and Bagger.

VASSILKOV, Yaroslav 1989–90. Draupadī in the assembly-hall, Gandharva-husbands and the origin of the Gaṇikās. *Indologica Taurinensia* 15–16: 387–398.

ViP = *Viṣṇupurāṇa. Śrīviṣṇupurāṇam: The Viṣṇu-Purāṇam (A System of Hindu Mythology and Tradition)*. Sanskrit text and English translation, with various notes derived from other texts, by H. H. Wilson, ed. and revised by K. L. Joshi. Delhi: Parimal Publications, 2002.

WITZEL, M. 1999. Substrate languages in Old Indo-Aryan (Ṛgveda, Middle and Late Vedic). *Electronic Journal of Vedic Studies* 5.1.

The 16,108 Wives of Kṛṣṇa in the *Harivaṃśa*

HORST BRINKHAUS

Kṛṣṇa is famous as the 'Divine Lover', and ever since Jayadeva's *Gītagovinda* it is the love between Kṛṣṇa and Rādhā in particular that has been a central theme in the Kṛṣṇa cult. Rādhā is regarded as Kṛṣṇa's mistress, not as his wife. In the older epic and Purāṇic literature, that is, in the *Mahābhārata* (Mbh), the *Harivaṃśa* (Hv), and in Kirfel's *Purāṇapañcalakṣaṇa*, and even in the *Kṛṣṇacaritas* of the *Viṣṇupurāṇa* and *Bhāgavatapurāṇa*, there is not a single mention of Rādhā's name anywhere. As a preview of things to come, as it were, 'Kṛṣṇa's dalliance with the herdswomen' is referred to for the first time in one passage of the *Harivaṃśa*.[1] This relates, however, only to Kṛṣṇa's life in Mathurā, when he was still a youth and, of course, unmarried. Later on, after moving to Dvārakā, he is universally described as an adult and a married man. As a husband, however, he was not living in monogamy, but was concurrently described in the epic and Purāṇic literature as married with thousands of wives. Very probably the oldest testimony of this state of affairs is to be found in the *Mahābhārata*. In the following I shall start with a short summary of all the information this epic provides about Kṛṣṇa's wives.

---

[1] *Harivaṃśa* 63.15–35; cf. Sheth 1984: 13–14.

MAHĀBHĀRATA

In the Ādiparvan it is reported that Kṛṣṇa, Baladeva, Pradyumna and other members of Vasudeva's family once came down to the earth as partial incarnations of Nārāyaṇa and other gods. On this occasion, 16,000 Apsarases were given along with Nārāyaṇa-Vāsudeva to serve as his wives on earth. On the other hand, Pradyumna, his son born by his wife Rukmiṇī, is well attested in the epic,[2] and thus we are not astonished to find in a couple of Northern manuscripts an interpolation which adds Rukmiṇī to the 16,000 anonymous wives. How this was done is quite remarkable: at first only Draupadī is mentioned in the critical text, as a partial incarnation of the goddess Śrī – a fact testified to elsewhere in the *Mahābhārata* – but the interpolation now identifies Rukmiṇī with Śrī and declares Draupadī to be an incarnation of Śacī (Mbh 1.61.93–95).[3] This shows a later tendency to tighten up the older text to a certain extent: the divine consorts Śrī and Śacī were assigned as incarnations on earth to the incarnations of their own heavenly partners.

Mention of the anonymous 16,000 wives and other single main spouses called by name does occur in several further passages of the *Mahābhārata*:

(1) In the Udyogaparvan, it is briefly narrated that Kṛṣṇa once defeated the Asura Naraka and freed and carried off the 16,000 girls the Asura had kept in prison (Mbh 5.155.8–9).[4]

---

[2] Cf. Sörensen 1904–25, s. v. 'Pradyumna' and 'Rukmiṇī'.

[3] *gaṇas tv apsarasāṃ yo vai mayā rājan prakīrtitaḥ* I *tasya bhāgaḥ kṣitau jajñe niyogād vāsavasya ca* II 93 II *tāni ṣoḍaśa devīnāṃ sahasrāṇi narādhipa* I *babhūvur mānuṣe loke nārāyaṇaparigrahaḥ* II 94 II *śriyas tu bhāgaḥ saṃjajñe ratyarthaṃ pṛthivītale* I

N3, V1, B5, D (except D3) insert: *bhīṣmakasya kule sādhvī rukmiṇī nāma nāmataḥ* I *draupadī tv atha saṃjajñe śacībhāgād aninditā* I *drupadasya kule kanyā vedimadhyād aninditā* II 95 II

[4] *nirjitya narakam ... ṣoḍaśa strīsahasrāṇi ... pratipede hṛṣīkeśaḥ ...*

(2) A decidedly sectarian passage of the Anuśāsanaparvan, where Kṛṣṇa is presented as a prominent worshipper of Śiva, offers the following account: Jāmbavatī, another main wife of Kṛṣṇa specified by name, jealously complains that Kṛṣṇa with the help of Śiva had begotten eight sons upon Rukmiṇī – whose names she pedantically enumerates (Mbh 13.14.16d–17).[5] Therefore she asks him to do the same now for her. Kṛṣṇa promises her a son, to be called Sāmba, and accordingly sets off to the Himālaya to worship Śiva. After a long period of ascetic practice, Śiva appears together with Umā and offers eight boons to him. Strangely enough, Kṛṣṇa now asks for 10,000 sons. Śiva agrees, but immediately afterwards Umā turns to Kṛṣṇa saying that after Śiva had granted him his son Sāmba (sic!), she also wants to do him a favour, so Kṛṣṇa once again asks for sons, this time a hundred. What the goddess grants, however, are 16,000 wives![6] – Even as Pradyumna is attested as the son of Rukmiṇī, so too Sāmba is attested as Jāmbavatī's son, elsewhere in the *Mahābhārata*.[7]

(3) In the Mausalaparvan, it is Arjuna who brings the news of Kṛṣṇa's death to his unsuspecting widows, at first to the 16,000 anonymous wives and afterwards to Satyabhāmā and Rukmiṇī (Mbh 16.6.6 and 6.13). Satyabhāmā, the daughter of Satrājit, is also mentioned in different places in the *Mahābhārata* as Kṛṣṇa's wife, though not as the mother of any of his children,[8] and she is even repeatedly called Kṛṣṇa's *mahiṣī priyā* (Mbh 3.180.11; 222.3 and 10; and 4.8.17).

---

[5] *... rukmiṇyā janitāḥ sutāḥ* || 16 || *cārudeṣṇaḥ sucāruś ca cāruveṣo yaśodharaḥ* | *cāruśravāś cāruyaśāḥ pradyumnaḥ śambhur eva ca* || 17 ||

[6] Mbh 13.16.2d: (Kṛṣṇa:) *vṛne sutānāṃ ca śataṃ śatāni* ||
16.5ab: (Umā:) *datto bhagavatā putraḥ sāmbo nāma tavānagha* |
16.6b: (Kṛṣṇa:) *śataṃ sutānām ...*
16.7c: (Umā:) *bhāryāsahasrāṇi ca ṣoḍaśaiva*

[7] Cf. Sörensen 1904–25, s. v. 'Çāmba'.

[8] Cf. Sörensen 1904–25, s. v. 'Satyabhāmā'.

(4) Two *adhyāyas* later in the Mausalaparvan, we are informed that several of Kṛṣṇa's widows, namely Rukmiṇī, Gāndhārī, Śaibyā, Haimavatī and Jāmbavatī, entered the funeral pyre, whereas Satyabhāmā and the other Devīs went into forests (Mbh 16.8.71–72).[9] I have already mentioned Rukmiṇī, Satyabhāmā and Jāmbavatī; Gāndhārī is mentioned in one other place in the epic as Kṛṣṇa's wife (Mbh 7.10.10), whereas Śaibyā and Haimavatī do not appear elsewhere in it. Probably what is meant by the 'other Devīs' is the 16,000 anonymous wives.

(5) Of these it is finally said, in the Svargārohaṇaparvan, that they entered the river Sarasvatī in order to become Apsarases again and join Vāsudeva in heaven (Mbh 18.5.21).[10]

To summarize these occurrences throughout the *Mahābhārata*: There are two groups of Kṛṣṇa's wives: On the one hand, there are 16,000 women called Kṛṣṇa's *parigraha*, who are alluded to only as an anonymous group; on the other, a handful of women are marked as Kṛṣṇa's 'main wives' (*mahiṣī*) and called by individual names; some of these *mahiṣīs* are mentioned only once, and some turn up several times in the text. Two of them are well known for being the mothers of famous sons of Kṛṣṇa. There is only one place in the epic where the two groups of Kṛṣṇa's wives are directly associated with each other. In one passage of the Anuśāsanaparvan Durvāsas is said to have rewarded Rukmiṇī for her loyal devotion and granted her the boon of being the foremost among the 16,000 wives of Kṛṣṇa (Mbh 13.144.42).[11]

---

[9] *rukmiṇī tv atha gāndhārī śaibyā haimavatīty api* I *devī jāmbavatī caiva viviśur jātavedasam* II 71 II *satyabhāmā tathaivānyā devyaḥ kṛṣṇasya saṃmatāḥ* I *vanaṃ praviviśū rājaṃs tāpasye kṛtaniścayāḥ* II 72 II

[10] *ṣoḍaśa strīsahasrāṇi vāsudevaparigrahaḥ* I *nyamajjanta sarasvatyāṃ kālena janamejaya* I *tāś cāpy apsaraso bhūtvā vāsudevam upāgaman* II 21 II

[11] *ṣoḍaśānāṃ sahasrāṇāṃ vadhūnāṃ keśavasya ha* I *variṣṭhā sahalokyā ca keśavasya bhaviṣyasi* II 42 II

HARIVAṂŚA

With these results on the *Mahābhārata* in mind let us turn now to the *Harivaṃśa*. The *Harivaṃśa* has two main functions as a supplement (*khila*) to the *Mahābhārata*: one is to embed the *Mahābhārata* events within the wider history of the world and to present the genealogy of its main heroes; the other is to illuminate the figure of Kṛṣṇa in particular, whose descent, as specified in the epic, is highly obscure (cf. Brinkhaus 2005: 375–376). The problem of his origin is treated in the Yādava passage of the *Vaṃśānucarita* section (Hv 23.134–163; cf. Brinkhaus 2005: 373–383), which ends by addressing the question of the future of Kṛṣṇa's family after the *Mahābhārata* events. Accordingly, the theme of Kṛṣṇa's wives and children is taken up briefly at the end of the *Vaṃśānucarita* section, but it is interrupted by a discussion of Kṛṣṇa's divine origin and a long retrospective on his childhood and youth spent in Mathurā. At the end of this flashback, the return to Kṛṣṇa as a grown-up and the head of his family is accomplished by reporting the change of residence of the Yādavas from Mathurā to Dvārakā. Kṛṣṇa is afterwards described as having settled down in his new town and begotten many children there upon his many wives. This whole passage after the flashback can be read as a continuation and conclusion of the Yādava part of the *Vaṃśānucarita* section, and, in fact, this chapter ends with the formulation: *iti vaṃśaḥ samāpyate* (Hv 98.27d).

First of all, let us take a closer look at the *Vaṃśānucarita* section preceding the *Kṛṣṇacarita*, which I will refer to as Passage I in the following. In this passage only single wives are mentioned by their names. An indirect hint of Rukmiṇī's existence is to be found in *adhyāya* 24, where a son of Kṛṣṇa is called *raukmiṇeya* (Hv 24.30: ... *raukmiṇeyo* ... *kanīyān* ...). In *adhyāya* 28, however, the first part of the Syamantaka episode, Kṛṣṇa's wives are accorded extensive treatment. This episode clearly refers to the Kṛṣṇa of the *Mahābhārata*: It is localized in Dvārakā and obviously endeavours to turn Kṛṣṇa's trickster image in the epic into one of an honest and reliable individual. Also reported are the circumstances under which Kṛṣṇa was given Jāmbavatī to wed by her father Jāmbavat, the king of bears (Hv 28.28).[12] Soon after this the Yādava king Satrājit also gave his daughter Satyabhāmā and two further daughters as wives to Kṛṣṇa (28.32–34).[13] Thus three wives of Kṛṣṇa are named in Passage I, all of whom are well known from the *Mahābhārata*; among them Rukmiṇī, it can be presupposed, was widely recognized as Kṛṣṇa's wife, whereas information is supplied about how Jāmbavatī and Satyabhāmā came to be married to Kṛṣṇa.

After a long interruption, the theme of Kṛṣṇa's wives is taken up again in *adhyāyas* 87–94, which I designate as Passage II in the following. In the long *Kṛṣṇacarita* inserted between Passages I and II, Kṛṣṇa is, of course, still unmarried as a child or youth, though it is remarkable enough that here too he is reported – at least in one passage – to have been very fond of dallying with many herdswomen all at the same time. What connection this has with Kṛṣṇa's many wives in his later days is a matter of speculation, since the text itself does not give any explicit hint in this regard.

Kṛṣṇa's youth ends with the – more or less compulsory – move to Dvārakā reported in *adhyāyas* 84–86. Kṛṣṇa is now a grown-up man, and the following Passage II focuses on his getting married and founding a family. Here is also reported in great detail how Kṛṣṇa won Rukmiṇī as his wife. Kṛṣṇa's abduction of Rukmiṇī, who had been promised to his archenemy

---

[12] *vāsudevas tu nirjitya jāmbavantaṃ mahābalam* । *lebhe jāmbavatīṃ kanyāṃ ṛkṣarājasya saṃmatām* ॥ 28 ॥

[13] *satrājitaḥ ... kumāryaś cāpi tisro ...* ॥ *satyabhāmottamā strīṇāṃ vratinī ca dṛḍhavratā* । *tathā padmāvatī caiva bhāryāḥ kṛṣṇasya tā dadau* ॥ 34 ॥

Śiśupāla, is occasionally hinted at in the *Mahābhārata*, but a thorough account of this event is only given in the *Harivaṃśa*. As I said before, the *Mahābhārata* contained an enumeration of eight sons of Rukmiṇī, and this list is extended in Hv 88 to ten sons and one daughter (88.36–39b).[14] Immediately afterwards seven other main wives are mentioned by name or descent (88.40–42).[15] This list is finally concluded by the remark that Kṛṣṇa had, moreover, won 16,000 wives, on whom he begot his many thousands of sons (88.43).[16] After this mention of Kṛṣṇa's 16,000 wives, the Naraka episode is told in a more detailed way than in the corresponding *Mahābhārata* account. Now the figure of 16,000 girls who had been kept in prison and were freed by Kṛṣṇa is corrected to the number 16,100 and it is said that these girls were the daughters of Gandharvas, gods, human beings and seven groups of Apsarases (Hv 91.12–13b).[17] Kṛṣṇa safely brought them to Dvārakā where he – with explicit permission from his relatives – gave generous presents to them (94.24–25). He then placed them all under the domestic control of Rukmiṇī (94.27cd)[18] and provided each of them with a separate house, garden and lotus-pond of her own (94.28). Earlier, however, the

---

[14] *sā [= rukmiṇī] hi tasyābhavaj jyeṣṭhā patnī kṛṣṇasya* ... | ... || 36 || *tasyām utpādayām āsa putrān daśa mahārathān* | *cārudeṣṇaṃ sudeṣṇaṃ ca pradyumnaṃ ca mahābalam* || 37 || *suṣeṇaṃ cāruguptaṃ ca cārubāhuṃ ca vīryavān* | *cāruvindaṃ sucāruṃ ca bhadracāruṃ tathaiva ca* || 38 || *cāruṃ ca balināṃ śreṣṭhaṃ sutāṃ cārumatīṃ tathā* |

[15] *mahiṣīḥ sapta kalyāṇīs tato 'nyā madhusūdanaḥ* | *upayeme mahābāhur guṇopetāḥ kulodbhavāḥ* || 40 || *kālindīṃ mitravindāṃ ca satyāṃ nāgnajitīm api* | *sutāṃ jāmbavataś cāpi rohiṇīṃ kāmarūpiṇīm* || 41 || *madrarājasutāṃ cāpi suśīlāṃ śubhalocanām* | *satrājitīṃ satyabhāmāṃ lakṣmaṇāṃ cāruhāsinīm* | *śaibyāṃ sudattāṃ rūpeṇa śriyā hy apsarasopamām* || 42 ||

[16] *strīsahasrāṇi cānyāni ṣoḍaśātulavikramaḥ* | *upayeme hṛṣīkeśaḥ* ...

[17] *gandharvāṇāṃ ca yāḥ kanyā jahāra narako balī* | *yāś ca devamanuṣyāṇāṃ sapta cāpsarasāṃ gaṇāḥ* || 12 || *caturdaśa sahasrāṇi ekaviṃśac chatāni ca* |

[18] *kuṭumbasyeśvarī tv āsīd rukmiṇī bhīṣmakātmajā* || 27 ||

text had described eight very lavish palaces as belonging to Kṛṣṇa's eight main wives (93.39–50).

On the whole, the treatment of Kṛṣṇa's wives in Passage II can be regarded as a continuation of Passage I, and both together can be interpreted as a considerable elaboration on the theme of Kṛṣṇa's wives beyond the passing treatment accorded it in different parts of the *Mahābhārata*. The circumstances and events surrounding Kṛṣṇa's marriages to his different wives are reported, the groups of both his main wives and his anonymous wives are slightly enlarged, and the number of Rukmiṇī's children, who constitute Kṛṣṇa's main line of succession, is also extended, – but all in all the details of Kṛṣṇa's wives and children in both the *Mahābhārata* and the *Harivaṃśa* largely agree.

With this, however, the theme of Kṛṣṇa's wives and children in the *Harivaṃśa* is not laid to rest. There is a Passage III in *adhyāya* 98 which contains a systematic enumeration of them, and particularly the children of the eight main wives of Kṛṣṇa. This passage starts with practically the same list of Kṛṣṇa's main wives as given in *adhyāya* 88 (and again in 93), and thereafter a complete roster of sons and daughters is provided for each of them (98.2–5a).[19] For Rukmiṇī, this means a virtual repetition of the list given in *adhyāya* 88. One possible reason for extending the list of Kṛṣṇa's sons may have been the feeling that his son Sāmba, who is mentioned several times in the *Mahābhārata* as having been born to Jāmbavatī, but is missing in the *Harivaṃśa* Passages I and II, was owed recognition of his rightful status; and another motive could have been the intention to finally ascribe children to Satyabhāmā who, after all, was the second-best known wife of Kṛṣṇa after Rukmiṇī, but had not been cred-

[19] *aṣṭau mahiṣyaḥ putriṇya iti prādhānyataḥ smṛtāḥ* I ... II 2 II *rukmiṇī satyabhāmā ca devī nagnajitī tathā* I *sudattā ca tathā śaibyā lakṣmaṇā cāruhāsinī* II 3 II *mitravindā ca kālindī jāmbavaty atha pauravī* I *subhīmā ca tathā mādrī raukmiṇeyān imāñ śṛṇu* II 4 II *pradyumnaḥ prathamo yajñe*

ited with any child of him in the whole of the *Mahābhārata* and in Passages I and II of the *Harivaṃśa*. On the other hand, almost all the many names of Kṛṣṇa's sons and daughters given in this list seem to have been newly invented; obviously those familiar with the text at the time seem not to have been familiar with these names. This, at least, is suggested by the particularly numerous variant readings given in the Critical Edition of the *Harivaṃśa*, and by the many differing forms of these names in the parallel passages of the early Purāṇic testimonies within the *Purāṇapañcalakṣaṇa* corpus. I shall come back to this shortly.

Let me first try to come to a conclusion about the pertinent *Harivaṃśa* passages treated before on the theme of Kṛṣṇa's wives and children: The fact that these three passages of text are scattered the way they are in the *Harivaṃśa* may be explained by historical reasons of textual development and growth. Possibly the whole *Kṛṣṇacarita* – at least, the part which is set in Mathurā – was interpolated as a sort of flashback to the divine origin and the childhood and youth of Kṛṣṇa, so that Passages I and II were radically separated from each other.

Passage III, for its part, is separated again from Passage II mainly by the detailed account of the Naraka episode, which could have been added as an elucidation and rectification of the mention of 16,000 wives at the end of *adhyāya* 88. Alternatively, Passage III could have been conceived later as a more systematic restatement of the theme presented in *adhyāya* 88, and thus have been attached at the end as a comprehensive summary of it, and at the same time as a fitting conclusion to the whole of the *Vaṃśānucarita*.

PURĀṆAPAÑCALAKṢAṆA

As was mentioned before, *adhyāya* 98 of the *Harivaṃśa* has a parallel in the Purāṇic testimonies of Kirfel's *Purāṇapañca-*

lakṣaṇa corpus,[20] which Kirfel himself, however, failed to notice. A comparison particularly with the parallel versions in the core of the *Brahmāṇḍa-* and *Vāyupurāṇas* (Bḍ-Vā) and that of the *Matsya-* and *Padmapurāṇas* (Mt-Pd), and in the *Viṣṇupurāṇa*, shows that these do not contain a full elaboration of the basic scheme, but only extracts of the more important parts of it. Thus the question arises as to what the text-historical relationship is between the *Harivaṃśa* text and its Purāṇic parallel: Was it the Purāṇic version that stimulated a *Harivaṃśa* redactor to add the extended and systematized new view to Passages I and II, which represent an older developmental stage seen in the *Mahābhārata* itself? Or was this more comprehensive version conceived within the *Harivaṃśa*, in the process of its growth, so that it could serve as a model for the Purāṇic parallel versions? To begin with a more general consideration, if my assumption is correct that the main motive behind adding Passage III to the *Harivaṃśa* was to round out the scheme of Kṛṣṇa's wives and children, then one could expect that this was done completely, and not only partially. The complete list, however, may have been rejected by the Paurāṇikas as being too pedantic and filled with too many unknown names, so that they replaced it with extracts containing the more important names. The reverse developmental assumption would mean that the empty positions in the scheme were filled at first partially, and later on in total. The second option is not impossible, but still the first one seems to me to be more plausible.

In addition to these considerations, there are some textual details, revealed by closer inspection, which seem to support my view. In the parallel which comes closest to the *Harivaṃśa* version, namely the one in the Bḍ-Vā core, the text not only offers a clear correspondence with the enumeration of Kṛṣṇa's main wives and children in Hv *adhyāya* 98, but in its first part bears

[20] Kirfel 1927, esp. 477.14–479.22 (Bḍ-Vā and Mt-Pd).

even greater similarity with the list of Kṛṣṇa's wives in Hv *adhyāya* 88.[21] The ensuing list of Kṛṣṇa's descendants in Bḍ-Vā corresponds to Hv 98.7–18 and contains parallels to two asterisked passages of the Critical Edition of the *Harivaṃśa*.[22] The relationship between the *Harivaṃśa* and the Bḍ-Vā in this chapter appears to me to be similar to that in another passage which I dealt with in a former article:[23] The redactor of the Bḍ-Vā version seems to have borrowed and at the same time refashioned the text of the *Harivaṃśa* in a relatively developed state of its textual history.

As for the parallel version in the Mt-Pd core, the text is clearly one step further away from the *Harivaṃśa* text than that of the Bḍ-Vā core; indeed the latter version can be assumed to have been the direct source of the Mt-Pd parallel. Similarly, the retelling in prose of the corresponding passage in the *Viṣṇupurāṇa* (ViP 4.15.19–20) very probably depends on the Bḍ-Vā version.[24] At the same time, however, another part of the *Viṣṇupurāṇa*, namely the *Kṛṣṇacarita* in *aṃśa* 5, is certainly based on the *Harivaṃśa* version, or more precisely, almost the whole of the middle part of the *Harivaṃśa* which previously went under

---

[21] The formulation and contents of e.g. VāP 96.233–235b are much nearer to Hv 88.41–43b than to Hv 98.2–4.

[22] The exact correspondences e.g. in the VāP are:

Hv 98.7c–f	—	VāP 96.239
*1102	—	240ab
8–11	—	240d–243
12–14	—	
15–17	—	246–249
*1105	—	244–245

(The two *passages are well-attested in N- and S-Mss. of the *Harivaṃśa*, though not extant in the five manuscripts Ś1, Ñ1 and M1–3 which are 'constitutive' for the Critical Text.)

[23] Cf. Brinkhaus 2005: 371–373 and 385–387, where the relationship between parts of Hv 21 (namely Appendices 6B and 7) and 23 and the parallels in VāP 91–92 / BḍP 2.66–67 was dealt with.

[24] Cf. also Vielle 2005: 543, n. 30: '... it is almost clear that the ViP is posterior, being often a mere rewriting (by paraphrazing or shortening) of VāP-BḍP passages.'

the name of Āścaryaparvan (*adhyāyas* 30–113).[25] This is the reason why there are as many as three passages in the whole of the *Viṣṇupurāṇa* where Kṛṣṇa's wives and descendants are treated: The first one has already been mentioned as being located within the *Purāṇapañcalakṣaṇa* part of *aṃśa* 4. The other two are to be found in the *Kṛṣṇacarita* of *aṃśa* 5: There is a parallel in ViP 5.28.1–5 to the enumeration of Kṛṣṇa's main wives in Hv 88, and these wives are mentioned again together with Kṛṣṇa's descendants in ViP 5.32.1–5 which is a parallel version of the corresponding part of Hv 98. Thus, to summarize and briefly restate the main result of the preceding observations: The text-historical development as presented here implies once again the view I have expressed elsewhere (Brinkhaus 2005: 373) that the *Harivaṃśa*, even in its developed form, still precedes the texts of the extant Purāṇas, and therefore should be viewed more as 'late epic' than 'early Purāṇic'.

One last remark in conclusion: The only article I have been able to find on the specific theme of Kṛṣṇa's wives in epic and early Purāṇic literature was published by Asha Goswami in 1974.[26] The title of the article, 'The Monogamist Kṛṣṇa', itself discloses the underlying trend of thought, namely that Rukmiṇī was traditionally seen as the only wife of Kṛṣṇa. My results cannot corroborate this somewhat arbitrary view, but the predominance of Rukmiṇī as Kṛṣṇa's wife in the epic in contrast to the later Purāṇic evidence is well attested in the *Mahābhārata*, and in the earlier textual stages of the *Harivaṃśa* as well.

---

[25] Cf. Brinkhaus 2002, esp. pp. 168–169.

[26] E.g. W. G. Archer (1957) is mainly concerned with 'the loves of Kṛṣṇa' in the *Bhāgavatapurāṇa* and later Kṛṣṇa poetry.

REFERENCES

ARCHER, W. G. 1957. *The Loves of Krishna in Indian Painting and Poetry*. London: George Allen & Unwin.

BRINKHAUS, Horst 2002. The division into parvans and the Bhaviṣyaparvan of the Harivaṃśa. In: Mary Brockington & Peter Schreiner (eds.), *Composing a Tradition. Concepts, Techniques and Relationships. Proceedings of the First Dubrovnik International Conference on the Sanskrit Epics and Purāṇas, August 1997*: 157–176. Zagreb: Croatian Academy of Sciences and Arts.

——— 2005. Duplicates in the Somavaṃśa account of the Harivaṃśa: early text-historical developments in the Harivaṃśaparvan. In: Petteri Koskikallio (ed.), *Epics, Khilas, and Purāṇas. Continuities and Ruptures. Proceedings of the Third Dubrovnik International Conference on the Sanskrit Epics and Purāṇas, September 2002*: 363–387. Zagreb: Croatian Academy of Sciences and Arts.

GOSWAMI, Asha 1974. The monogamist Kṛṣṇa. In: *Charu Deva Shastri Felicitation Volume*: 443–450. Delhi: Charu Deva Shastri Felicitation committee.

Hv = *The Harivaṃśa. Being the Khila or Supplement to the Mahābhārata*. For the first time critically edited by P. L. Vaidya. 2 vols. Poona: Bhandarkar Oriental Research Institute, 1969–71.

KIRFEL, Willibald 1927. *Das Purāṇapañcalakṣaṇa. Versuch einer Textgeschichte*. Bonn: Kurt Schroeder Verlag.

Mahābhārata, Critical Edition. Ed. by V. S. Sukthankar, S. K. Belvalkar, P. L. Vaidya *et al*. 19 vols. Poona. Bhandarkar Oriental Research Institute, 1933–66.

SHETH, Noel 1984. *The Divinity of Krishna*. New Delhi: Munshiram Manoharlal.

SÖRENSEN, Sören 1904–25. *An Index to the Names in the Mahābhārata with Short Explanations and a Concordance to the Bombay and Calcutta Editions and P. C. Roy's Translation*. London: Williams and Norgate.

Vāyupurāṇa. The Vāyumahāpurāṇam. Edited by V. Caturvedi. Delhi: Nag Publishers, 1983 (Reprint of the Venkatesvara edition of AD 1895).

VIELLE, Christophe 2005. From the *Vāyuprokta* to the *Vāyu* and *Brahmāṇḍa Purāṇas*: preliminary remarks towards a critical edition of the *vāyuprokta Brahmāṇḍapurāṇa*. In: Petteri Koskikallio (ed.), *Epics, Khilas, and Purāṇas: Continuities and Ruptures. Proceedings of the Third Dubrovnik International Conference on the Sanskrit Epics and Purāṇas, September 2002*: 535–560. Zagreb: Croatian Academy of Sciences and Arts.

Viṣṇupurāṇa. Viṣṇupurāṇam. Critically edited by M. M. Pathak. 2 vols. Vadodara: Oriental Institute, 1997–99.

'Surprise, Surprise!' Authors' Stratagems and Audiences' Expectations in the *Rāmāyaṇa*

MARY BROCKINGTON

For perhaps 2,500 years, audiences have been captivated by the tale of Rāma Dāśarathi continually retold and adapted in a large number of languages and every possible medium. Out of the many reasons for its continued appeal this article will address only one, the most fundamental: that is, the narrative skill of the original tellers and some of their earliest successors. The developed forms of the story are now so well-known that it is difficult for the modern reader to imagine the impact the earliest recitations must have had on an audience who were hearing the *Rāmāyaṇa* for the first time and had no idea of how the plot was going to unfold, but research on epic language and manuscript transmission published since the appearance of the Baroda Critical Edition has enabled scholars to define the broad outline of the story in its earlier forms. Several years spent producing a translation of those parts identified as relatively early have helped me to appreciate the timeless strategies employed by its authors to arouse, maintain, or re-awaken their listeners' interest at all points of a taut, coherent, fast-moving narrative.[1] At

1 Brockington & Brockington 2006 (passages identified for translation are listed on pp. 373–380). Page references in this article are to this translation; text references are to the Baroda Critical Edition.

this stage, while the story was still unfamiliar, we can assume that it would have to have been recited in its entirety, even if the performance was extended over several days or nights.[2] The unsuspecting audience was presented with a series of reversals of plot and surprises of characterisation, with contrasting and paradoxical situations, with abrupt changes of scene, with lively, varied diction, and occasionally with tantalising delaying tactics. The tellers knew how to capture and how to manipulate their audience; how to avoid boredom, and also how to exploit it.

It is not the purpose of this article to develop the debate about oral versus written *composition* of ancient Indian texts, but to evaluate the effects of the *presentation* of a text, however composed, to a largely if not exclusively non-literate audience. The challenges and opportunities that confront an oral performer with a multiple audience differ from those faced by the writer whose work will be read privately by a single individual. Each performance happens only once, so it is crucial to create an immediate impact and to maintain interest over a long period; if members of an audience once start to drift away they will not return, unlike the individual reader, who may resume his reading whenever he feels like it, or has nothing better to do. Conversely, the listener, unlike the reader, has no opportunity either to skip boring passages, or to turn back the pages to check details for accuracy. The dynamics of a multiple audience as opposed to an individual are also important, for enthusiasm and dissatisfaction are alike contagious, as any demagogue knows. The oral performer is both more powerful and more vulnerable than the writer.

Narratorial tricks such as starting the story *in medias res* with no preamble, then filling it with surprise twists of plot, leaps of scene, and tense cliff-hanger situations before a predictable

---

[2] The English translation in Brockington & Brockington 2006 has been estimated to take at least twenty hours to perform.

happy ending are but the stock-in-trade of the third-rate hack. The composers of the *Rāmāyaṇa* certainly do use all these well-worn techniques, but if they did no more than that it is unlikely that Rāma's story would have achieved the status of the *ādi-kāvya*. Careful advance planning is a marked characteristic of the original author (M. Brockington 1999), and his surprises are no exception. Throughout the work he repeatedly teases the audience by deliberately building up their expectations in a misleading direction, often merely by subtle hints, only to dash these legitimate expectations with a sudden reversal of plot or characterisation to which this misleading preparation gives an even greater shock value.

The original opening of the story can no longer be identified with any precision, but the audience appears to have been presented with an agitated Rāma bursting into Sītā's apartments to tell her the shock news of his sudden exile; this is an intimate conversation in which neither character need be made to hide personal feelings beneath a front of propriety and restraint (2.23.18–31). The impression given by Rāma is that Daśaratha has acted in a high-handed manner and that Bharata may be expected to act harshly towards his displaced brother's wife, so it comes as a surprise shortly afterwards that the king is introduced in person not as the conventional capricious tyrant but as a frail, pathetic old man; his first words incite Rāma to depose him, and he continues to make affectionate suggestions for alleviating the rigours of the exile, his only strength now being his refusal to revoke his pledged word (2.31.13–34.8). Any idea that Rāma's initial indignation will be translated into action against his father is immediately overturned by his adamant refusal to compromise his father's integrity.

Bharata is absent from Ayodhyā at this point and is not introduced in person until after Daśaratha's death (2.62); had he been present all the time, the narrator would have had to contrive some other way of preventing him refusing the kingship at too early a stage in the story. His continued absence after Rāma leaves may seem surprising to an audience who expect Kaikeyī to send for her son immediately to be installed as *yuvarāja*, but it allows the narrator to focus exclusively on the departure of the exiles and the king's death while not contradicting the initial impression that Bharata will be a harsh ruler. The motif of the stepmother whose jealousy sends a tale's hero away from home to begin his adventures is so all-pervasive and so natural a feature of traditions the world over that this rare example of a son siding with his mother's victim must have shattered their expectations.[3] It is unfortunate that the episode where he learns of his father's death and angrily rejects Kaikeyī's scheming has been so heavily reworked that it is impossible to identify exactly how it was first presented (2.65–69).

The composer was too skilful and too sensitive not to understand that some things are better left unsaid, and he passes over the death of Daśaratha in dignified silence (2.59.6–7). With similar reticence he does not explain how Rāma learns the news, or how he reacts; anomalies of this nature can be the more easily concealed in a fast-moving, orally presented text such as this. A lesser poet might have included the scene, but Rāma would surely have to have been assigned a display of emotion surpassing the already extravagant grief which sends Bharata and Śatrughna writhing about on the ground after the funeral (2.70–71); such a competition in grief could seem inappropriate and cheap. The alternative reaction, impassivity, could have been misinterpreted as indifference or even understandable resentment. The quiet pathos of the scene where the widowed queens discover Rāma's poverty-stricken *badam*-flour offering, implying

---

[3] For the possible influence of this variation on the traditional hostility of stepchildren upon the unrelated international tale 'The Twins or Blood-Brothers' (ATU 303, also known as 'The Two Brothers') see M. Brockington 1998: 141–143.

without overtly stating all the audience need to know, is far more effective, especially as it is so unexpected.[4]

The next scene is Bharata's futile attempt to persuade Rāma to return to Ayodhyā. Again, there is slightly less drama here than might be expected. The audience have been prepared for Rāma's gracious welcome of Bharata by his rebuke of Lakṣmaṇa's excitable suspicions (2.90–91), and it is debatable whether they ever really expected Bharata's attempt to succeed. The surprise is all Bharata's, and the tension is varied by a little gentle comedy as his avowed allies all desert him. Awed by the power of Rāma's presence, Sumantra does not even hear the order to spread him a *kuśa*-grass seat and Bharata succumbs to the indignity of doing it himself (2.103.12–15); then the citizens who had been so delighted at the news of an expedition to bring back Rāma reply to Bharata's plea for support:

> We acknowledge that Rāma Rāghava is talking correctly to Bharata Kākutstha, for the distinguished prince is standing by his father's promise, so we certainly cannot deflect him. (2.103.20–21; trans. pp. 65–66; cf. 2.76.22)

The audience are convinced that Bharata has been overcome by the unexpected reversal of his confident expectations of success, and lose sight of him as he prepares to re-enter the desolate Ayodhyā, bearing the sandals that symbolise his authority to rule (2.105.24).

The tension is now relaxed as the exiles resume their forest idyll, visiting a succession of sages until the next surprise, the episode that abruptly reverses the whole direction and tenor of

---

4 Rāma's restrained reaction to the news of his father's death is the explicit focus of the *Dasarathajātaka* adaptation, where, more logically, the news is brought by Bharata, but the portrayal is far less convincing. The consummate poet Kālidāsa also has the news brought in straightforward fashion by Bharata, but does not mention the reaction of either prince to the news in his brief summary of the episode, thus wisely avoiding any feeling of competition in their grief (*Raghuvaṃśa* 12.13–15).

the narrative. This is Rāma's first display of military power as he effortlessly slaughters Khara and his whole army of *rākṣasas*, convincing Rāvaṇa that to abduct Sītā safely he will have to adopt decoy tactics. Startling though it is, this episode risks inducing complacency in the audience: if Rāma is so powerful, why should he have any difficulty defeating Khara's brother? The narrator will have difficulty in maintaining the tension of his story, a problem he resolves by giving the audience the biggest surprise of the whole narrative when Rāvaṇa appears at the hermitage. His conventional introduction of himself to Sītā hits the audience with the force of a bombshell:

> That was the opportunity Daśagrīva had been waiting for; in the guise of a mendicant he quickly strode up to Vaidehī.
>
> 'Fair-skinned lady, I am half-brother to Vaiśravaṇa; my name is Rāvaṇa, if you please, the mighty ten-necked one! ... That half-brother of mine, Vaiśravaṇa, had some quarrel with me, but I attacked him in my anger and defeated him in battle. It was fear of me that drove him, Naravāhana, in distress from his own wealthy home to live on excellent Mount Kailāsa.' (3.44.2, 46.2, 4–5; trans. pp. 104–105)

The hero's adversary is no ordinary monster, but the paternal half-brother of the god of wealth himself, also called Kubera, leader of the *yakṣas*, his fearsome demonic image well known to the audience; Vaiśravaṇa is skulking in the Himālaya, it now appears, to escape from the younger but more mighty Rāvaṇa. Rāma may be able to kill 14,000 fierce *rākṣasas* single-handed and simultaneously, but will this new threat be too much even for him?

From this point on the technique changes: outright reversals of plot are now replaced by setbacks in progress towards the inevitable killing of Rāvaṇa. The audience are repeatedly thrilled and frightened, but scarcely surprised, as tension mounts towards the long-hoped-for, long-expected climax and each new peril is prepared and surmounted; the question tantalising them is not so much, 'Will Rāma be defeated?' but 'How will the author resolve this latest problem?' Many of these setbacks can be regarded as the narrator's delaying tactics, adding colour and excitement to his tale (episodes on the search party's journey, for instance, or Hanumān's repeated delays in returning from Laṅkā, and the disabling injuries to Rāma and Lakṣmaṇa on the battlefield) but others are more carefully integrated into the plot: Kabandha and Saṃpāti prove not to be the dangerous enemies they at first appear, but in fact help the heroes in their quest (3.65–69; 4.55–57). Problems are solved in a variety of unexpected ways, such as the unheralded appearance of Garuḍa to free the brothers enmeshed by Indrajit's magic (6.40.33–57), or Hanumān's fetching the healing herbs – indeed the whole healing-herb mountain – to cure Lakṣmaṇa's wounds (6.89.13–24), and sometimes by a complete reversal of what the audience have been deliberately led to expect. Sugrīva's jealous ownership of the Madhuvana is several times emphasised: surely the exultant *vānaras* have not ruined the whole enterprise at the very moment of success (5.59.7–61.26)? The resolutions of individual episodes continue to tease and surprise, but the overall outcome is never in serious doubt.

Then, with each new threat successfully overcome, just when characters and audience alike are relaxing after victory and happy reunion, the author springs his final surprise: Bharata, last heard of about to enter the desolate Ayodhyā to take on the role of king at Rāma's insistence, is now revealed by divine announcement to have been sharing his brothers' hardships all along and living as an ascetic, news which is such a shock to Rāma that it sends him hurrying back to Ayodhyā without waiting long enough even to have a bath (2.105.24; 6.108.16, 109.4–7). Once again, this carefully-prepared twist has a more complex function than merely surprising the audience and providing a neat ending to the tale. It reinforces the original composer's decision to create, not an epic involving the fate of a nation, but a romance centring on his single hero. Rāma's intention had always been to punish Rāvaṇa and rescue Sītā, not to conquer the *rākṣasa* nation as a whole and rule Laṅkā himself. He had Vibhīṣaṇa consecrated king before the beginning of the war (6.13.7–8, ratified at 6.100.9–18) and the immediate return to Ayodhyā avoids any suggestion that he might remain, even as adviser to the new *rākṣasa* ruler.

Not all these surprise reversals can be accounted successes: it would surely take an extremely sentimental and gullible audience to be convinced by Vālin's dying repentance and Sugrīva's heartbroken reaction to his brother's death. This scene may be supposed to represent a wrongheaded attempt to ascribe to Sugrīva and also to Aṅgada (by extension from his father Vālin), the moral grandeur thought appropriate both to Rāma's sworn ally and to one of the leading *vānara* warriors, inserted as a relatively late contribution to the composition of the early material;[5] when at moments of crisis Aṅgada's relationship to his father is stressed, the reference is to his military prowess, not to his morals (e.g. 4.38.25, 63.14; 6.54.3).

Surprises are supplemented by pseudo-surprises, where the disingenuous author switches abruptly from one situation to its direct opposite without detailing the intervening steps. The climax of Mārīca's vehement and increasingly panic-stricken refusal to co-operate with Rāvaṇa in abducting Sītā by transforming himself into a golden deer to decoy Rāma is followed in the next verse by his appearance trotting dutifully around the hermitage:

> '... If you carry off Sītā from the hermitage with my help, then that will be the end of you, me, Laṅkā and the *rākṣasas*.'
>
> The shapely lady was picking flowers when she caught sight of him with his golden, silver-adorned flanks ... (3.39.19–41.1; trans. pp. 99–100)

[5] The reference forward to Aṅgada's role in slaughtering the *rākṣasas* is typical of later material: see below.

Mārīca's capitulation raises a laugh, but does it really surprise anyone, least of all Mārīca? The technique is familiar to all comedians, but here Mārīca's prophecy of doom not only for himself but for Rāvaṇa, Laṅkā and all the *rākṣasas* is also a clever instance of audience manipulation operating on two levels. It creates the expectation of inevitable victory for Rāma – no great surprise there; the surprise is that that expectation is to be only partially fulfilled when Rāma leaves the city and its surviving inhabitants in the capable hands of a *rākṣasa* ruler, his ally Vibhīṣaṇa.

This narrator is no pedant; information that can be inferred is passed over, unless it can be used effectively. Abrupt changes of scene keep the story moving along, so that interest does not flag. Accordingly there is no account of Hanumān's reception by the *vānaras* on his return from Laṅkā: he is in Laṅkā, then he is back among his exultant companions, and that is all the audience need to know. They all heave a sigh of relief, only to be immediately tantalised by the perilous Madhuvana episode already mentioned (5.59.7–62.24). More spectacularly, not a word is wasted in telling Rāvaṇa's death: after a carefully contrived and protracted build-up, *rāvaṇaṃ hatvā* (6.97.19a) shocks by its very suddenness and completeness. No dying speeches this time, no unconvincing deathbed repentance from the arch-villain, no forgiveness from his resentful brother (Vibhīṣaṇa is unwilling so much as to perform the funeral ritual that is his elder brother's due; 6.99.30–42). The audience have never seriously doubted that Rāma would ultimately triumph, but this accomplished narrator still contrives to make victory come as a surprise.

Paradoxical situations provide an effective variant on narrative surprises; of the many examples that could be explored, I mention only one. When Rāma's troubles have reduced him to begging for help from a monkey, Lakṣmaṇa concludes his account to Hanumān in pathetic fashion:

> The man who once upheld the world, dispensing wealth and earning glory unparalleled, seeks the aid of Sugrīva, in the expectation that Sugrīva and his lords will be gracious to grief-stricken, heartbroken, supplicant Rāma. (4.4.15–16; trans. pp. 140)

Audience boredom is an ever-present danger to a live performer who wishes to safeguard his professional reputation and fee, but he can also exploit it to his benefit at points in the narrative where he is confident that his listeners are desperate to know what happens next, and will not just wander away: if he is sufficiently skilful, he may even persuade them to bribe him to carry on. At two of the high points in the tale of Rāma, the narrator teases his captive audience unmercifully. As Rāvaṇa is hurrying off from the hermitage with Sītā (will Rāma and Lakṣmaṇa hear her screams and rush back to the rescue?) the sleepy, ancient Jaṭāyus intervenes but, before attacking Rāvaṇa, launches into a maddeningly lengthy, dreary sermon that must have set audiences baying at the performer to get on with the story (3.48). Similarly, when all the most senior *rākṣasa* warriors are already dead and Rāvaṇa at long last takes the field against Rāma, the *rākṣasa* lord is suddenly provided with three more subordinates to enter the fray alongside him, and the audience are forced to endure the accounts of their battles and defeats before they are finally rewarded with the fight they really want to hear (6.83–86). If they were reading this story written in a book, they would have turned to the last chapter long before.

Sometimes, however, the narrator takes the audience into his confidence, and it is not they but the characters who are surprised. We know that Rāvaṇa's account to Sītā of a night-battle in which Rāma and all the leading *vānara*s have been killed is all a lie and the severed head of her husband with which she is presented is not a grisly trophy but an illusion conjured up by Vidyujjihva (6.22.6–42). Similarly, when Indrajit immobilises Rāma and Lakṣmaṇa and goes back to Laṅkā convinced that they are dead, Vibhīṣaṇa immediately reassures *vānara*s and audience alike that the heroes will recover from their unconsciousness; this time not only Sītā but all her enemies are taken in (6.36.11–41.17). On the third occasion it is Sītā who is presented as having been killed by Indrajit, but again the audience already know that this Sītā is an illusion of *rākṣasa* magic before Vibhīṣaṇa explains the deception to the distraught Rāma and Hanumān (6.68.1–71.9). This variation of the narrator's usual technique increases the sense of audience participation, enabling them to enjoy and mock the *rākṣasas*' delusion and to identify with grief-stricken Sītā and Rāma in their distress without themselves being deluded into leaving the performance early because they think the story is over.

Surprises of characterisation are more subtle and perhaps more effective than these surprises of plot. Even the steadfast Rāma is not portrayed as completely static. Particularly at the beginning he is made to let fall a few hints that his personal feelings conflict with his acceptance of his public duty. He does not lightly give up his pleasures, such as his love of hunting in the Sarayū forest (2.43.13–14); but the negative feelings he harbours towards Bharata and Daśaratha are largely focused not on his own suffering but on the welfare of his wife and birth-mother: he warns Sītā not to antagonise her new king (2.23.23–31), and is quietly scathing when he begs his father to look after Kausalyā's interests:

> You can grant boons; you have a particular responsibility towards her. (2.33.18; trans. pp. 9)

The implication is not lost on the conscience-stricken king, who swoons at the respectful taunt.

Perhaps it is no accident that the story of Rāma is framed by incidents where the otherwise inflexible hero is persuaded to countermand his clearly-expressed instructions: he gives in to the pleas of Sītā and Lakṣmaṇa to accompany him in his exile (2.26.1–28.11) and allows all his allies to go with him back to

Ayodhyā for a visit after he has told Vibhīṣaṇa and his ministers to remain to rule Laṅkā and the *vānaras* to return to Kiṣkindhā (6.110.11–20). Presumably we are meant to see these gracious climb-downs as in fact reflecting his own inner desires: after all, he's only human.

Sheldon Pollock draws attention to the quality of Rāma's grief at the loss of Sītā:

> The problem in this episode is that the hero contradicts virtually everything the poet, up to this point, has encouraged us to believe about him. ... If there is any single virtue that characterizes the hero's conduct through the first two and a half books of the poem, it is his equanimity. ... Yet what we are presented with in this deeply moving passage of the third book negates this hitherto consistently drawn portrait. This seems less the exploration of another side of his character than a complete reversal. (Pollock 1991: 55)

The intensity of the passion that overmasters Rāma is touchingly portrayed at, for example, 3.59; 4.1, 4.6.12–22, and 4.29.21, 32–33, although the wilder excesses of the passage Pollock has in mind (3.60.36–52) can be identified as later elaborations. Yet the surprising discovery that the hero's emotional attachment to his wife can conquer even his rigid self-control is but one stage in the progressive decline in Rāma's fortunes, a decline that not only threatens the hero's resolve but undermines the balance of his mind. Any romantically-inclined listeners who may have been lulled into the belief that the *Rāmāyaṇa* is a love-story are eventually jerked back to reality when Rāma reaches the very nadir of his fortunes and is convinced that Lakṣmaṇa has met his death in his service. Loss of social position, loss of wife are nothing compared to loss of his cherished brother.

> What's the use of Sītā, what's the use of my life now I can see my brother lying defeated in battle? If I searched the world I could find another wife as good as Sītā, but not a brother as good as Lakṣmaṇa, my comrade-in-arms. If that growing joy to Sumitrā has met his end, then I will give up my life too in the sight of the *vānaras*. ... It's because of me that Lakṣmaṇa there is lying lifeless, fallen on to a bed of arrows. ... I feel real burning shame about my vain rhetoric:

> Vibhīṣaṇa has not been made king of the *rākṣasas*. (6.39.5–7.12.22; trans. pp. 289–290)
>
> I shouldn't be bothering with fighting or my life or with Sītā when I can see my brother Lakṣmaṇa slain in the dust of battle. What do I want with kingship, what do I want with life? I can see no reason in fighting when Lakṣmaṇa here lies killed on the battlefield. (6.89.7–8; trans. pp. 338–339)

The centrality of this theme to the plot is underlined by its repetition, and even more by Rāvaṇa's similar lament for Kumbhakarṇa, the culmination in the progressive disintegration of the villain's personality. Exemplified in varying degrees by the three widely different sets of brothers (Rāghavas, *vānaras* and *rākṣasas*), fraternal loyalty is presented in this *kṣatriya* tale as the supreme virtue, a virtue which, at moments of crisis, threatens to overwhelm even the *kṣatriya* ideal, the pursuit of personal honour, military prowess and kingship, until all that Rāma values most is successively restored to him. Rāma's laments may come as a surprise, but they are not a contradiction.

Rāvaṇa is less one-dimensional than might be expected, and his character shows a gradual development. His initial bombastic self-confidence has been subtly undermined by doubt even before Rāma reaches Laṅkā (6.6.1–3); his worries grow when he sees the *vānara* army (6.20.1–13) and his foremost warriors are successively defeated. He dare not express these worries before subordinates but valiantly keeps up a bold front, until he too can find relief by confiding his fears to his grumpy younger brother Kumbhakarṇa in a refreshingly intimate scene (6.50). His grief at the death of Kumbhakarṇa is his first and only show of genuine affection for anyone (he will react to his son's death with obstinate resignation) and the *rākṣasa* who has usurped one of his brothers and provoked another to defect mourns his remaining brother in terms that recall Rāma's laments for Lakṣmaṇa:

> Kingship's nothing to me. What do I want with Sītā? I've no pleasure in life now Kumbhakarṇa's been torn from me. I must be the one to kill my brother's killer, the Rāghava, in battle; if not, death would

> be better for me than this purposeless life. This very day I shall follow where my brother's gone. (6.56.12–14; trans. pp. 309)

For Rāvaṇa, self-doubt is a new and surprising experience, and he can find no way to dispel it; his only recourse is to ignore it, even after the deaths of Kumbhakarṇa and of his last hope, his son Indrajit, when he has no option but to take the field himself (6.83).

At an early stage in the story the audience have been disabused of their supposition that Bharata is to function as the villain of this tale, but it is not until the final scene (6.108.16) that they fully realise that his character is an exact counterpart of Rāma's, with all his elder brother's determined, upright self-sacrifice; all he lacks is the seniority of birth that would enable him to put his wishes into effect. He must obey Rāma, but only as long as Rāma is in a position to impose his will. Left to himself, he finds a way of sharing the hardships he has been unable to spare his brother.

The biggest, and most carefully crafted, surprise of characterisation is the development in the character of Sītā: initially portrayed as an inexperienced young girl, she matures instantly after her capture, now that she must rely on herself, not on Rāma's protection. She surprises Rāvaṇa, and she surprises the audience. The young princess can repeat accurately but parrot-fashion the rules of behaviour she has been brought up to follow (2.34.23–27), but has not yet had these precepts tested in earnest; she has romantic notions that she would like to live the life of an ascetic in the forest, but when the time comes, she does not even know how to put on an ascetic's clothes (2.26.11–13, 33.10–11). Her determination and her fervent devotion to Rāma are apparent from the start, but they are expressed in terms bordering on the hysterical (2.26; 3.43.1–34). Rāma and Lakṣmaṇa must hunt deer for their food, but she makes pets of all the forest animals (3.41.9–11, 59.5). She has all a young girl's frivolity: Rāma and Lakṣmaṇa cling desperately to the hope that she is playing a trick on them when they find the hermitage deserted (3.59.4, 15). Daśaratha recognises her love of finery and insists on gratifying it with jewels and ornaments to be worn in the forest (2.34.14–18), and Rāvaṇa plays on this trivial streak when he orders the decoy to be a golden deer studded with jewels (3.34.17, 38.15, and 3.41).[6] The stratagem works, and it is Sītā's very love of finery and the fervour of her devotion to Rāma in sending away Lakṣmaṇa in another hysterical outburst (3.43) that bring about her capture. The portrayal is not unsympathetic: Agastya and Rāma himself comment on her frail physique (3.12.2–3, 56.12); and both Anasūyā and Agastya commend her determination and devotion to her husband (2.109.21–28; 3.12.3–8). Her immature behaviour is not condemned but excused as only what is to be expected of an inexperienced young wife who has led the sheltered life of a princess, but how will this tender, frail girl withstand the rigours and temptations of captivity? The audience fears for Rāma: will he be robbed of his wife's loyalty as well as her presence?

Almost immediately they are reassured, as Sītā instantly grows up and her determination matures from hysteria into unshakable resolution. The earlier portrayal has not been totally misleading: the frail girl who found life in exile physically difficult, endured those hardships with no recorded complaint, exhibiting the first signs of the emotional and bodily toughness that enable her to survive some eleven months living in the open air constantly threatened in the most gruesome fashion (Rāma and Lakṣmaṇa at least have a comfortable cave in which to spend the four months of the rains). She now becomes self-reliant and quick-witted, as her craving for finery deserts her and she abandons her ornaments without hesitation in the hope that the *vānaras* will somehow be able to save her (3.52.1–4). Gone, too, is

---

[6] The symbolism of Sītā's jewels has been examined by Alf Hiltebeitel (1980–81).

her hysteria: throughout her captivity she retains her dignity, never demeaning herself, never pleading with her captors, but assuming a position of moral and social superiority, refusing Rāvaṇa's blandishments with imperious disdain alternating with patient explanation (3.46.10–22, 51.3–24, 52.14–16, 53.17–54.19; 5.18–22). The stark realism of her quiet despair as she clings tenaciously to life makes a sharp contrast with the exaggerated threats of suicide which she had initially thought would persuade Rāma to let her accompany him in his exile (5.23.17); her comportment does not disguise her distress, but rather enhances it. Rāvaṇa's threats cannot sway this timorous girl; his wealth holds no attraction for this lover of finery; and the position of chief queen of the *rākṣasas* cannot seduce the princess whom we first met looking forward to her role as wife to the *yuvarāja* of the Rāghavas. The end of the poem leaves her beginning to take on the role of consort when she refuses Hanumān leave to wreak vengeance on her *rākṣasī* tormentors: like Rāma, like Daśaratha, she subordinates personal considerations to higher values; unlike Rāma, unlike Daśaratha, she does not promise to grant a request before learning what it is (6.101.23–39). She has learned to think, and the audience are now sure that she will make a fitting, wise queen.[7]

Composition of the *Rāmāyaṇa* was not a once-for-all process: from the very beginning each new teller was free to add his own touches, and it is no longer possible to retrieve the original text with absolute certainty.[8] A number of its minor inconsistencies should not be classed as deliberate authorial surprises, but as

---

[7] The contrasts depicted both within Sītā's character, and between her portrayal and the later idealised conception of her, are examined by Sally J. Sutherland (1989).

[8] Episodes which appear to have been incorporated towards the later part of the first stage of composition include one if not both explanations of the curse on Kabandha (3.67.1–17); one if not both versions of the Dundubhi story (4.11.7–42, 45.3–15); and the first battle between Vālin and Sugrīva and their reconciliation (4.12.14–32 and 4.22).

the anomalies characteristic of a developed text, some of them specifically of an orally-composed text, where authors' errors and the problems of transmission should not be subjected to over-literal interpretation.[9] Later reworkings can be identified by their language, and reveal not only the differing values of a later society, but the narratorial problems confronting the tellers of a tale that was by then well known and well loved.[10] Audiences were thoroughly familiar with the plot and characters, so the surprises no longer surprised. New tactics were needed if the story were not to lose its appeal and atrophy: problems had to be turned into opportunities.

But that's another story.[11]

---

[9] The identity of Timidhvaja's son (2.39.11) is an example of a question better left unexplored. The anomaly that Lakṣmaṇa speaks as if Mārīca is already known to him and Rāma (3.41.4–5) may result from problems of transmission: the next two verses are insufficiently supported by the manuscript evidence and should be excised from the text. Rāma's reference to Mārīca by name (3.55.7–8) sounds like nothing more significant than a slip by an excited narrator. A glaring problem is that from the moment of capture to the moment when Hanumān brings her news of Rāma, Sītā (who is supposed to be convinced that her husband has been killed or at least badly injured by Mārīca) has been consistently behaving as if she believed him to be still alive; this anomaly probably masks another of the earliest additions to the text (3.47.20–35, 51.10–23, 54.4–17; 5.19.17–29, 24.12–25; at 5.19.26 a different, simpler explanation is given for the brothers' absence from their hut – they are out looking for food; M. Brockington 2002; 2004). On the other hand, the manliness attributed alike to *vānaras*, to Saṃpāti the vulture, and to *rākṣasas* (e.g. *pauruṣa*, 4.8.37d, 55.2d, and 6.99.23b respectively) may result not from thoughtlessness, but from a desire to create a deliberate effect: neither realistic humans nor realistic animals/monsters, they are the composite creatures familiar from traditional tales.

[10] It comes as a considerable surprise to modern audiences and readers to discover that many features or incidents they have always assumed are fundamental to the story are not in fact present in the early layers. A minor example is that Śatrughna is consistently called Lakṣmaṇa's younger brother rather than his twin (*lakṣmaṇānuja* 2.72.1b, 5b, 23b). The incorporation of the birth story, detailed c. 1st–3rd century AD (1.8–17), made it necessary for all four brothers to be born more or less simultaneously, but audience resistance would have made it impossible to create a fourth mother at this late stage.

[11] It is the issues raised in this story (that is to say, the early tale with the later material superimposed, better thought of as a set of successive sto-

REFERENCES

ATU = *The Types of International Folktales. A Classification and Bibliography*. 3 vols. By Hans-Jörg Uther. (Folklore Fellows Communications, 284–286.) Helsinki: Academia Scientiarum Fennica, 2004.

BHATT G. H. & U. P. SHAH (crit. eds.) 1960–75. *The Rāmāyaṇa of Vālmīki*. 7 vols. Baroda: Oriental Institute.

BROCKINGTON, John & Mary BROCKINGTON (trans.) 2006. *Rāma the Steadfast. An Early Form of the Rāmāyaṇa*. London: Penguin.

BROCKINGTON, Mary 1998. The relationship of the *Rāmāyaṇa* to the Indic form of 'The Two Brothers' and to the Stepmother redaction. In: Lauri Honko, John Miles Foley & Jawaharlal Handoo (eds.), *The Epic. Oral and Written*: 139–150. Mysore: Central Institute of Indian Languages.

—— 1999. The art of backwards composition: some narrative techniques in Vālmīki's *Rāmāyaṇa*. In: Mary Brockington & Peter Schreiner (eds.), *Composing a Tradition. Concepts, Techniques and Relationships. Proceedings of the First Dubrovnik International Conference on the Sanskrit Epics and Purāṇas, August 1997*: 99–110. Zagreb: Croatian Academy of Sciences and Arts.

—— 2002. The rise and fall of Mārīca: stages and transitions in the portrayal of the Rāmāyaṇa's golden deer. In: Mary Brockington (ed.), *Stages and Transitions. Temporal and Historical Frameworks in Epic and Purāṇic Literature. Proceedings of the Second Dubrovnik International Conference on the Sanskrit Epics and Purāṇas, 1999*: 177–192. Zagreb: Croatian Academy of Sciences and Arts.

—— 2004. Who was the golden deer? Narrative inconsistency in the abduction of Sītā. In: Renata Czekalska & Halina Marlewicz (eds.), *2nd International Conference on Indian Studies. Proceedings*. (Cracow Indological Studies, 4/5): 73–84. Kraków: Institute of Oriental Philology, Jagiellonian University. [First read at 12th International *Rāmāyaṇa* Conference, Leiden, 28–30 August 1995.]

---

ries) that are explored by most modern commentators, e.g. Dimmitt (1986), Goldman & Goldman (1996: 57–62) and Pollock (1986: 48–73). I explore the techniques adopted by later contributors and compilers in M. Brockington 2007.

BROCKINGTON, Mary 2007. Another Story: re-composing the *Rāmāyaṇa*. *Rocznik Orientalistyczny* 60.2 (= *Theatrum Mirabiliorum Indiae Orientalis: A Volume to Celebrate the 70th Birthday of Professor Maria Krzysztof Byrski*): 68–81.

Dasarathajātaka 1901. *The Jātaka or Stories of the Buddha's Former Births*, IV.461; trans. by W. H. D. Rouse: 78–82. Cambridge: Cambridge University Press.

DIMMITT, Cornelia 1986. Sītā: fertility goddess and *śakti*. In: John Stratton Hawley & Donna Marie Wulff (eds.), *The Divine Consort. Rādhā and the Goddesses of India*: 210–223. Boston: Beacon Press (first edition: Berkeley: Graduate Theological Union, 1982).

GOLDMAN, Robert P. (gen. ed.) 1984– . *The Rāmāyaṇa of Vālmīki. An Epic of Ancient India*. Princeton: Princeton University Press.

GOLDMAN, Robert P. & Sally J. Sutherland GOLDMAN (trans.) 1996. *Sundarakāṇḍa*. (Goldman 1984– , vol 5.)

HILTEBEITEL, Alf 1980–81. Sītā *vibhūṣitā*: the jewels for her journey. *Indologica Taurinensia* 8–9: 193–200.

POLLOCK, Sheldon I. (trans.) 1986. *Ayodhyākāṇḍa*. (Goldman 1984– , vol. 2.)

——— (trans.) 1991. *Araṇyakāṇḍa*. (Goldman 1984– , vol 3.)

SUTHERLAND, Sally J. 1989. Sītā and Draupadī: aggressive behavior and female role-models in the Sanskrit Epics. *Journal of the American Oriental Society* 109: 63–79.

'Then in His Warlike Wrath Rāma Bent His Bow': Weaponry of the Early *Rāmāyaṇa*

JOHN BROCKINGTON

It is indeed for very good reasons that in the later tradition Rāma Dāśarathi is identified as Rāma with the bow, since this is an aspect of his portrayal that goes right back to the earliest layers of the tradition.[1] In this article I look again at the weaponry employed in the early *Rāmāyaṇa*. I previously examined this topic, as you are no doubt aware, in my *Righteous Rāma* (1985: 140–149), in which I cite the older literature on the subject. But the preparation of the translation which Mary Brockington and I made of an early form of the *Rāmāyaṇa* (trans. Brockington & Brockington 2006) prompted me to look at the topic again for two reasons. One of these is that translating the text necessarily involved another close examination of the exact nature of the weapons mentioned and the other is that, in the process of translation, we refined the criteria for the text translated. This underlying text is slightly modified from that identified in *Righteous Rāma* as the first stage of development (when the analysis was based essentially on complete *sargas*), mainly by taking into account the textual evidence – the degree of manuscript support – to remove some further single verses or

[1] The title of this article quotes Rām 6.95.10ab.

small blocks of verses accepted into the Critical Edition despite their inadequate support from the Northern recension, but also sometimes by slight adjustments to the boundaries between first and second stage passages, and very occasionally for other reasons also. Since an article which presents the arguments for these minor modifications has already been published (J. L. Brockington 2006), I will not go further into the matter now. So far as translating the various terms is concerned, we aimed at using one English term for each Sanskrit one but necessarily had to modify this principle quite often in the interests of producing a readable translation; for example, while the *Rāmāyaṇa* uses several terms for arrows, that term is the only one still in any sense alive in English.

As I noted previously (1985: 140), terms for weapons in general are more frequent in later stages of the text compared to the terms for specific weapons, obviously reflecting their greater distance from the military background of the early epic. Though much more frequent in later stages, so much so that less than a quarter of its occurrences are in the text we translated, *astra* is the most frequent general term for weapon, for which we usually adopted the translation 'missile', because of the nuances of especially powerful, because magical weaponry (found especially in the area of its greatest concentration, 6.66–96), although we used 'weaponry' on an appreciable number of occasions and sometimes simply 'weapon'. On the other hand, the other three terms used, *āyudha*, *praharaṇa* and *śastra*, we mostly translated as 'weapon', except where they occurred in a compound (e.g. *śastrapāṇi* or *parighāyudha*), when 'armed' or 'armed with' was usually more appropriate.

Arrows are much the most frequently mentioned weapon and, along with the bows and quivers needed to launch and to store them, are nearly three times as commonly mentioned as all other specific weapons. The most frequent term by far is *śara*, which with over 250 occurrences accounts for over half the total and is more than twice as frequent as *bāṇa*. We translated these two, along with the relatively infrequent terms *iṣu*, *mārgaṇa* and *sāyaka*, normally as 'arrow', with occasional use of 'shafts' or even 'reed' for variety or for euphony, while we reserved 'bolt' for *nārāca*, commonly and probably correctly held to be an iron arrow, although the one direct reference to iron arrows uses the term *śara* (*pañcāyasaḥ* ... *śarāḥ*, 5.44.20). All these terms are very much more frequent in our translated text than elsewhere in the Critical Edition text, confirming their predominance in early epic warfare. Other terms are very rare: *bhalla* occurs just four times (and once translated as 'quarrel' because of its juxtaposition to other terms for arrow), *nālīka* and *vikarṇin* three times (both in a set phrase with *nārāca*), the crescent-headed *ardhacandra* and the razor-sharp *kṣura* twice each, and the eared *karṇin* and the stone-ground *śilīmukha* once each; their significance is marginal at best.[2] Indeed, the last (at 3.49.9d) is probably better treated as adjectival rather than as a substantive, comparable to the frequent description of arrows as *niśita*, 'sharpened' or 'whetted'. There are also a number of references to the notch, *puṅkha*, which is several times said to be made of gold, and occasional allusions in general to the feathering and in particular to the use of heron-feathers; the term *viśikha*, occurring four times (3.19.20a; 6.34.23a, 78.9d, 17b) to denote arrows, is somewhat ambiguous, but we decided to translate it as 'well-feathered' rather than unfeathered.[3] The frequency of references

---

[2] Another term noted in *Righteous Rāma* (p. 141) as occurring once in the first stage (*pṛṣatka* at 6.78.7a, also 6.47.95d [l.v.]) is in a passage, 6.78.6ef–7, not sufficiently supported by manuscript evidence. The terms *añjalika*, *ardhanārāca* and *vatsadanta* found only at 6.35.23ab were excluded from our translation for the reasons suggested in *Righteous Rāma* (p. 142, n. 26). Incidentally, the term *margaṇa*, though rare (11 times in the text we translated) is predominantly early (17 times in total in the CE text), whereas *sāyaka* is commoner later (just 33 in the text we translated out of 84 overall) against the general trend.

[3] This represents a change from my view in *Righteous Rāma*, where I took *viśikha* to mean unfeathered arrows (p. 142); however, there is no other

to 'networks of arrows' (*śarajāla*, *bāṇajāla*), 'showers of arrows' (*śaravarṣa*, less often *śaravṛṣṭi*, *bāṇavarṣa*) and 'floods of arrows' (*śaraugha*, *bāṇaugha*) indicate that a premium was set on speed of discharge above accuracy of aim, shown also indirectly by occasional allusions to warriors bristling with arrows or as lying on a bed of arrows (6.36.4d, 39.12d, 15a, 40.3cd).

Despite their practical importance in archery, quivers (*tūṇī*, once *kalāpa*) are rarely mentioned and even the bowstring (for which the only term is *jyā*) is infrequently named, while the bow itself is named not much more than a third as often as arrows. Indeed, it is only because the bow was normally unstrung when not in use, and so stringing it (*sajjyaṃ* + √*kṛ*) was a warlike act, that the bowstring is mentioned much at all. The term *dhanus* is much the most frequent term, more than twice as frequent as *cāpa*, itself almost twice as common as *kārmuka*, with the compound term *śarāsana* being infrequently used.[4] None of these terms tell us anything much about the type of bow meant, although the term *kārmuka* to some extent implies a longbow, since it indicates that the bow is made of the wood of the *kṛmuka* tree. The main clue is in fact in the way that the bow is handled, with the arrow drawn back to the ear (e.g. 6.75.26b and 78.32ab), indicating that long arrows were the norm; the rare mention in later stages of short arrows points in the same direction. Possible contrary evidence consists of two mentions of the bow as triply bent or curved: *trivinataṃ cāpam* at 3.42.2a (where the commentators take it to be made of horn, so presumably a compound bow) and *kārmukam ... triṇatam* at 6.35.24bd, where however the use of the term *kārmuka* rather negates the implication that

---

mention of the use of unfeathered arrows but some, noted above, of feathering.

[4] The cognate of *dhanus*, *dhanvan*, is rare (only found at 3.24.6a, 49.5a, 63.23a), while the derivative *dhanvin* is infrequent, occurring just 11 times (the same frequency as *śarāsana*).

the reflex curvature must indicate a compound bow.[5] But the bulk of the evidence points to a long bow held vertically.

The next most frequently mentioned category of weapons, after bows and arrows, are spears, javelins and the like, with a little over one hundred mentions in all. Distinctions between types of weapons seems to be based more on size than on whether they were hurled or thrust (the usual distinction in European warfare); we tended to use the translation 'spear' for both *śakti* and *śūla*, although when the two occurred in close proximity we used the more exact, but more archaic, 'pike' for *śūla*, which is in fact the most commonly mentioned of this group. being almost twice as frequent as *śakti*. There are a certain number of references to the *śakti* being thrown but the weapons in this group seem often to be used for thrusting; however, since they occur mainly in lists of weapons, their precise nature and use are difficult to define with precision. The *paṭṭasa / paṭṭiśa*[6] (for which we adopted the translation 'javelin') is especially characteristic of the early parts of the text, although the remaining weapons in this group tend to be somewhat more common than the average in developed parts of the text. The other terms which do also occur in the early parts are *tomara* (usually 'lance', 10 times), *prāsa* ('dart', 8 times), *bhiṇḍipāla* ('dart' or 'javelin', three times) and *ṛṣṭi* ('lance', once – a lingering survival from the Vedic texts).

Clubs and similar weapons are only a tenth as common as bows and arrows (and two thirds as common as spears, lances and javelins). Both the *gadā* (usually 'club') and the *parigha*

---

5 For some discussion of this issue see Emeneau 1953 and for instances in early sculpture Auboyer 1956, who notes the first of these occurrences at p. 180. There is also a reference to a triply-curved bow in the second stage at 6.59.17a.

6 The different forms are simply the result of changing editorial policy in the CE, with *paṭṭiśa* being adopted uniformly in the Bāla to Sundara *kāṇḍa*s and *paṭṭasa* in the Yuddha and Uttara *kāṇḍa*s.

(usually 'mace') seem to be large and occasionally made of iron, while the *gadā* is once described as spiked (*gadāṃ tāṃ bahukaṇṭakām*, 6.42.33b), but the precise difference between them is not clear. The other terms in this group are *musala* ('pestle', found ten times), *kūṭamudgara* ('cudgel', found six times and from its etymology presumably a concealable, so smaller, version of the *mudgara*), *mudgara* ('hammer', also found six times). Staves (*daṇḍa* or *yaṣṭi*, found four times) should perhaps also be included here. More problematic is the *śataghnī*, found a couple of times in the text we translated; for the occurrence at 6.83.25c the context indicates that it is hand-held and so probably a particularly fearsomely spiked club (for which we simply adopted the translation 'bludgeon') but at 6.3.22a it is mentioned as part of the defences of Laṅkā next to siege engines (*yantra*, also mentioned earlier in this passage and at 4.14.5c – three occurrences in all) and so either a catapult or ballista is intended or possibly a large version of a caltrop.[7] Presumably the descriptive aspect of the term was still dominant, which enabled it to be applied to these very different types of weapon, but in any case it is of marginal significance.

It is noteworthy that swords play a relatively small part in the action, mentioned just under fifty times in the text we translated. The two main terms, almost equally frequent, are *asi* and *khaḍga*, with *nistriṃśa* occurring just three times. While we normally translated all three terms as 'sword' we also used 'scimitar' and 'broadsword' once each for variety. The sheath, *kośa*, is also mentioned once, when Indrajit unsheathes his sword to behead an illusory Sītā (*kṛtvā vikośaṃ nistriṃśaṃ*, 6.68.14c),

---

[7] Thus we translate 6.83.25 as 'many were the weapons in their hands – swords, javelins, pikes, clubs, pestles, ploughs, sharp-edged spears, great cudgels, darts, bludgeons and other fine weapons' and 6.3.22 as 'Furthermore, evil Rāvaṇa's city of Laṅkā is adorned with beams and catapults and siege engines of all types.' It is worth noting that a considerable number of N mss lack 6.83.25.

as is the hilt, *tsaru*, when Rāma picks up his golden-hilted sword and his bow to go after the golden deer (3.42.1).

In addition to the various types of weapon discussed so far, there are occasional references to battle-axes (a dozen times in all, mostly *paraśvadha*, once *paraśu*), nooses (*pāśa*, 5 times) and discuses (*cakra*, 4 times). Although battle-axes are by their nature normally hand-held, at 3.24.23 *rākṣasas* hurl spears, swords and battle-axes at Rāma, which seems rather out of place (which is no doubt why most Northern manuscripts have alternative readings eliminating the oddity). Again, we find at 6.88.7 that

> With frightful force wily Daśagrīva dispatched from his bow discuses, sunlike and huge, hurtling to every quarter and landing everywhere around,

where Rāvaṇa's *cakras*, despite their description as 'sun-like' implying the disk shape, come from his bow.

This completes the tally of weapons used by human warriors in the early parts. There are also occasional references to the gods using their characteristic weapons, for example Indra hurling his hundred-forked lightning bolt (*vajreṇa śataparvaṇā*, 3.67.10b) at Kabandha and driving his head and legs right into his body. But rather more significant are the quite frequent allusions to the *vānaras* in particular using branches, whole trees or boulders to fling at their opponents, as well as using their teeth, nails and fists, or pulling hair and ears. These natural weapons are by no means confined to the *vānaras* – even Rāvaṇa resorts to rocks and trees in his final duel with Rāma (6.95.17–18) – but they use them predominantly.

The concentration on the warriors' prowess in the use of their weapons, whichever one they are using at that moment, is underlined by the paucity of mention of defensive armour of any kind (and similarly the limited interest in the defences of cities). The commonest term for armour, *kavaca*, occurs a mere fourteen times (four of which are in the form *kavacin*), with just two occurrences of *varman* and one of *saṃnāha*, and shields are mentioned only eleven times (*carman* ten times and *śarāvara* once),[8] while helmets are mentioned just four times (in *saśirastrāṇa* three times, as *śirastra* once). In contrast, chariots are very frequently alluded to because they belong with the exercise of skill in warfare (chariots themselves nearly two hundred times, charioteers however much less often, and the component parts of chariots less than ten times). Unusually, there is a fairly full description of a chariot's parts at 6.45.24–26:

> It was ready and waiting, harnessed to swift-paced steeds and driven by an expert charioteer; it rumbled like huge clouds, it dazzled with the brilliance of moon and sun; it was protected by its serpent-standard and safeguarded by its stout rail; its frame was sturdy, its lattice-work was golden, and it seemed to be exulting in its splendour.

* * *

To sum up, the picture is overwhelmingly one of warriors mounted on chariots, using long bows to shoot large numbers of arrows at their opponent as quickly as possible, the speed of discharge being more valued than the accuracy. Great warriors are described as fighting in the duels which receive much greater attention than mass fighting. Next most frequent are the other group of weapons that can be used from a chariot, the spears, lances, javelins and the like, which in that situation would no doubt usually have been thrown; their use in the manner of medieval European jousting cannot be ruled out but seems unlikely. Clubs and similar weapons and swords, which belong to hand-to-hand combat when dismounted, are much less frequent

---

[8] It is worth noting that not only does the term *carman* indicate that the shield is made of leather but it is specifically described as made of cowhide at 6.84.21a (another mention at 5.1.22c belongs to the second stage of growth). Also, many mentions are linked with use of a sword, probably indicating that the shield was not worn while using a bow.

and indeed there is a definite trend towards their use solely in the closing stages of a duel, after one or both warriors' chariots have been destroyed. Other weapons, though mentioned occasionally, are in reality of very limited significance. The only major qualification to this basic pattern, but one that in fact serves rather to reinforce it, is that the *vānaras* regularly fight from the ground and use rocks, trees and anything else to hand.

REFERENCES

AUBOYER, Jeannine 1956. L'arc et la flèche dans l'iconographie ancienne de l'Inde. *Artibus Asiae* 19.3–4: 173–185

BHATT G. H. & U. P. SHAH (gen. eds.) 1960–75. *The Rāmāyaṇa of Vālmīki*. 7 vols. Baroda: Oriental Institute.

BROCKINGTON, J. L. 1985. *Righteous Rāma. The Evolution of an Epic.* Delhi: Oxford University Press.

——— 2006. Some *Rāmāyaṇa* textual issues. In: Danuta Stasik & Anna Trynkowska (eds.), *India in Warsaw / Indie w Warszawie*: 202–212. Warszawa: Elipsa.

BROCKINGTON, John & Mary BROCKINGTON 2006. *Rāma the Steadfast. An Early Form of the Rāmāyaṇa.* London: Penguin Books.

EMENEAU, M. B. 1953. The composite bow in India. *Proceedings of the American Philosophical Society* 97: 77–87.

Nikumbhilā's Grove: *Rākṣasa* Rites in Vālmīki's *Rāmāyaṇa*

SALLY J. SUTHERLAND GOLDMAN

In Vālmīki's *Rāmāyaṇa*, the word *nikumbhilā* occurs only in the three final *kāṇḍa*s,[1] where it is used either as the name of the grove in which the *rākṣasa*s perform rites or as the name of the goddess to whom those rites are offered. The reference in either case seems to be to a mother goddess cult and/or some type of *abhicāra*[2] worship associated with her. The word is consistently associated with either the *rākṣasī* attendants of Sītā or with Indrajit, the most famous son of Rāvaṇa. The term *nikumbhilā* appears to be unique to the *Rāmakathā*, and the epic's marking of this figure and/or her grove as belonging to the *rākṣasa* world raises questions as to the epic's rationale for such a juxtaposition between a mother-goddess figure and the *rākṣasa*s.[3] The Sundara kāṇḍa, as I have argued elsewhere, locates itself in the feminine,

---

[1] This paper focuses on the use of the term *nikumbhilā* primarily in the Sundara and Yuddha *kāṇḍa*s. See footnote 47 on the Uttarakāṇḍa's much more limited use of this word. See specifically 5.22.41; 6.60.18, 69.23, 72.10–14; and 7.25.2. All references, unless otherwise noted, are to the critical edition of the *Vālmīkirāmāyaṇa*, and all translations and notes are referenced to the Princeton translation. See R. P. Goldman 1984; Pollock 1986; 1991; Goldman & Goldman 1996; R. P. Goldman *et al.* 2008.

[2] The word *abhicāra* normally carries with it a negative semantic connotation, and its range of meanings in English includes, 'exorcising, employment of magical spells for malevolent purposes, *or* the employment of magic'. The use of 'magical spells' traditionally is considered an *upapātaka*, or 'lesser sin' (*abhicāreṣu sarveṣu kartavyo dviśato damaḥ*, *Manusmṛti* 9.290).

[3] See, for example, Seely 2004.

and uses gender and the exotic as two strategies with which to depict Rāvaṇa's demonic realm.[4] On the other hand, the Yuddhakāṇḍa locates itself in the masculine and constructs the demonic world as hauntingly similar to that of the imagined brāhmaṇic one. Yet the use of the term *nikumbhilā* marks the two *kāṇḍa*s as possessing a site of worship that is aligned with both the demonic and the feminine.

Vālmīki populates much of his epic with a number of fearsome and terrifying female figures, both human and non-human. Thus the audience is made aware of the threatening and dangerous nature of the feminine through the introduction of such figures as the gargantuan *rākṣasī* Tāṭakā, the beautiful and beguiling Kaikeyī, and her deformed maid-servant Mantharā, and the aggressive and infatuated sister of Rāvaṇa, Śūrpaṇakhā,[5] whose very name calls to mind her sexual excess. But it is not until the action moves to Laṅkā in the Sundarakāṇḍa that the depth and obsessiveness of Vālmīki's fascination with and fear of the feminine is thoroughly revealed. Moreover, it is here that the feminine is clearly and systematically linked with the demonic.[6]

What appears to be of far less concern to Vālmīki's epic is the figure of the goddess. In the later *Rāmāyaṇa* tradition, Sītā herself is considered both an incarnation of Śrī or Lakṣmī and to be born from Pṛthivī, the Earth goddess. However, in Vālmīki's epic this identification, partly necessitated by the logic of the narrative, is, as we shall see, largely backgrounded or simply inferred.[7]

---

[4] R. P. Goldman 2000; S. J. S. Goldman 2003.

[5] The name Śūrpaṇakhā is a *bahuvrīhi samāsa*, meaning 'having nails like winnowing baskets' (*śūrpāḥ eva / iva nakhāḥ yasyāḥ sā*). Long nails are seen as signs of sexual excess, as indicated by the poetic convention of telltale nail marks left on one's lover after lovemaking. See Ingalls 1965: 199–200.

[6] After all, Vālmīki expresses his distrust of the feminine through the action in the court at Ayodhyā, where Kaikeyī is allowed to hold sway over her love-sick husband. See Sutherland 1992.

[7] Sītā once or twice is compared to Śrī, but not explicitly identified with her, unlike the male figures Rāma and Lakṣmaṇa who are both explicitly

In the Bālakāṇḍa, Sītā is said to spring up from the earth (1.65.14–15), but, it must be noted that the inclusion of these verses in the critical text is seriously suspect and that they were admitted only for reasons of 'higher criticism'.[8] Here, King Janaka tells the circumstances of his daughter's birth. She rose up as he was plowing a field (*atha me kṛṣataḥ kṣetraṃ lāṅgalād utthitā mama*). Twice we are told that she is 'risen up from the surface of the earth' (*bhūtalād utthitā*), and that she is 'not born of a womb' (*ayonijā*).[9]

This association of Sītā with the earth is, however, reiterated in the Ayodhyākāṇḍa, where Sītā relates to Anasūyā the story of her birth and marriage. Here it is quite clear that Sītā's origins are understood to be 'otherworldly'. She tells Anasūyā that her father Janaka, discovered her after 'she had broken through the earth' (*bhittvā jagatīm*)[10] and, after he had taken her, that a disembodied voice, 'a human voice, but unlike any ever heard',[11] verified that she was no ordinary infant. Eventually, Janaka, worried about the marriage prospects for such an unusual daughter, since she, 'was not born of a woman's womb', gives the order for her *svayaṃvara*.[12]

---

understood to be *avatāras* or *aṃśāvatāras* of Viṣṇu. See, for example, 5.14.6, 20.42, 36.2. See R. P. Goldman *et al.* 2008 (Introduction to the Yuddhakāṇḍa, The Major Characters of the Yuddhakāṇḍa: Rāma). See, too, the note to Yuddhakāṇḍa 28.28–29 on the divinity of Rāma, and Pollock 1991: 15–54.

[8] See notes to 1.65.14–15.

[9] 1.65.14–16: *atha me kṛṣataḥ kṣetraṃ lāṅgalād utthitā mama* | *kṣetraṃ śodhayatā labdhvā nāmnā sīteti viśrutā* || 14 || *bhūtalād utthitā sā tu vyavardhata mamātmajā* | *vīryaśulketi me kanyā sthāpiteyam ayonijā* || 15 || *bhūtalād utthitāṃ tāṃ tu vardhamānāṃ mamātmajām* | *varayām āsur āgamya rājāno munipuṃgava* || 16 ||

[10] 2.110.27: *ahaṃ kilotthitā bhittvā jagatīṃ nṛpateḥ sutā.*

[11] 2.110.30: *antarikṣe ca vāg uktāpratimā mānuṣī kila.*

[12] 2.110.36: *ayonijāṃ hi māṃ jñātvā nādhyagacchat sa cintayan.* Note the parallels in the story of Sāvitrī in the *Mahābhārata*, where her father Aśvapati, worries about finding a husband for his daughter, Sāvitrī, who was given to him as a boon by the goddess Sāvitrī, and becomes distraught (Mbh 3.277.31).

At Yuddhakāṇḍa 104.15, Sītā once again makes mention of her birth from the earth:

> Since my name [Jānakī] is derived from Janaka, you failed to take into account the fact that I was born from the earth itself, nor, though you are an expert judge of conduct, have you given due consideration to my virtuous conduct.
>
> *apadeśena janakān notpattir vasudhātalāt* |
> *mama vṛttaṃ ca vṛttajña bahu te na puraskṛtam* || 15 ||

A literal translation would be, 'by the name from Janaka, the birth from the surface of the earth is not, nor is my conduct properly put forward by you, o knower of conduct'.[13] The syntax of the verse as it appears in the critical edition and the texts and printed editions of Vivekatilaka, Govindarāja, and Maheśvaratīrtha is awkward and apparently elliptical. The translation follows the interpretation of Vivekatilaka and Govindarāja, who understand the participle *puraskṛtam*, 'put forward', to also govern the feminine nominative *utpattiḥ* in *pāda* a. Thus they supply a feminine form, *puraskṛtā*. This reading enables them to then account for both the instrumental *apadeśena*, 'by name', and negative particle *na* in *pāda* a. However, despite the difficulty of the reading, Sītā's point here is quite clear. Regardless of her epithets such as Jānakī, Maithilī, etc., she is reminding Rāma that she is not the actual child of the human king Janaka, but rather that she was born from the surface of the earth (*vasudhātalāt*).[14]

There is thus little doubt that the epic understands Sītā's birth to be non-human and that she had emerged from the [surface of] earth. However the passages do not explicitly identify her as either a goddess or as born from a goddess and associate, or at least contain, the locus of her birth within a very masculine

---

[13] See R. P. Goldman *et al.* 2008, notes to Yuddhakāṇḍa 105.15.

[14] See 1.65.14–17 and notes on Sītā's birth. See, too, 7.88, where the goddess Pṛthivī fetches Sītā back. See Sutherland 1989 and S. J. S. Goldman 2006.

space, the field (*kṣetraḥ*) (1.65.14; 2.110.27). While the word *kṣetra* literally refers to a 'field' or 'fertile land', and is masculine in gender, it comes to be used also as a term for a 'wife', as the wife is conceptualized as not only a husband's possession, but the 'fertile' locus in which offspring are produced.[15]

In the Uttarakāṇḍa, there is little doubt that Sītā's birth is from not just the surface of earth, but the goddess herself. At 7.88.12, Dharaṇī, the earth goddess, arrives in divine form and takes back the hapless queen after she has called upon her to provide testimony to her faithfulness to her husband (7.88.12–14 GPP).[16] In this passage, the earth Dharaṇī is called *devī*, 'goddess', and is depicted rising up out of the surface of the earth (*bhūtalād utthitam*) on a divine throne being carried on the heads of *nāgas* (*divyaṃ siṃhāsanam ... dhriyamāṇaṃ śirobhis ... nāgair*). Note however, that the locus of the event is still masculine, occurring at the *aśvamedha* sacrifice of Rāma (7.83ff.). Nevertheless, although there is a close connection between the earth and Sītā, the critical text of the *Vālmīkirāmāyaṇa*, outside of the Uttarakāṇḍa does not appear to explicitly identify the earth as a divinity and nowhere does it appear to explicitly identify Sītā as the daughter of Pṛthivī.

The only clear identification of Sītā as a goddess identifies her as Lakṣmī in a passage at the end of the Yuddhakāṇḍa (6.105.25), where Brahmā reveals to Rāma who he truly is:

> Sītā is Lakṣmī; you are god Viṣṇu. You are Kṛṣṇa. You are Prajāpati, lord of creatures. It was in order to slay Rāvaṇa that you entered a human body here. (6.105.25)

---

[15] The motif of non-human birth is common in the epics, and allows, as I have argued elsewhere, the patriarchy to claim control of what is essentially a biological function of the female. See S. J. S. Goldman 1996.

[16] 7.88.11cd–13ab: *bhūtalād utthitaṃ divyaṃ siṃhāsanam anuttamam* || 11 || *dhriyamāṇaṃ śirobhis tan nāgair amitavikramaiḥ* | *divyaṃ divyena vapuṣā sarvaratnavibhūṣitam* || 12 || *tasmiṃs tu dharaṇī devī bāhubhyāṃ gṛhya maithilīm* |

sītā lakṣmīr bhavān viṣṇur devaḥ kṛṣṇaḥ prajāpatiḥ I
vadhārthaṃ rāvaṇasyeha praviṣṭo mānuṣīṃ tanum II 25 II

The passage is well attested and has been admitted to the critical text on solid text editing principles, and appears to be the only explicit reference to Sītā in Vālmīki's epic as either a goddess or an incarnation of Lakṣmī.

References to other female divinities are similarly rare. The Bālakāṇḍa mentions Pārvatī (Umā), Gaṅgā, and the Earth (Pṛthivī) in its narration of the birth of Kārtikeya and the descent of the Gaṅgā, along with a few other female divinities,[17] but these roles are mostly secondary, especially compared to the roles of these same figures in variant versions.

There is one important exception to this. One of the more intriguing stories of the Bālakāṇḍa concerns a goddess aligned with the darker powers, and about whom we hear little in the larger tradition. The story of Diti, the mother of the *daityas*, and Indra is fairly prominent in the first *kāṇḍa*, occupying the better part of three *sargas* (1.44.14–46.9). Of particular note here, is that the story, which I have discussed elsewhere in its critical structural role in the Bālakāṇḍa, establishes a strong link among divinity, the feminine, and the demonic. Outside of these references, the figure of the goddess is relatively unknown, with a few notable exceptions (S. J. S. Goldman 2004).

It is in the Sundarakāṇḍa, with its focus on both the feminine and *rākṣasa* world, that we clearly see not just the juxtaposition

---

[17] For example, 1.33 mentions Satyavatī, sister of Viśvāmitra, who follows her husband, and is transformed into a river (Kauśikī). She is called a goddess at 1.33.8–12. At *sargas* 34–36, we have a sequence of Śaiva based stories that make mention of goddesses. We are told the story of Gaṅgā, Umā and Rudra, which occurs in a very abbreviated form at 1.34.12–21; at *sarga* 35 we are told of the birth of Kārtikeya, but again, it is very abbreviated and there are only a few references to Umā (6, 7, 10, 13, 19, 21, 22, 23); at *sarga* 36, we are told of the birth of Kārtikeya, where the Ganges is not able to bear Śiva's semen (11–17); at Bālakāṇḍa 40–43, the descent of the Ganges is narrated. In addition, there is the story of Indra and Diti. Although Aditi is mentioned, the story mainly concerns Indra's killing Diti's sons, and then cutting up Diti's fetuses (1.44–46). See S. J. S. Goldman 2004.

of gender and exotica, but the virtual superimposition of one onto the other from the very outset of the action. As Hanumān leaps over the ocean, he confronts two monstrous, devouring female figures, Surasā and Siṃhikā, respectively. In sailing past Mt. Maināka,[18] Hanumān has symbolically passed beyond the range of the Āryan world of Rāma and Lakṣmaṇa into the unknown. It is no accident, I would argue, that Vālmīki next puts Hanumān to a test, nor is it accidental that the test involves both the feminine and at least in form, demonic. Surasā, the first of the two female figures encountered by Hanumān, is specifically called *devī* (5.1.134), arguably a multivalent term, and is called 'the mother of the serpents' (*nāgamātaram,* 5.1.130). The gods and other assembled divinities request her to take on the form of a *rākṣasī* and create an obstacle for Hanumān,

> ... she stood in the middle of the ocean, taking on the deformed and hideous appearance of a *rākṣasa*, terrifying to all. (5.1.134cd–135ab)
>
> *samudramadhye surasā bibhratī rākṣasaṃ vapuḥ* ॥ 134 ॥
> *vikṛtaṃ ca virūpaṃ ca sarvasya ca bhayāvaham* ।

Moreover, Surasā claims that she must devour Hanumān and that the gods have ordained him as her food. In fact, the gods have said nothing to Surasā about her dietary habits. They have only requested that Hanumān should be tested:

> For we wish to ascertain his strength and his valor. Will he foil you through some stratagem or will he give way to despair? (5.1.133)
>
> *balam icchāmahe jñātuṃ bhūyaś ca parākramam* ।
> *tvāṃ vijeṣyaty upāyena viṣādaṃ vā gamiṣyati* ॥ 133 ॥

The figure of Surasā is both a goddess, or at least a *devī* with close ties to the divine, and a *rākṣasī*, and is clearly a threat

---

[18] One should not forget that prior to this, Hanumān's first encounter is with the phallic father figure, Mount Maināka, whom, we are explicitly told, is a barrier against the hordes of *asuras* who inhabit the underworld known as Pātāla. He is urged to rise up out of the ocean and offer assistance to Hanumān (5.1.79–129, esp. verse 81).

to the masculine. As she seeks to devour Hanumān, her ever-increasing size threatens to consume the entire world if not held in check.[19] After Hanumān contracts his size to 'no bigger than a thumb' (*aṅgulamātrakaḥ*, 5.1.151), he enters her, and then exits, having allowed himself to be eaten symbolically. He then pays homage to her, because she is a goddess, and a mother, and worthy of such respect:

> Homage to you, Dākṣāyaṇī. I have entered your mouth. Now I shall go where Vaidehī is. May your words ever prove true. (5.1.153)[20]
>
> *praviṣṭo 'smi hi te vaktraṃ dākṣāyaṇi namo 'stu te* I
> *gamiṣye yatra vaidehī satyaṃ cāstu vacas tava* II 153 II

Here, Hanumān addresses Surasā by her patronymic, *Dākṣāyaṇī*.[21] The term normally refers to any daughter of Dakṣa, including Aditi, Diti, or Kadrū. Since Surasā has also been described as *nāgamātaram*, it is not unlikely that the reference is to Kadrū, who is said to be the mother of the snakes and a figure used by Vālmīki elsewhere (see 6.40.49 and note).[22] The commentators only mention the genealogical connection. Thus Govindarāja understands that she is the descendant of Dakṣa (*dakṣasyāpatyaṃ dākṣāyaṇī*), while Nāgojibhaṭṭa understands that she belongs to the lineage of Dakṣaprajāpati (*dākṣāyaṇī dakṣaprajāpatisaṃtāne*). Hanumān appears to realize Surasā's real nature, addressing her as Dākṣāyaṇī, before she resumes her true shape (5.1.133–134); however the episode calls to mind the maxim: *nārī pratyakṣarākṣasī*.[23] Vālmīki's introduction of the character of

---

[19] See S. J. S. Goldman 2003. Cf. too the role of Kumbhakarṇa in the Yuddhakāṇḍa.

[20] See note to 5.1.153.

[21] *dākṣayaṇī*: *dakṣiṇa*, not uncommonly an epithet of Śiva or Viṣṇu, conveys a semantic range of meanings such as, 'capable, able, right, southern'.

[22] For the story of Kadrū and Vinatā, see Mbh 1.14ff.

[23] Despite the phallic gist of the passage, Vālmīki has set the tone of the *kāṇḍa* to come. To reinforce this, he introduces yet another horrific female figure, Siṃhikā. The passages seem very similar in nature, with Hanumān

Surasā establishes a strong link among the feminine, the demonic, and divinity. Here importantly, the three are located at the margins of society in that liminal region between us and them, as it were.

Subsequently, in the critical edition,[24] Hanumān's search centrally takes him through the *antaḥpuram* of Rāvaṇa, an area wherein the feminine is located, albeit possessed and dominated by a male, and it is in the *antaḥpuram* of the *rākṣasa* lord that the demonic is most strikingly aligned with the feminine. But it

---

in each confronting a female figure who is trying to destroy him. The second passage is relatively short (only 14 verses, compared with 26 verses in the Surasā passage). Although one's temptation is to discount the second as a poor imitation of the first, it has been admitted with strong textual support to the critical text. Unlike her predecessor, however, Siṃhikā is inherently a *rākṣasī*, marking, in fact, a boundary between the first and second episodes. Hanumān has entered the world of the *rākṣasa*s at this point in the story, and the figures need no longer 'assume' *rākṣasa* shapes and forms to 'test' the hero, but are, in fact, *rākṣasas*. In keeping with her *rākṣasa* nature, the sole purpose of Siṃhikā's aggression is to sate her insatiable hunger. She has not eaten in a long time, and wishes to capture and devour Hanumān (verse 167). No gods or other divine figures request her to do so, she does not change form. The conflict is quite brief, no challenges verbal or physical occur between the two. Hanumān's forward progress is stopped as he realizes that his shadow has been caught by the demoness Siṃhikā. He expands to a huge size, in an instant contracts, and enters into her guts. There he slashes her inner 'vital organs', killing her. He exits her body, and is praised by the gathered hosts of gathered celestials (only first mentioned here), and resumes his venture (166–179).

While the two episodes at first appear remarkably similar, I would argue that they serve different purposes. Both are part of the transitional nature of the first *sarga* of the *kāṇḍa* that leads Hanumān from the male Āryan world to that of the *rākṣasa* realm. Both Surasā and Siṃhikā mark the space as female, dangerous, and threatening. The confrontation with Surasā reminds us that, as we leave the known world of *āryavarta*, dangerous and threatening females lurk all about, ready to devour us. Like all females, these creatures are powerful and threatening and can appear in many guises. If properly controlled and/or propitiated, some are not unequivocally destructive. Others, however, are like Siṃhikā, dangerous and evil through and through, *rākṣasas* to the core. These creatures are deserving of little attention and are in need of aggressive counter-action.

[24] In the GPP and other southern versions of the *Vālmīkirāmāyaṇa* (NSP, VSP, KK, Gita Press), Hanumān confronts the demoness of the city, Lāṅkā, as he enters, providing yet another encounter with a demonic female/ liminal figure.

is in the *aśokavana*, which is located in the midst of a garden attached to the *antaḥpuram*, the site of sexual pleasure, the very heart, as it were, of the *rākṣasa* world, where Vālmīki sets the vast majority of the *kāṇḍa*'s action. Here we find Sītā surrounded by hideous and frightful *rākṣasī* guards, who stand in stark contrast to the lovely women that inhabit the *antaḥpuram*. The density and detail of the description is in itself telling:

> Straining to see Vaidehī, Hanumān, son of Māruta the wind god, saw – a short distance away – frightful-looking *rākṣasa* women.
>
> One had but a single ear, while another's ears covered her entire upper body. One had no ears at all. Another had pointed ears, while yet another had her nose on her forehead.
>
> One had a huge head; another's neck was long and thin. One had hardly any hair, while another was completely bald. Yet another's hair covered her like a woolen blanket.
>
> One had pendulous ears and a bulbous forehead; another's breasts and belly hung down. One's lips protruded, while another's lips were on her chin. Yet another's mouth hung open, while still another's knees stuck out.
>
> One was very short; another very tall. One was hunchbacked, another was deformed, while yet another was a dwarf. One was snaggletoothed; another had a sunken face. One had yellow eyes, while another's features were distorted.
>
> The best of monkeys stared at those *rākṣasa* women, whose appearance caused his fur to bristle. They were deformed, yellowish, black, wrathful, fond of quarrels, and armed with spears, mallets, and war-hammers of black iron. Some had the faces of boars, deer, tigers, buffalo, goats, or jackals. Some had feet like those of elephants, camels, and horses. Others' heads were sunk into their bodies. Some had only one hand; others one foot. Some had the ears of donkeys, horses, cows, elephants, or monkeys. Some had no noses; others had huge ones. Some had crooked noses or noses without nostrils. Some had noses like elephants' trunks, while others had nostrils in their foreheads. Some had the feet of elephants, some had enormous feet, some had feet like cows' hooves, and some had hairy feet. Some had enormous heads, necks, breasts, and bellies. Some had enormous mouths and eyes, while others had long tongues and nails. Some of those dreadful-looking *rākṣasa* women had the faces of goats, elephants, cows, sows, horses, camels, or donkeys. They held lances and war-hammers; and they were wrathful and fond of quarrels. Those *rākṣasa* women were snaggletoothed. Their hair was like smoke, and their faces were hideous. Addicted to meat and wine, they were constantly guzzling strong drink. Their bodies were smeared with flesh and blood, for such was their food.

> They were seated all around the mighty trunk of that massive tree. Beneath it fortunate Hanumān observed that lady, Janaka's daughter, the blameless princess. She was lusterless and wracked with sorrow; her hair was matted with dirt. (5.15.4–19)[25]

The *rākṣasī*s are thus deformed, act like men [they hold weapons], are quarrelsome, and are hideously ugly. Like all of their race, they are fond of raw flesh and liquor (S. J. S. Goldman 2003). The portrait is one of horror and disgust, but at the same time the very 'thickness' and detail of description indicates a fascination with these women and reveals, it would seem, a voyeuristic pleasure in their very grotesqueness.

The *rākṣasī*s are once again introduced into the narrative at *sarga* 20 of the *kāṇḍa*, when Rāvaṇa, having repeatedly tried to woo Sītā, is again rejected. Frustrated but still infatuated, the king of the *rākṣasa*s, enlists the aid of the *rākṣasī* guards, with their various means of persuasion, to convince Sītā that she should submit to her captor. These women are again described in

---

[25] *didṛkṣamāṇo vaidehīṃ hanūmān mārutātmajaḥ* I *sa dadarśāvidūrasthā rākṣasīr ghoradarśanāḥ* II 4 II *ekākṣīm ekakarṇāṃ ca karṇaprāvaraṇāṃ tathā* I *akarṇāṃ śaṅkukarṇāṃ ca mastakocchvāsanāsikām* II 5 II *atikāyottamāṅgīṃ ca tanudīrghaśirodharām* I *dhvastakeśīṃ tathākeśīṃ keśakambaladhāriṇīm* II 6 II *lambakarṇalalāṭāṃ ca lambodarapayodharāṃ* I *lamboṣṭhīṃ cibukoṣṭhīṃ ca lambāsyāṃ lambajānukām* II 7 II *hrasvāṃ dīrghāṃ ca kubjāṃ ca vikaṭāṃ vāmanāṃ tathā* I *karālāṃ bhugnavaktrāṃ ca piṅgākṣīṃ vikṛtānanām* II 8 II *vikṛtāḥ piṅgalāḥ kālīḥ krodhanāḥ kalahapriyāḥ* I *kālāyasamahāśūlakūṭamudgaradhāriṇīḥ* II 9 II *varāhamṛgaśārdūlamahiṣājaśivāmukhāḥ* I *gajoṣṭrahayapādāś ca nikhātaśiraso 'parāḥ* II 10 II *ekahastaikapādāś ca kharakarṇyaśvakarṇikāḥ* I *gokarṇīr hastikarṇīś ca harikarṇīs tathāparāḥ* II 11 II *anāsā atināsāś ca tiryaṅnāsā vināsikāḥ* I *gajasaṃnibhanāsāś ca lalāṭocchvāsanāsikāḥ* II 12 II *hastipādā mahāpādā gopādāḥ pādacūlikāḥ* I *atimātraśirogrīvā atimātrakucodarīḥ* II 13 II *atimātrāsyanetrāś ca dīrghajihvānakhās tathā* I *ajāmukhīr hastimukhīr gomukhīḥ sūkarīmukhīḥ* II 14 II *hayoṣṭrakharavaktrāś ca rākṣasīr ghoradarśanāḥ* I *śūlamudgarahastāś ca krodhanāḥ kalahapriyāḥ* II 15 II *karālā dhūmrakeśīś ca rākṣasīr vikṛtānanāḥ* I *pibantīḥ satataṃ pānaṃ sadā māṃsasurāpriyāḥ* II 16 II *māṃsaśoṇitadigdhāṅgīr māṃsaśoṇitabhojanāḥ* I *tā dadarśa kapiśreṣṭho romaharṣaṇadarśanāḥ* II 17 II *skandhavantam upāsīnāḥ parivārya vanaspatim* I *tasyādhastāc ca tāṃ devīṃ rājaputrīm aninditām* II 18 II *lakṣayāmāsa lakṣmīvān hanūmāñ janakātmajām* I *niṣprabhāṃ śokasaṃtaptāṃ malasaṃkulamūrdhajām* II 19 II

hideous terms in a passage reminiscent, and in some instances identical with, the earlier, although not quite as lengthy. That they dwell in the women's quarters as guards of Sītā seems physically to interrupt the intended loveliness of the place and its inhabitants. In addition, their grossness and deformity contrast vividly with Sītā's frailness and loveliness, as in the same manner they had earlier served as a foil to the lovely women of the harem. The *rākṣasīs* entreat, cajole, and finally threaten Sītā. If she does not give in to Rāvaṇa's wishes, they say, then she will suffer. The threats are explicit and go to the heart of this discussion.

Their threats allow us, the audience, to glimpse the imagined world of these women. They are, we learn, like *rākṣasas* elsewhere, consumers of human flesh. Such consumption can be understood as an expression of sexual desire, and excessive or inappropriate libidinal urges (S. J. S. Goldman 2000; 2003). The specificity with which these women detail their cravings as well as the locus of the cravings mark the passage as particularly significant.

> 'Or else, Maithilī, if you do not do exactly as I have told you, I shall tear out your heart and eat it!'
>
> Then another fierce-looking *rākṣasa* woman named Caṇḍodarī said these words as she brandished a huge spear:
>
> 'The moment I saw this woman that Rāvaṇa has brought here, her eyes darting like a doe's and her breasts trembling with fear, I conceived a tremendous craving.'
>
> 'I conceived a desire to eat her liver, spleen, *utpīḍa*, her heart with all its veins and arteries, her intestines, and her head.'
>
> Then a *rākṣasa* woman named Praghasā said, 'What are we waiting for? Let's strangle this heartless woman!'
>
> 'We will tell the king that the woman is dead, and without a doubt, he will tell us, "You may eat her!"'
>
> Next a *rākṣasa* woman named Ajāmukhī said, 'Butcher her; then cut her all up into equal, bite-sized pieces.' (5.22.32–38)[26]

[26] *utpāṭya vā te hṛdayaṃ bhakṣayiṣyāmi maithili* | *yadi me vyāhṛtaṃ vākyaṃ na yathāvat kariṣyasi* || 32 || *tataś caṇḍodarī nāma rākṣasī krūradarśanā* | *bhrāmayantī mahaccchūlam idaṃ vacanam abravīt* || 33 ||

The precise anatomical detail indicates a familiarity with body parts that could only arise from previous experience. The *rākṣasīs* are clearly intended to be portrayed as practiced at dissection; their scientific approach adds a gruesome calculatedness to the undertaking that combined with the passion of their cravings creates a scenario of true terror. The demonic, the consumptive, and the feminine are linked yet once more, as they were in the Surasā episode. Here, I would argue that the *rākṣasīs*' words are of more than thematic, emotional, and/or narrative import, for they provide yet another homo-erotic reference in the Sundarakāṇḍa (cf. *sarga* 7.57–59). This time the eroticism is masked in consumptive terms and its subjects are not the beautiful women of the harem, but the hideous women guards.

At this point the *rākṣasīs* introduce a rationale, it seems, for their blood-thirsty behavior. It is not merely to delight in the partaking of a tasty snack, but rather to engage in the worship of the goddess Nikumbhilā.

> [Ajamukhī continues:]
> 'Then we can all have a share. I hate arguments! So quickly bring lots of things to drink and all kinds of garlands.'
> Then a *rākṣasa* woman named Śūrpaṇakhā said, 'I agree with what Ajāmukhī just said.'
> 'So quickly bring wine, the banisher of every sorrow! Let us eat human flesh and dance before the goddess Nikumbhilā.' (5.22.39–41)[27]

---

imāṃ hariṇalokākṣīṃ trāsotkampapayodharām I *rāvaṇena hṛtāṃ dṛṣṭvā daurhṛdo me mahān abhūt* II 34 II *yakṛtplīham athotpīḍaṃ hṛdayaṃ ca sabandhanam* I *antrāṇy api tathā śīrṣaṃ khādeyam iti me matiḥ* II 35 II *tatas tu praghasā nāma rākṣasī vākyam abravīt* I *kaṇṭham asyā nṛśaṃsāyāḥ pīḍayāmaḥ kim āsyate* II 36 II *nivedyatāṃ tato rājñe mānuṣī ṣā mṛteti ha* I *nātra kaścana saṃdehaḥ khādateti sa vakṣyati* II 37 II *tatas tv ajāmukhī nāma rākṣasī vākyam abravīt* I *viśasyemāṃ tataḥ sarvān samān kuruta pīlukān* II 38 II

[27] *vibhajāma tataḥ sarvā vivādo me na rocate* I *peyam ānīyatāṃ kṣipraṃ mālyaṃ ca vividhaṃ bahu* II 39 II *tataḥ śūrpaṇakhā nāma rākṣasī vākyam abravīt* I *ajāmukhyā yad uktaṃ hi tad eva mama rocate* II 40 II *surā cānīyatāṃ kṣipraṃ sarvaśokavināśinī* I *mānuṣaṃ māṃsam āsādya nṛtyāmo 'tha nikumbhilām* II 41 II

The actual site of Nikumbhilā's worship is unclear – either in the garden of the *antaḥpuram* itself or at another site – but the *rākṣasīs* associated with her worship are located in the garden of the *antaḥpuram*, the site of sexual activity, and their activities are associated with this locus. That the rite of *rākṣasī* women is situated here serves multiple functions in the text, the most obvious of which is a narrative device through which the hapless Sītā can be terrorized into submission, but it can also be seen as a mechanism through which the *rākṣasa* world is once again aligned with a threatening, consumptive/sexualized, female, who also is understood in fact to be a goddess.

The identification of Nikumbhilā is problematic. The name is, to the best of my knowledge, not known outside of the *Rāmāyaṇa* story. The commentators, anticipating the events of the Yuddhakāṇḍa, understand Nikumbhilā here to be the name of or a manifestation of Bhadrakālī, worshiped at a shrine in Laṅkā. Govindarāja understands her shrine to be located at the western gate of the city (*nikumbhilā nāma laṅkāyāḥ paścimadvārapradeśavāsinī bhadrakālī*), while Nāgojibhaṭṭa and Śivasahāya understand her shrine to be located in the western section of Laṅkā (*paścimabhāgavartī*). The name Bhadrakālī, however, is unattested in the critical edition.[28]

While the introduction of the figure of the *rākṣasīs* into the text at this point in the epic serves a narrative function, the graphic descriptions allow an opportunity for the author to express a deep-seated anxiety concerning women and sexuality, reinforced by the demonic and gustatory proclivities of the women. But similar attitudes and depictions, as noted above, are seen elsewhere. What makes this passage unique is the interjec-

[28] However, the critical edition does know Kālarātrī (6.23.15, 34.15, 58.31). It is also of interest to note that several *rākṣasas* have names that include the word *kumbha*, 'pot', e.g. Kumbha, Nikumbha, and Kumbhakarṇa. (Kumbhahanu 6.46.15, 19; Kumbha 6.47.20, 62.37, and 6.63; Kumbhakarṇa *passim*; Nikumbha 6.62.37 and 6.64), although no connection is made in the text or by the commentators of any relationship.

tion into the narrative of the ritual to a mother-goddess figure and the locating of this figure in the demonic and feminine. Vālmīki's anxieties seem to extend to include a mother goddess cult and her worship.

The threat of the feminine, while very much of concern in the first five books of the epic, seems to fade in the Yuddhakāṇḍa. While the Sundarakāṇḍa focuses on the feminine, locating itself in the *antaḥpuram* of Rāvaṇa, the sixth book the Yuddhakāṇḍa locates itself solidly in the masculine. The very name of the *kāṇḍa*, the Yuddha, or 'Book of the Battle', leaves little doubt as to its dominant theme. But the text is complex, with struggles over technicalities of *nīti* and *dharma* held in court, in camp, and on the battlefield. Yet these struggles reflect the world of men, or rather of males. The use of the term *nikumbhilā* in the Yuddha reflects this change of locus and emphasis. The word is specifically linked to the site of the rituals of Indrajit. There is no goddess referred to, and, in fact, the term is used here exclusively for the shrine or grove in which the sacrifice of Indrajit is offered and, which is, it seems, located at the western gates of Laṅkā.[29] The identification of the location is probably based on the assignment of Indrajit to the western gate (6.28.11–12, 68.3) by his father Rāvaṇa and the fact that Indrajit resorts to the grove during battle. The actual source of Indrajit's powers, especially that of invisibility, is problematic and somewhat overdetermined in the Yuddhakāṇḍa, being said to derive from both his thrice-performed ritual and from a boon (6.60.18–25, 67.4–10, 69.23–24, 72.12).[30]

The ritual performances held at the Nikumbhilā shrine are at once familiar yet clearly belong to 'the other'. The first occurrence, and the more detailed of the rites, is described at 6.60.17–

---

[29] See notes to 6.60.18, 19 and 6.69.23, 71.13; 5.22.41.

[30] 6.36.10 and 7.30.10–13. Cf. text and/or notes to 6.13.5, 34.28–29, 35.15, 40.4, 47.15, 67.27, 72.12–13, 32.

28. Perhaps of special note for our discussion is that the text places the shrine of Nikumbhilā in the possession of the *rākṣasa* Indrajit, and locates it on or near the battlefield.

> Addressed in this fashion by the *rākṣasa* lord and having received his fulsome blessings, the hero proceeded swiftly to the shrine of Nikumbhilā in his horse-drawn chariot.
>
> Upon reaching the battleground, that immensely powerful tamer of his foes stationed the *rākṣasas* all around his chariot. (6.60.18–19)
>
> *tathokto rākṣasendreṇa pratigṛhya mahāśiṣaḥ* ।
> *rathenāśvayujā vīraḥ śīghraṃ gatvā nikumbhilām* ॥ 18 ॥
> *sa saṃprāpya mahātejā yuddhabhūmim ariṃdamaḥ* ।
> *sthāpayāmāsa rakṣāṃsi rathaṃ prati samantataḥ* ॥ 19 ॥[31]

What follows is a somewhat detailed description of a sacrifice, but with a number of subtle inversions, indicating the '*abhicāra*' nature of the ritual.

> Then to the accompaniment of sacred *mantras* and in accordance with the ritual prescriptions, the foremost of *rākṣasas*, whose splendor was like that of Agni, god of fire, eater of oblations, offered oblations to Agni, god of fire, eater of oblations.
>
> The valorous *rākṣasa* lord offered oblations to Agni, the purifier, god of fire, along with ritual offerings of parched grain accompanied by flowers and sandalwood paste. (6.60.20–21)
>
> *tatas tu hutabhoktāraṃ hutabhuksadṛśaprabhaḥ* ।
> *juhuve rākṣasaśreṣṭho mantravad vidhivat tadā* ॥ 20 ॥
> *sa havir lājasaṃskārair mālyagandhapuraskṛtaiḥ* ।
> *juhuve pāvakaṃ tatra rākṣasendraḥ pratāpavān* ॥ 21 ॥

Of note here is that the ritual, carried out by a male, replicates, although not perfectly, a brahmanical one. The sacrifice is not to Nikumbhilā, as it was in the Sundarakāṇḍa, but to the Vedic god Agni. It has the familiar requisites of the Vedic sacrifice, such as parched grain, etc., and it, much in the same vein as the Vedic sacrifice, seeks specific gains for the sacrificer. The change in tenor of the sacrifice comes in verses 22 and 23.

---

[31] Note that these verses should probably be considered suspect (see note to 6.60.16).

Weapons served as the *śarapatra* grass, myrobalan wood[32] was the kindling. His garments were red and his ladle of black iron.

Having strewn the fire altar there with weapons, including iron cudgels, in place of *śarapatra* grass, he seized the throat of a live pure black goat. (6.60.21–22)

śastrāṇi śarapatrāṇi samidho 'tha vibhītakāḥ I
lohitāni ca vāsāṃsi sruvaṃ kārṣṇāyasaṃ tathā II 21 II
sa tatrāgniṃ samāstīrya śarapatraiḥ satomaraiḥ I
chāgasya sarvakṛṣṇasya galaṃ jagrāha jīvataḥ II 22 II

While the interpretation of the passage is somewhat problematic, what is important for the discussion at hand is the striking familiarity of the ritual and yet the introduction of elements that disturb that familiarity. The translation follows the interpretation of the commentators in understanding that the first and third articles mentioned in verse 22, i.e., the weapons and myrobalan wood, serve in the roles of the second, the *śarapatra* grass, and the fourth, the kindling, which are regular elements of any sacrificial performance. The commentators make this judgment on the grounds that these substitutions are mandated in the case of black magical (*abhicāra*) or violent (*raudra* – Maheśvaratīrtha) rites (*atra tv abhicāraviśeṣe tatsthāne śastrāṇi kṛtānīty bodhyam* – so Govindarāja). One should note that the *Mānavaśrautasūtra* 5.1.9.17–18 prescribes the use of arrows instead of kindling sticks, *samidh*, in a ritual designed to gain the upper hand in a battle.[33]

According to Nāgojibhaṭṭa, Śivasahāya, Govindarāja, and Maheśvaratīrtha, *śarapatra* grass is a type of *kuśa* grass. This grass (*Saccharum spontaneum*) is generally used for mats and roofs, and is personified together with *kuśa* grass (*Poa cynosu-*

---

[32] Myrobalan wood (*vibhītakāḥ*) is the *Terminalia bellerica*, one of the Myrobalans. The tree is known from the 10th book of the *Ṛgveda* (ṚV 10.34.1), as its nuts are said to be used as dice for gambling.

[33] All translations consulted take the term, mistakenly, we believe, in its most common sense of, 'fire'. Surely it is the altar that is being spread with weapons, and not the fire itself. Compare 1.13.22 for Vālmīki's use of *agni* in this sense and see Apte (s.v. *agni*). See 6.60.22 and note.

roides) as one of Yama's attendants. Typically the sacrificial grass is spread over a stretch of ground to serve as the arena for the sacrifice. This very common tall grass is used among other rites specifically in rituals for the deceased.[34] The garments of a Vedic practitioner would normally be white. The commentators note that red garments (*lohitāni ... vāsāṃsi*) are probably meant to be associated here with black magic.[35] On the other hand, Govindarāja in his comments to 6.67.5 notes that the Vedas specify red turbans for *ṛtvik* priests. The ladle made of black iron (*sruvaṃ kārṣṇāyasam*) is considered inauspicious. For example, at 1.57.9, Triśaṅku's royal ornaments turn to iron when he is cursed to become a *cāṇḍāla*.[36] Normally, of course, the ladle for the Vedic sacrifice would be made of acacia or cutch wood (*Acacia catechu*). The pure black goat (*chāgasya sarvakṛṣṇasya galaṃ jagrāha*), as the commentators explain, is to be used by Indrajit as the sacrificial offering (*homārtham iti śeṣaḥ* – so Tilaka). The significance of a pure black goat is not made clear by the commentators, however a female black goat is used in the crematory ritual[37] and the color black is associated with things that are inauspicious as we see, for example, at 5.25.18, where in Trijaṭā's prophetic dream Rāvaṇa is said to be wearing black garments.[38]

Indrajit performs a similar sacrifice again at 6.67.4–10, where many of the verses repeat in part or in full those found in *sarga*

---

[34] See *Baudhāyanapitṛmedhasūtra* 1.14.9 as quoted in the *Śrautakośa* 1.2, p. 1096. Monier-Willliams (s.v.) identifies *śarapatra* only as *Tectona grandis* (teak), clearly an unsuitable meaning in this context, where instead of *śarapatra* grass, weapons are used; the red garments and black iron are used to indicate an association with the *abhicāra* ritual. See notes to 6.60.22.

[35] See 5.25.20 where red garments are tokens of ill omens. See 6.59.21, 58.45–46, 67.5 and notes. See, too, *Agnipurāṇa* 229.14.

[36] See, too, *Agnipurāṇa* 230.1–4.

[37] See *Āśvalāyanagṛhyasūtra* 4.2.7; see also *Śrautakośa* 1.2, p. 1071. The color black is associated with things that are inauspicious, see 1.57.9 and 5.25.18.

[38] See 1.57.9.

60. However, there are a few notable additions. At the onset of the sacrifice we are told:

> As Rāvaṇi began his oblation into the sacrificial fire, *rākṣasa* women, bear. 'g red turbans, came in haste to where he stood. (6.67.5)
>
> *juhvataś cāpi tatrāgniṃ raktoṣṇīṣadharāḥ striyaḥ* |
> *ājagmus tatra saṃbhrāntā rākṣasyo yatra rāvaṇiḥ* || 5 ||

The presence of the women at the pseudo-Vedic sacrifice is problematic for both commentators and translators. According to Nāgojibhaṭṭa and Kataka, these women are sacrificial attendants (*homaparicārikāḥ*). Govindarāja and Śivasahāya note that they bring the turbans for the officiating priests to wear (*ṛtvigdhāraṇārthaṃ raktoṣṇīṣāṇy ānayantya ity arthaḥ*). Govindarāja further substantiates this idea by quoting an untraced Vedic passage in which *ṛtvik* priests are described as wearing red turbans (*lohitoṣṇīṣā ṛtvijaḥ pracarantīti śruteḥ*).[39] A number of the existing translations understand that the women themselves are wearing the turbans, but this seems unlikely.[40] Nevertheless, the dissonance created between the women carrying (wearing?) red turbans and the sacrifice is intentional on the part of the composer, indicating to the audience once again that the sacrifice is not 'typical' and its goals are in keeping with those of the 'Other'. While alterity is clearly marked as demonic, the introduction of the *rākṣasīs* reminds us that it is in addition marked as feminine. The reintroduction of *rākṣasa* women into the text thus reinforces the connection among the demonic, the divine or sacred, and the feminine.

---

[39] Remember that at *sarga* 60 and here as well, Indrajit is said to be wearing red garments.

[40] Pagani thus, for example, renders, 'coiffées de turbans rouges', (1999: 1089). In her note to this passage (p. 1670) she indicates that the commentators understand the *rākṣasa* women to be carrying out the duties of the sacrificers ('Ces rākṣasī au turban rouge font, d'après les commentaires, office de sacrifiants'). This, however, as noted above, is not quite correct.

As if to strengthen the connection with the demonic, an additional verse is added to the passage, wherein the sacrifice is no longer to please only Agni, but:

> Once he had made his oblations into the sacrificial fire, and thus gratified the gods, *dānavas*, and *rākṣasas*, he mounted his excellent and radiant chariot, which could be rendered invisible. (6.67.10)
>
> *hutvāgniṃ tarpayitvātha devadānavarākṣasān* |
> *āruroha rathaśreṣṭham antardhānagataṃ śubham* || 10 ||

Otherwise, the sacrifice is virtually identical to that previously described. The weapons again serve as the *śarapatra* grass, the myrobalan wood as the kindling, in addition Indrajit's red garments, the iron ladle, and the black goat are all described virtually verbatim.

A third sacrifice is referred to in *sarga* 69, wherein Indrajit, after slaying an illusory Sītā, returns once again to the Nikumbhilā shrine and offers oblations.[41] The description here is brief, only three verses, but they are telling, and the text leaves little doubt that the ritual being performed belongs to the demonic. The sacrifices earlier both made reference to the sacrifice of an animal, a black goat, and although the significance of this particular animal is not made clear in the text, sacrifice of similar animals is not unknown in the Vedic tradition, and certainly Vālmīki is familiar with animal sacrifices.[42] But here, we are told:

---

[41] 6.69.23. 'Now when Indrajit saw that Hanumān was heading back to where Rāghava waited, he retired to the Nikumbhilā shrine, eager to make oblation in the sacred fire, purifier of all things.'

[42] According to the commentators, the reference is to the shrine, sanctuary, or sacred grove dedicated to the goddess Nikumbhilā. Tilaka adds as an alternative that it may refer to a [sacred] banyan tree (*nikumbhilāṃ caityā tadākhyādevālayā vaṭavṛkṣā vā*). See 5.22.41 and note and 6.60.18, 72.13 and notes. According to several commentators, Nikumbhilā is said to be the name of a manifestation of Bhadrakālī, worshiped at a shrine in western Laṅkā (*daśamyāṃ pūrvāhne nikumbhilāgamanam*). Tilaka understands that this visit of Indrajit to Nikumbhilā's shrine takes place on the morning of the tenth day of the battle.

> The sacred fire, purifier of all things, blazed up fiercely, as, in accordance with the sacrificial injunctions, the *rākṣasa* poured oblations of blood into it there on the sacrificial ground.
>
> Glutted with the oblations of blood and swathed in flames, that fierce fire blazed up like Āditya, the sun god, at twilight. (6.69.24–25)
>
> *yajñabhūmyāṃ tu vidhivat pāvakas tena rakṣasā* |
> *hūyamānaḥ prajajvāla homaśoṇitabhuk tadā* || 24 ||[43]
> *so 'rciḥpinaddho dadṛśe homaśoṇitatarpitaḥ* |
> *saṃdhyāgata ivādityaḥ sa tīvrāgniḥ samutthitaḥ* || 25 ||

Unlike the earlier two, these verses make clear reference to the offering of blood into the sacrificial fire, an action associated with *rākṣasas* and with the polluting of the Vedic sacrifice.[44] The offering of blood and flesh, as well as its consumption, was also associated with worship of Nikumbhilā in the Sundarakāṇḍa.

At *sarga* 71 we are told of Indrajit's intention to perform yet another sacrifice this time to make himself unassailable in battle (verses 13–14). Here the name *nikumbhilā* is used in apposition to the term *caitya*, 'shrine',[45] clearly denoting the term as a referent to the locus rather than the object of worship although these need not be mutually exclusive. In the following two *sargas* we are informed by Vibhīṣaṇa that the only way to defeat Indrajit is to interrupt his sacrifice (6.71.14, 72.10–14, 74.2–6) before it is completed. This is done, and at 6.73.13–14, Indrajit emerges from the grove with his sacrifice incomplete. The final reference in the critical edition of the Yuddhakāṇḍa to the grove of Nikumbhilā is found at the opening of *sarga* 74. Here, however, the name Nikumbhilā is not used; rather it is called simply a *mahad vanam*, 'vast grove'.

---

[43] The translations follows Govindarāja in reading the compound *homaśoṇita-* as *paranipāta* for *śoṇitahoma-*, 'oblations of blood'. A number of manuscripts and the texts and printed editions of Govindarāja and Maheśvaratīrtha read *māṃsa-*, 'flesh', for *homa-*, 'oblation', yielding the sense, 'eater of flesh and blood'.

[44] See for example 1.18.5, 29.10–21, where Mārīca and Subāhu pollute the sacrifice of Viśvāmitra with blood.

[45] Compare 5.22.41; 6.60.18, 69.23 and notes.

> Proceeding for a short distance, Vibhīṣaṇa entered a vast grove and there pointed out to Lakṣmaṇa the sacrificial ground. (6.74.2)
>
> *avidūraṃ tato gatvā praviśya ca mahad vanam* |
> *darśayāmāsa tat karma lakṣmaṇāya vibhīṣaṇaḥ* || 2 ||

Here, following the commentators once more, the phrase, *tat karma* is interpreted as a reference to the site of the sacrifice, rather than the sacrifice itself, since, at this point the sacrifice has not commenced (*homakarmasthānam* – so Tilaka). The shrine is specifically described as a fearsome looking banyan tree (*nyagrodhaṃ bhīmadarśanam*, 6.74.3)[46] that resembles a black storm cloud (*nīlajīmūtasaṃkāśaṃ*). The connection with the demonic is again made very clear here, in that the oblations are no longer said to be offered to Agni alone, but,

> It is only after offering oblations to malignant spirits at this spot that the mighty son of Rāvaṇa goes forth to battle. (6.74.4)
>
> *ihopahāraṃ bhūtānāṃ balavān rāvaṇātmajaḥ* |
> *upahṛtya tataḥ paścāt saṃgrāmam abhivartate* || 4 ||

The term *bhūtānām*, is rendered here as, 'malignant spirits', a meaning that Vālmīki employs elsewhere. Moreover, in the context of Indrajit's *abhicāra* ritual, it appears that the term must refer to ghosts and/or associated malignant spirits. Nāgojibhaṭṭa and Kataka understand the reference to be to the hosts or class of beings or spirits in whom the passionate and turgid qualities, that is to say the *rajas* and *tamas guṇas*, predominate (*rajastamaḥpradhānabhūtaguṇānām*).[47]

---

[46] There Rāvaṇa's powerful brother pointed out to Lakṣmaṇa a fearsome-looking banyan tree that resembled a black storm cloud. (*nīlajīmūtasākāśaṃ nyagrodhaṃ bhīmadarśanam* | *tejasvī rāvaṇabhrātā lakṣmaṇāya nyavedayat* || 3 ||) Śivasahāya understands the reference to be Nikumbhilā's banyan tree (*ficus indicus*) (*nikumbhilāvaṭam*). See notes to 6.72.13, 69.23.

[47] Compare 6.4.14, 47.9, 59.12 and notes. The Uttarakāṇḍa uses the word *nikumbhilā*, once only, at 7.25.2, where Daśagrīva is said to have gone into 'that vast grove in Laṅkā, named Nikumbhilā' (*tato nikumbhilā nāma laṅkāyāḥ kānanaṃ mahat ... praviveśa*). There he sees his son, Indrajit, engaged in a vow of silence and Uśanas Kāvya, the *purohita* of the demons,

That the Yuddhakāṇḍa understands the term *nikumbhilā* to refer to a site, rather than a divinity, given the masculine focus of the *kāṇḍa*, with its pseudo-Vedic sacrifice, its martial exploits, and focus on male relationships, can be understood as contextually apposite and not necessarily reflective of an inconsistency on the part of the composer. Rather the use of the term in the Yuddhakāṇḍa may be seen instead as a clever strategy that allows the feminine to infiltrate its masculine world. That the feminine word *nikumbhilā* is clearly identified earlier in the text as a goddess, marks the field, the shrine, and/or the grove associated with the name as belonging to her. In addition to the word *nikumbhilā* itself, the interjection of the *rākṣasīs* into Indrajit's sacrifice, the pouring of blood into the Vedic sacrifice, and the use of implements that are reflective of *abhicāra* practices, all bring to mind the rite associated with the goddess Nikumbhilā in the Sundarakāṇḍa and thus situate the Vedic sacrifice within the demonic and feminine locus of this goddess. Moreover the interruption and destruction of Indrajit's sacrifice suggests the impending defeat not just of the *rākṣasas*, but of the goddess, and of the feminine world, with its manifold dangers, that she represents. Thus juxtaposition of the exotic marked by the demonic and feminine with the familiar represented by the

---

speaks for him. We are told that Indrajit Rāvaṇi is performing not just one sacrifice, but six, all of which are Vedic sacrifices: The *agniṣṭoma*, the *aśvamedha*, the *bahusuvarṇaka*, the *rājasūya*, the *gomedha*, and a sacrifice propitiating Viṣṇu (*vaiṣṇavas tathā*) (7.52.5–8) The purpose of this elaborate ritual undertaking is, to receive boons from Śiva. Specifically we are told that he is to receive: a heavenly car, and the Tāmasī Māyā that would allow him to be invisible to his enemies, a power brought about by creating a darkness that surrounds the enemy rendering the possessor of the spell's movements undetectable (10–11). In addition he is said to receive inexhaustible quivers, and a bow (12).

The passage is intended to provide a rationale for Indrajit's possession of the boon of invisibility as well as his prowess in battle, evidenced earlier in the epic. The grove is once again the site of a Vedic sacrifice; but now it has been clearly associated with not the mother goddess but with Śiva and to some extent Viṣṇu.

pseudo-Vedic sacrifice of Indrajit, can be seen as intentional, a mechanism through which Vālmīki continues to articulate attitudes about and discomfort with the goddess, represented here as Nikumbhilā, – a figure that he envisions as no less a threat than that represented by the demonic world – along with the rites associated with her as well as the feminine world in general.

REFERENCES

Editions of the *Vālmīkirāmāyaṇa*

Rāmāyan of Vālmīki (1914–20). With three commentaries called *Tilaka*, *Shiromani*, and *Bhooshana*. Ed. by Shastri Shrinivas Katti Mudholkar. 7 vols. Bombay: Gujarati Printing Press. (GPP)

The Rāmāyaṇa of Vālmīki (1930). With the commentary (*Tilaka*) of Rāma. Ed. by Wāsudeva Laxmaṇ Śāstrī Paṇaśīkar. 4th rev. edition. Bombay: Nirṇayasāgar Press. (NSP)

Śrīmad Vālmīki-Rāmāyaṇa (1969). 3 vols. Gorakhpur: Gita Press. (Gita Press)

Śrimadvālmīkirāmāyaṇam (1911–13). [Also called the Kumbakonam Edition.] Ed. by T. R. Krishnacharya & T. R. Vyasacharya. 7 vols. Bombay: Nirṇayasāgar Press [reprint 1930]. (KK)

Śrīmadvālmīkirāmāyaṇam (1935). With the commentaries of Govindarāja, Rāmānuja, and Maheśvaratīrtha and the commentary known as *Taniślokī*. Ed. by Gaṇgāviṣṇu Śrīkṛṣṇadāsa. 3 vols. Bombay: Lakṣmīveṅkaṭeśvara Mudraṇālaya. (VSP)

The Vālmīki-Rāmāyaṇa. Critical Edition (1960–75). General editors G. H. Bhatt & U. P. Shah. 7 vols. Baroda: Oriental Institute.

General References

Agnipurāṇa. Agnipurāṇam. (Ānandāśrama Sanskrit Series, 41.) Poona: Ānandāśrama Press, 1900 (reprint 1957).

APTE, V. S. 1957–59. *The Practical Sanskrit-English Dictionary*. 3 vols. Poona: Prasad Prakashan.

Āśvalāyanagṛhyasūtra. Āśvalāyanagṛhyasūtram. (Ānandāśrama Sanskrit Series, 105.) Poona: Ānandāśrama Press, 1938.

Baudhāyanadharmasūtra (Electronic Text). Online version based on E. Hultzsch's [1922] and Pandeya's [The Kashi Sanskrit Series

1972] versions. Göttingen Register of Electronic Texts in Indian Languages (GRETIL), Archive of E-Texts in Unicode (UTF–8), http://www.sub.uni-goettingen.de/ebene_1/fiindolo/gret_utf.htm#DhSutra.

GOLDMAN, Robert P. (ed. & trans.) 1984. *The Rāmāyaṇa of Vālmīki. An Epic of Ancient India*, vol. 1: *Bālakāṇḍa*. Introduction by Robert P. Goldman. Annotation by Robert P. Goldman and Sally J. Sutherland. Princeton: Princeton University Press.

—— 2000. Rāvaṇa's kitchen: a testimony of desire and the Other. In: Paula Richman (ed.), *Questioning Ramayanas. A South Asian Tradition*: 105–116, 374–376. Delhi: Oxford University Press – Berkeley: University of California Press.

GOLDMAN, Robert P. & Sally J. Sutherland GOLDMAN (eds. & trans.) 1996. *The Rāmāyaṇa of Vālmīki. An Epic of Ancient India*, vol. 5: *Sundarakāṇḍa*. Introduction and annotation by Robert P. Goldman and Sally J. Sutherland Goldman. Princeton: Princeton University Press.

GOLDMAN Robert P., Sally J. Sutherland GOLDMAN & Barend A. van NOOTEN (eds. & trans.) 2008. *The Rāmāyaṇa of Vālmīki. An Epic of Ancient India*, vol. 6: *Yuddhakāṇḍa*. Introduction by Robert P. Goldman and Sally J. Sutherland Goldman. Princeton: Princeton University Press.

GOLDMAN, Sally J. Sutherland 1996. Soul food: eating, conception, and gender in the literature of pre-modern India. Paper delivered at Annual Meeting of the Association of Asian Studies, Honolulu, Hawaii, April 1996. (A revised version was presented at the Annual Conference on South Asia. Madison, Wisconsin, October 1996.)

—— 2000. The voice of Sītā in Vālmīki's *Sundarakāṇḍa*. In: Paula Richman (ed.), *Questioning Ramayanas. A South Asian Tradition*: 223–238. Berkeley: University of California Press.

—— 2003. Re-siting Sītā: gender and narrative in *Vālmīki's Sundarakāṇḍa*. *Purāṇa* 45.2: 115–135.

—— 2004. Gendered narratives: gender, space, and narrative structures in Vālmīki's *Bālakāṇḍa*. In: Mandakranta Bose (ed.), *The Ramayana Revisited*: 47–85. New York: Oxford University Press.

—— 2006. Illusory evidence: *māyā* in the Yuddhakāṇḍa of Vālmīki's *Rāmāyaṇa*. Paper presented at the annual meeting of the American Oriental Society, Seattle, Washington, March 2006.

INGALLS, D. H. H. 1965. *An Anthology of Sanskrit Court Poetry. Vidyākara's Subhāṣita-ratnakoṣa* (Harvard Oriental Series, 44). Cambridge, Mass.: Harvard University Press.

Mahābhārata, Critical Edition. Ed. by V. S. Sukthankar, S. K. Belvalkar, P. L. Vaidya *et al.* 24 vols. Poona. Bhandarkar Oriental Research Institute, 1933–71.

Manusmṛti. Manusmṛti with the commentary *Manvarthamuktāvali* of Kullūka. Ed. by N. R. Archarya. 10th edition. Bombay: Nirṇayasāgar Press, 1946.

PAGANI, Brigitte 1999. *Le Rāmāyana de Vālmīki*, Canto 6: *La Guerre*. Ed. by Madeleine Biardeau. Paris: Gallimard.

POLLOCK, Sheldon (trans.) 1986. *The Rāmāyaṇa of Vālmīki. An Epic of Ancient India*, vol. 2: *Ayodhyākāṇḍa*. Ed. by Robert P. Goldman. Introduction and annotation by Sheldon Pollock. Princeton: Princeton University Press.

——— (trans.) 1991. *The Rāmāyaṇa of Vālmīki. An Epic of Ancient India*, vol. 3: *Araṇyakāṇḍa*. Ed. by Robert P. Goldman. Introduction and annotation by Sheldon Pollock. Princeton: Princeton University Press.

Ṛgvedasamhitā. Ṛgveda-Samhitā with the commentary of Sāyaṇāchārya. 5 vols. Poona: Vaidic Samshodhan Mandal, 1933.

SEELY, C. 2004. *The Slaying of Meghanada – A Ramayana from Colonial Bengal*. New York: Oxford University Press.

Śrautakośa. Śrautakośa: Encyclopedia of Vedic Ritual. Comprising the two complementary sections, namely, the Sanskrit section and the English section. 2 vols. Poona: Vaidika Saṃśodhana Maṇḍala, 1958.

SUTHERLAND, Sally. J. (Goldman) 1989. Draupadi and Sītā: aggressive behavior and female role-models. *Journal of the American Oriental Society* 109.1: 63–67.

——— 1992. Seduction and counter-seduction: bedroom politics and the Indian Epics. *Journal of Indian Philosophy* 20: 243–251.

A Comparative Study of Polity in the *Nītiprakāśikā* and the *Rāmāyaṇa* (Bālakāṇḍa and Ayodhyākāṇḍa)

URMI S. SHAH

The word *rājanīti* is made of two terms: *rājan* derived from the root √*rāj* 'to shine' or √*rañj* 'to gladden', therefore denoting the king as he is the radiant one who gladdens his subjects; and the word *nīti* giving the sense of *nīyante saṃlabhyante upāyādayaḥ aihikāmuṣmikārthā vāsyāmānayā vā* (*Śabdakalpadruma*, part 4), that is, means or techniques to obtain/fulfil the goals of the material world leading to happiness/success here and in the life after.

Thus this compounded word refers to the science of polity by which the king achieves the welfare of the state through prudent, judicious and discreet use of the means on hand, and thereby following his *dharma*/duty achieves gain even in the next world. In other words, *rājanīti* denotes royal policy, statesmanship or polity, which is the *dharma* of the king as the leader of the state. This implies that the king as the leader of the state is supposed to have the fourfold knowledge of philosophy (*ānvīkṣikī*) in order to have the knowledge of direct perception and spiritual authority, Vedas (*trayī*), the science of economics (*vārtā*) and polity and military science (*daṇḍanīti*). The word *arthaśāstra* fundamentally denotes how peaceful acquisition and maintenance of material wealth became an intrinsic part of polity; while *daṇḍanīti* primarily referring to chastisement, secondarily as army, symbolizes the coercive authority and power of the state/ king which in turn presupposes the function of the state to maintain law and order. Both together ensure social and political stability accompanied chiefly by economic growth by which the state prospers. Thus

> *Daṇḍanīti*, which comprises the principles relating to the coercive authority of the temporal ruler, seems to denote the art of government, rather than the science of Polity (Dandekar 1998: 336).

Therefore *daṇḍanīti* also means *rājanīti*. Thus *nīti* / *rājanīti* / *daṇḍanīti* come to mean – a) politics and state administration, and b) military organization and operation. Towards the end of the *Nītiprakāśikā* (NP), Vaiśaṃpāyana calls this instruction *nītiśāstra* (NP 8.99).

Kauṭilya's *Arthaśāstra* has been the most authoritative treatise on which the successive thought of Indian polity rests. However, there are some treatises on polity which are not much known. Vaiśaṃpāyana's *Nītiprakāśikā* is one such text. It may be noted that hardly any reference has been made to its date (it is once referred to as a work on *dhanurveda*) but Gustav Oppert takes the *Nītiprakāśikā* to be contemporary with the *Mahābhārata* in the introduction to his critical edition. It is commented upon by Sītārāma in the *Tattvavivṛti* (TV), which Oppert assigns to approximately the 16th century AD. In my further study of the texts, which has just commenced, I hope to find some conclusive data about the dating of both.

The *Nītiprakāśikā*, as the title suggests, is 'the expounder of / the book throwing light on polity' which is to be followed by the king as the head of the state. However, the first and the last of its eight *sargas* refer to *rājanīti* / *rājadharma*, while the other six (that is, cantos two to seven) discuss the military science (*dhanurveda*) as a part of *rājanīti* at length. When Vaiśaṃpāyana went to Takṣaśilā to meet Janamejaya, the king informed him about his awareness of the *rājadharmas* as already propounded in the *Mahābhārata* (NP 1.4–9). However he had a doubt about carrying them out as the world was then being gripped by the age of Kali, there was a degradation of *dharma*, emphasis on materialism and predominance of the baser instincts (NP 1.10–17). In fact kings themselves behaved like *dasyus* (NP 1.17d). In this situation Janamejaya desired to know the *rājanīti* in detail to control the society and the state along with the knowledge of *dhanurveda* (NP 1.18). Vaiśaṃpāyana assured him that he had come there to impart to him *rājadharma*. Thus the *Nītiprakāśikā* is primarily an instruction on *rājanīti* by Vaiśaṃpāyana to the King Janamejaya.

The sage was the last teacher of the original tradition on the subject and was imparting the most abridged edition of the *rāja-śāstra* (NP 1.21–28). Its first exponent Brahmā gave the original, unabridged science to Pṛthu Vena; he protected his subjects in accordance with *dharma* and protected them by overpowering death (NP 1.30). Hence his subjects were happy and held Pṛthu in high esteem (NP 1.31–35). Therefore Brahmā revealed himself to give him a boon (NP 1.36). Pṛthu requested him to impart the knowledge of *dhanurveda* with all its secrets and sections. Hence Brahmā proceeded to narrate the significance of *dhanurveda* as *daṇḍanīti* being a part of *rājanīti*. In this context it is said that the purpose of *dhanurveda* was to enable the king to look after the subjects as if they were his own children – *sarvāḥ pālaya dharmeṇa prajāḥ putrān iva-aurasān* (NP 1.48).

This clearly indicates that following *dharma*, that is, taking proper care of the subjects was the duty of the king and that was *rājadharma*. The commentary *Tattvavivṛti* elucidates that Brahmā visualized that the person who deserved to receive the knowledge of *dhanurveda* was the deserving recipient of polity as well (NP 1.47). Therefore the exposition of the different aspects of *rājanīti*/*rājadharma* in this text through the *Nītiprakāśikā* fundamentally aims at expounding *dhanurveda*. This proves that both *rājanīti* and *daṇḍanīti* are synonymous terms.

It must be noted that the *Tattvavivṛti* on *Nītiprakāśikā* 1.48–49 has an elaborate commentary of Sītārāma along with quotations from the *Rāmāyaṇa*, Kāmandaka, Nārada and elsewhere; but it is apparently surprising to find that he never refers to the *Mahābhārata*. The reason behind this could be found in the reference made in Janamejaya's query that though Vaiśampāyana had elaborated the *rājadharma*s in the *Mahābhārata* he needed to be instructed again due to the influence of the age of Kali.

Oppert edited the *Nītiprakāśikā* (it being the first edition) in 1882. He noted that it was invaluable from the point of view of the information on *dhanurveda* but simultaneously on *rājanīti* too. While editing the *Nītiprakāśikā*, Oppert observes in the introduction,

> In its contents, the *Nītiprakāśikā* coincides in many points with the *Rāmāyaṇa*, especially with its first two books (Oppert 1882: 1),

although he admits that the reader is often reminded of the *Mahābhārata* and the *Harivaṃśa* which are not mentioned at all by Sītārāma. This calls for the comparative study of the two texts on *rājanīti*.

Though the *Rāmāyaṇa* is a living tradition of India, its incidental references to *rājanīti*/*rājadharma* have not been of much significance in *Rāmāyaṇa* study. Traditionally the ideal state has always been termed as *Rāmarājya* and it has influenced the Indian psyche to a great extent. The *Rāmāyaṇa* itself refers to the *Rāmarājya* towards the end of the opening canto of the Bālakāṇḍa. It is interesting to note that when Mantharā approaches Kaikeyī to remind her to make her legitimate demands from Daśaratha as per his promise, the queen's first reaction is that of the confidence in the establishment of an ideal state under Rāma (Rām 2.8.8). Moreover, having the theme of the royal family, the entire *Rāmāyaṇa* has direct and indirect references on polity. Incidentally we may remember Sheldon Pollock's observation here regarding the influence *Rāmāyaṇa* had on the medieval India, namely,

> the tale Rām. deeply influenced the literary imagination of the people since very early times but it exerted this effect not before 12th century A.D. under very special historical circumstances of India. It was this period when its political reflections started being dwelt upon and *Rāmarājya* became the paradigm every king was to strive to realize (Pollock 1995: 141–142).

However, the present paper restricts its scope to the first two *kāṇḍa*s in view of Oppert's observation as related to the *Nītiprakāśikā*. Consequently the main episodes of relevance for the present paper in order of importance are two episodes that occur in the Ayodhyākāṇḍa and form the pivot of the theme from which springs the action of the entire theme. However the Bālakāṇḍa does have some stray references to *rājanīti* as the epic is fundamentally the theme of the powerful royal family of Ayodhyā and its key role in the then existing environment.

The main episodes are:

(1) Daśaratha's decision to consecrate Rāma as the crown prince and its repercussions from Kaikeyī

(2) Bharata's meeting with Rāma on Citrakūṭa.

This episode of the dialogue and discussion between Rāma, Bharata, Vasiṣṭha and Jābāli in the presence of the officials and citizens of Ayodhyā is not just illustrative of the love of the brothers and love for *dharma*, they contain lively discussions on *rājanīti* and *rājadharma* as well. Of this group of *sarga*s, the 94th *sarga* can be regarded as entirely containing *rājadharmopadeśa*, since the informal query of Rāma to Bharata focuses on the proper administration in accordance with the tradition of the solar dynasty. The characteristic feature of this *sarga* is that each question Rāma asks Bharata begins with the interrogative particle *kaccit* which means 'I hope' (translated as 'I trust' by

Pollock 1986). Hence this *sarga* is often called *kaccitsarga*. P. V. Kane notes that

> The word 'Kaccit' is employed when the person asking the question desires to have a favourable reply, as the *Amarakośa* says 'Kaccit kāmapravedane' (Kane 1968: 400).

RĀJANĪTI IN THE *NĪTIPRAKĀŚIKĀ* AND *RĀMĀYAṆA*

Origin of the king

NP 8.3–7 gives the divine origin of the king almost exactly as given in *Manusmṛti* 7.5.7–9, 11 which says that the king was created to remove the state of anarchy that prevailed in the society.

The *Rāmāyaṇa* refers to the origin of the king by describing the kingless or pre-government state in the context of the main story and asserts that it was not acceptable. After Daśaratha's funeral the 'king-makers' (*rājakartāraḥ*, the leading and highly qualified brahmins like Mārkaṇḍeya, Maudgalya, Vāmadeva, Kaśyapa and others headed by the royal priest Vasiṣṭha) got together in the assembly (*sabhā*) along with the ministers (*amātyas*) to nominate the king and avert the evils of the kingless condition in the state (*arājakatva*) (Rām 2.61.1–3, 5–25). A later similar situation is described by Jābāli to convince Rāma to return to Ayodhyā and take up the reins of the state (2.100). In this context Ayodhyā is compared with a woman whose husband is away (*ekaveṇīdharā*, 2.100.8) and it is also said that the earth is a widow (*vidhavā*, 2.97.11).

Rājaguṇas or the virtues required by the king

These are specifically referred to in both the texts under discussion.

The *Nītiprakāśikā* says that the king must have a thorough knowledge of statesmanship and bad policy (*naya* and *apanaya*), so that the citizens of the state do not hesitate in revealing their prosperity before him (NP 8.8). The other virtues of the king who can be a *vijigīṣu* are referred to in connection with the external state policy – diplomatic relations with other kings, as the king whose coercive strength is sound can affect the equations of power within the *rājamaṇḍala*. The *Nītiprakāśikā* mentions that he must have qualities such as eloquence and the ability to convince others (*vaktā*), be in possession of logical and meaningful speech etc. (*Tattvavivṛti* on NP 1.49). He must have a thorough knowledge of and therefore the capacity of taking decisions of implementing peace, agreements and war (*sandhivigrahatattvajñaḥ*, NP 1.48a).

In the *Rāmāyaṇa* the qualities that are referred to are avoiding procrastination of duties and accomplishing tasks properly (Rām 2.94.14–15), extending due recognition to the learned men in the kingdom (2.94.17–18), nobility and purity, doing away with theft (bribe?), being above greed and thus being clean (2.94.47), honouring the elders, children, and the leading doctors with donations/grants/awards (money), warm heart and speech (2.94.51), paying proper and courteous respect to teachers, elders, ascetics, at temples, all those whose desires were already accomplished and the brahmins (2.94.52), being on the constant watch to avoid confusion between *dharma*, *artha* and *kāma* (2.94.53) and fulfilling each of them in his routine (2.94.54).

The king must not entertain atheist brahmins (*lokāyatika*) as they are shallow, clever in inspiring others towards wrong decisions, considering themselves to be scholars (2.94.32), shifting people's focus from the standards prescribed by the secular and moral laws evolved through ages (2.94.33).

As the king is a householder too, towards the end of the *kaccitsarga* Rāma reminded Bharata of his duty of not partaking of food alone but offering it to his friends who had good feelings for him (2.94.59).

Moreover, the king must remain in direct touch with the people and respond to them equally with love due to which Rāma was loved by all the *prakṛtis* (Rām 2.2.25–26, 3.8–9, 13, 28, 4.16.). Actually these were some of the most important qualities of Rāma which made the whole of the assembly, all the subjects including *mlecchas*, *āryas* and the foresters to unanimously support and endorse Daśaratha's proposal to designate him as the Crown Prince (2.2.14, 18 and 2.3.9). Further, the qualities of the king are referred to very specifically in relation to the virtues of Daśaratha (1.6.6–16) and Rāma (1.1.1–3, 12–18; 2.1.14–26, 31–32 and 2.2.18ff.; 2.3.9, 13; 2.16.46cd, 31.24, 35.17, 42.1 etc.).

As the prospective successor to the throne, Rāma was trained at the practical level in state and civil administration; he did this so as to benefit the subjects (2.1.12–13). Daśaratha sent him on military campaigns (2.1.24), also returning from which he would meet people personally, which further evoked their good will towards him (2.2.24). In other words, the epic emphasizes the king's direct contact with his subjects to be his most favourable quality.

It is worth noting that Sītā is called *abhijñā rājadharmāṇām* (2.23.4). It indicates that women/princesses were probably trained in this field or at least were well-conversant with the science of polity and the royal duties.

Prakṛtisampat **or** ***tīrthas*****:** **the elements constituting the state**

The *Nītiprakāśikā* does not mention the seven elements of the state; rather they are to be understood. However, Vaiśampāyana specifically says that the following eighteen *tīrthas* or heads of different departments of the state administration should be on his side, especially fifteen of them (*Tattvavivṛti* on NP 1.51). These are:

(a) *amātya* = ministers

(b) *purohita* = the chief officiating priest

(c) *yuvarāja* = the crown prince

(d) *bhūpati* = the tributary king
(e) *dvārapāla* = the main door-keeper (lord chamberlain)
(f) *antarvāsika* = commandant of the palace guards
(g) *kārāgārādhikārī* = warden of the prison
(h) *dravyasañcayakṛt* = chief treasurer
(i) *kṛtyākṛtyeṣv artheṣu viniyojaka* = chief of staff or secretary
(j) *pradeṣṭṛ* = commander-in-chief
(k) *nagarādhyakṣa* = the mayor of the city
(l) *kāryanirmāṇakṛt* = planning commissioner (?)
(m) *dharmādhyakṣa* = chief justice (who should be a brahmin according to NP 8.15)
(n) *sabhādhyakṣa* = president of the assembly (speaker?)
(o) *daṇḍapāla* = head magistrate or police commissioner
(p) *durgapāla* = commander of the forts
(q) *aṭavīpāla* = chief forest officer
(r) *anta* = governor of the frontiers.

All this *prakṛtisampat* should be constantly under the watch of the king through the espionage system, whether they are his own or those of the enemy. However the ministers, the crown prince and the priest of the king himself need not be watched which makes the number fifteen. By knowing their talk and behaviour the king should look after his subjects, provide the condition of fearlessness to the afflicted and also offer them wealth and other attractions. These acts would increase his people's love and dedication for him which would enable him to lead the state on the path of progress, *vṛddhi* (*Tattvavivṛti* on NP 1.51).

It is also said here that of these *prakṛtis*, the *mantrin* = advisers, *kośa* = treasury and *bhṛtyas* – ministers who serve him with sincerity and dedication constitute his threefold strength (*trividha bala*). The *Rāmāyaṇa* refers to these as a part of the story at appropriate places.

(a) The king – He was the head of the state and the fountain-head of all the state activities. He possesses all the elements of the state. The *Nītiprakāśikā* shows the king's qualities through Janamejaya and Pṛthu as noted above. However, polity of that time also shows that the king was not an absolute monarch. He could not take decisions regarding the state as he liked. To decide upon Rāma's consecration as the crown prince, Daśaratha called the council of ministers (*pariṣad*) to take its permission and consent (Rām 2.1.34). He placed the proposal before it and in view of Rāma's worthiness it was approved very easily. There is a mention of yet another incident besides the king being loved by his subjects and that being the sole merit, namely, that of the Prince Asamañja who was disowned by his father Sagara for torturing the children of the kingdom (Rām 1.37; 2.32).

(b) *Amātyas* – The ministers are called *sacivas* in the *Nītiprakāśikā* where it is said that they must be seven or eight in number, having specific qualities and must be experts in *nītiśāstra* (NP 8.18).

In the *Rāmāyaṇa* it is said that they must be appointed after properly checking their qualifications (Rām 2.94.21). (The commentary Tilaka explains the word *vicakṣaṇa* in 2.100.24 as *abhyastanītiśāstra*, that is, the minister must be an absolute expert in *rājanīti*. Similar is the meaning given by Mādhavayogī as *avadheyaviṣayaviseṣa*, that is, those who have grasped their subject well – in this case it is *rājanīti*.) They were the executive members of the cabinet (Dharma 1989: 65). Among them too due recognition was to be given to the most efficient ones (Rām 2.94.19). It is interesting to know that Rāma did not forget that the employees needed to be treated with personal feelings. He suggested to Bharata that he must take note of and proper care of the ministers in personal difficulties, as otherwise they were capable of bringing harm at the personal and state levels (2.94.49–50).

(c) *Bhṛtyas* – Assignment of proper duties to the ministers (*bhṛtyas*) was an important task. Rāma advised Bharata that they must be assigned appropriate duties by the king (Rām 2.94.20).

(d) *Mantrins* – Advisors/counsellors should have specific qualities. According to the *Nītiprakāśikā*, they must be two or three in number (NP 8.19). Meetings with them and their advice are so important that they form the root of victory for the kings, *vijayamūlam*. The same is said in the *Nītiprakāśikā* too. It proves their key role in advising the *ṣāḍguṇyanīti* of the king, his status in the *rājamaṇḍala* and maintenance of his *śakti* as well as *vṛddhi* as well as the state of emergency. They were probably the elders who helped the king in all kinds of decision-making as they had no personal interest in polity thus forming the advisory body.

(e) *Purohita* – The office of the *purohita* was very important. It was held by the highly spiritual sage Vasiṣṭha (Rām 2.94.7–8) who was *rājadharmavit* (2.75.8b). He was also the guru, the spiritual guide and the leading advisor to the king. Hearing Rāma's angry retort to Jābāli, Vasiṣṭha defended him and presented before Rāma the tradition of succession by primogeniture in the solar dynasty (2.102.2–31). Further in his capacity as the royal *purohita* Vasiṣṭha reminded Rāma that since in his capacity of being his preceptor (*ācārya*) he had given him the training to think rightly, he could not be overruled. Moreover he had the support of the members of the assembly (*pariṣad*), members of the governing bodies of the capital and other towns etc (*naigama*) and the members of the guilds (*śreṇī*) which could not be turned down. Further, it appears that when the state was kingless, the *purohita* led all the 'king-makers' (*rājakartāraḥ*) before the assembly (2.61.3) and had the power to solve the situation of interregnum, to summon the nominated king (here, Bharata) and then to offer him to assume the royal throne (2.62).

(f) *Pariṣad* – This was probably the assembly consisting of the cabinet of *amātyas*, *mantrins*, *rājakartāraḥ* or the learned and spiritual brahmins like Vāmadeva, Bharadvāja and others, *śreṇīs* or guilds of merchants and tradesmen, *naigamas* or the corporations governing the cities, *paurajānapadāḥ* or the people from both the urban and the rural areas, who would help the king in administration of the state as can be seen in Daśaratha seeking its approval to consecrate Rāma as the crown prince (2.2).

(g) *Vaidya* – The appointment of the doctor was very significant both for the times of peace and war. The *vaidya* had to be extremely loyal. Hence his qualifications were his expertise in medication, perfect knowledge and utilization of the four *upāyas* and *rājanīti*, close relations with the king, brave, and utmost loyalty to him so that he would never be tempted to kill him (2.94.23).

(h) *Kulaputrāḥ* – Great care was to be taken about the bodyguards of the king; they had to be hereditary so that they were loyal to him; when need arose they should be ready to lay their lives for him (2.94.28).

(i) *Upādhyāyas* – Teachers and priests belonged to different fields. To illustrate, the *kaccitsarga* refers to two:

– The religious preceptor (2.94.5)

– The teacher of the military science or *dhanurveda* (2.94.9) (while commenting in this *sarga* on 2.100.14, Tilaka notes that this *upādhyāya* named Sudhanvan was *arthaśāstraviśārada*, that is, *rājanītiśāstraṃ tatra viśāradaṃ paṇḍitam* = master of polity as also of *dhanurveda*).

(j) *Jānapada* – The citizens living outside the capital (*jānapadas*) were to be learned (literate), clever and full of understanding (2.94.29ab), as the *paurajānapadāḥ* together formed a major part of the assembly assisting the king to govern the state. They probably constituted the majority of the subjects and the *vaiśya* community contributing the largest share in the economic prosperity (Dharma 1989: 49–54). They were to be properly cared for (2.94.37–39), and the farmers and herdsmen of all types were to be carefully and lovingly protected so that they were happy (2.94.40). They were to be provided with protection and redemption from taxes to give them financial backing (2.94.41ab).

Here we have a very concrete statement of the chief duty of the king as the leader of the state, namely, protection and care of all the subjects according to *dharma* (*rakṣyā hi rājñā dharmeṇa sarve viṣayavāsinaḥ*, Rām 2.94.41cd).

The *Nītiprakāśikā* does not refer to this aspect from this point of view.

Further as a part of the subjects Rāma showed his concern about the women, as well as simultaneously cautioned Bharata regarding them. He suggested that women were to be provided emotional and physical comfort, as also protection (2.94.42ab); but they were not to be entirely trusted nor secrets be divulged to them (2.94.42cd). The *Nītiprakāśikā* does not refer to this aspect.

(k) *Daṇḍa* – The term refers to both justice/punishment and the army. In other words, coercive force for the maintenance of law and order are implied by this term.

The *Nītiprakāśikā* briefly refers to the former aspect as compared to the latter which is treated in great depth in *sargas* 2–7. It says that the king should carry out justice in a judicious manner along with the *amātyas*, the *purohita* and learned and wise brahmins (NP 8.22–23). Further, it shows the importance of *daṇḍa* as does Manu in 7.25 (NP 8.30).

It further describes and lays down the kinds of criminals and thieves and how they must be punished (NP 8.31–43).

This is not found in the Bālakāṇḍa and Ayodhyākāṇḍa.

Army and military administration/organization

Fourteen factors as mentioned in the *nītiśāstra* treatises were to be checked as: a) *deśa* = country, region, b) *durga* = forts, c) *ratha* = chariot, d) *hastin* = elephants, e) *vājin* = horses, f) *yodha* = warriors, g) *adhikāriṇaḥ* = officers in charge of different government departments, h) *antaḥpurānta* = harem and the interiors of the palace, i) *gaṇana* = the numbers of the animals, chariots etc. for use in war, j) *śāstra* = officer in charge of scriptures, k) *lekhya* = accounts section, l) *dhana* = finance and m) *āsava* = concealed enemies (*Tattvavivṛti* on NP 1.48–49). These fourteen needed to be constantly checked as they formed the strengths and weaknesses of the king. Of these the c–f refer to the structure of the army. Vaiśaṃpāyana says that the army constitutes of eight limbs and four sections; it should be led by well-trained and experienced commanders and military officers as the aim is to subdue the enemy army (NP 1.55).

The *Nītiprakāśikā* refers to the military campaign and its proper time especially with a view to conquer the enemy. Vaiśaṃpāyana notes that the king should check his threefold *bala* and speedily attack the enemy who is under the influence of the addictions/vices (*vyasana*) (NP 1.51). The *Tattvavivṛti* notes that these are ten in number: *strī* = women, *dyūta* = gambling, *mṛgayā* = hunting, *madya* = drinking liquor, *nṛtya* = dancing, *gīta* = music, *aṭana* = wandering, *vādya* = playing upon instruments (of music), *nindā* = being engaged in censure of people and *divāsvāpa* = sleeping during daytime.

Moreover, the *trividha bala*, that is, *mantrin* = counsellors, *kośa* = treasury and *bhṛtya* = ministers who serve him with sincerity and dedication also need to be regularly checked besides his threefold *śakti* = strength, namely, *prabhu* = control over all his elements of state, *mantra* = right counsel and *utsāha* = enthusiasm, courage for all undertakings for the state welfare. In this context the king ought to know thoroughly each of his *rājamaṇḍala* (*Tattvavivṛti* on NP 1.53). Vaiśaṃpāyana also refers here to the calamities (*vyasana*) of two kinds: natural (*daiva*) and those created by man (*mānuṣa*).

Thereafter he says that establishing himself firmly, that is, his own kingdom, he must attack the enemies, make efforts to defeat them and then keep them under control by giving them gifts etc. In any case they should not be allowed to become independent (NP 1.55).

After this is discussed *dhanurveda*, which forms the bulk of the *Nītiprakāśikā*. We find detailed narrations of various weapons, training, military organization and operations. The reason for this is that the *Nītiprakāśikā* aims at the control of law and order as the most vital part of the *rājanīti* in the age of Kali as noted earlier.

The *Rāmāyaṇa* too contains references to the army and military organization and operations. The entire superstructure of the foreign relations (*ṣāḍguṇyanīti*) depends on the strength of the army besides other *prakṛtis*. So Rāma indicates that the commander-in-chief must be bold, happy with the king, brave, patient, intelligent, pure, noble, faithful and resourceful (Rām 2.94.24). The other highly placed and chief officers in the army have to be physically strong, expert in the art of warfare, able to anticipate calamities and are adventurous. They must be honoured and felicitated (2.94.25).

The army must be given proper daily meals, monthly salary and share (in the booty captured after the victory in war), as also whatever was to be given (gift, bonus etc.) at the appropriate times (to keep them motivated and happy) (2.94.26); if the king tarried in this matter and the right time passes away, he would surely invite displeasure of the soldiers and others in the army which would amount to ushering a great calamity (2.94.27).

The king being in charge of the army must inspect whether the secret forests with elephants are properly maintained, and be on a constant guard against surprise enemy attack. Therefore as noted earlier in the daily routine of the king, he must always check the forts and fortifications. The king had to be a great and capable warrior skilled in weaponry and other aspects related to it like Viśvāmitra (1.20.11–14).

The army is referred to in various ways. Throughout both the *kāṇḍa*s we come across references to it such as: its expanse in the terms of *akṣauhiṇī* (1.19.3, 50.21), warrior tribes like Kāmboja, Pahlava and others which were warrior races (1.54.2–3), the important units of the four-sectioned army (*caturangiṇī senā*) made up of the infantry, elephants, cavalry and chariots (1.53.12, 54.4cd, 68.3, 6 and 73.9b), the king as the chief commander of the army (1.19.3; 2.76.22), the qualities of warriors as being brave/courageous, skilled in using all types of weapons – and missiles (1.19.4ab) and so on. We also find the description of the army in Ayodhyākāṇḍa (2.77.3–5, 7 and 2.76.24). The army, for any expedition, consisted of all kinds of manpower and equipment for it such as all types of artisans, geographers, engineers and others who construct roads etc. for the march (2.73.13, 74).

Usually Śiva is associated with *dhanurveda*; but there is a distinct variation, as in the *Nītiprakāśikā* where Brahmā passes it on to Pṛthu, and also another longer version given in the first canto under *rājanīti* (NP 1.21–28).

In the *Rāmāyaṇa* Viśvāmitra pleases Śiva with penance and obtains *dhanurveda* to fight against Vasiṣṭha to avenge his defeat at the hands of the army created by the *kāmadhenu* in the latter's hermitage (1.54.16–18).

Dharmayuddha or war according to the laid down rules/norms was the custom of the day. But communities like the *rākṣasa*s used to resort to the so-called 'foul' means (*kūṭayuddha*) to win the war (1.19.7). Such a reference comes even in the *Nītiprakāśikā* where adoption of similar tactics against those who use them are strongly recommended (NP 5.6–58).

The *Rāmāyaṇa* acknowledges the superiority of *brahmabala* (= power of Brahman / spirituality over that of *kṣatra* / physical and merely relying on the authority of power of kingship) (1.53.14, 55.4).

An interesting feature of the *Rāmāyaṇa* not found in the *Nītiprakāśikā* is that women seemed to be participating in war. Kaikeyī's incident is too well-known to be repeated here (2.9.9). She was Daśaratha's assistant in war against Śambara for which she was granted boons (1.1.19). Further when Bharata goes to meet Rāma at Citrakūṭa with his army, there is a reference to *yodhāṅganās* (2.76.23); this may refer to women warriors.

(l) *Pura* – This has the reference to the capital city. It is not found in the *Nītiprakāśikā*. But the *Rāmāyaṇa* has fond references to and descriptions of Ayodhyā, the capital of Kosala and the centre of all the political and economic activities. To illustrate, in the ***kaccitsarga*** Rāma asks Bharata whether he carefully protected this well-planned and beautiful city (1.5, 2.94.34–36).

(m) *Kośa* – This refers to the royal treasury. The *Nītiprakāśikā* specifies that the income of the state must be more than its expenditure (NP 8.28), whereas the ***kaccitsarga*** mentions it in the form of query by Rāma to Bharata. As an astute statesman, Rāma asked whether royal treasury (***kośa***) ever remained full, having lesser expenditure than income, wealth not going into the hands of the undeserving (2.94.45); on the other hand his indirect advice was that wealth could be given away (to an extent and for the benefit of the kingdom) to worship of gods, ancestors, maintaining the brahmins, welcoming the guests, and to warriors and friends (2.94.46).

(n) Civil administration – The *Nītiprakāśikā* refers to this at length with all the units of administration in a sequence, the smallest unit being a *grāma* (NP 8.49–54).

This is not found in the Bālakāṇḍa and Ayodhyākāṇḍa.

(o) Taxation – It is said in the *Nītiprakāśikā* that taxation should be fair (NP 8.56–57); this is not referred to specifically in the Bālakāṇḍa and Ayodhyākāṇḍa.

Rājamaṇḍala

This refers to the circle of the kings surrounding the *vijigīṣu* and his relations with them. This theory is not discussed in the *Nītiprakāśikā* but the references to the *vijigīṣu*, *pārṣṇi*, *ari*, *mitra* and so on show the assumption of the author that the reader is conversant with this Kauṭiliyan theory of the inter-state relations (NP 1.53).

Neither this theory nor such references are to be found in the Bālakāṇḍa and Ayodhyākāṇḍa. However, India during the *Rāmāyaṇa* period seemed to be having several states/kingdoms ruled by independent and vassal kings. Being a powerful kingdom, Kosala had good relations with all the kings. Daśaratha named the kings who were to be invited to the *putreṣṭiyāga* (Rām 1.12.17–24). Therefore although the kings of Ayodhyā were referred to as having conquered all enemies, it appears that they, especially Daśaratha had no enemies.

Ṣāḍguṇyanīti or the policy of using the six qualities/ factors/techniques in inter-state relations

The *Nītiprakāśikā* notes that the king must be an expert in the application of the six *guṇas*, namely, *sandhi*, *vigraha*, *yāna*, *āsana*, *dvaidhībhāva* and *āśraya* (NP 1.48).

Here the *Tattvavivṛti* notes that each of the two, *sandhi* and *vigraha*, are twofold:

> *tatra sandhivigrahau pratyekaṃ dvividhau* । *prabaladurbalayoḥ vijigīṣāvatoḥ samabhūmau yuddhaprasaṅgādyo vigraham icchati* । *itaras tu tena sahālpataragraheṇāpi sandhim evecchati* । *tayā ca durbaladvārakau sandhivigrahāv anyau* । *prabaladvārakāv anyau* । *yad idam uktam rāmāyaṇe 'dviyonī sandhivigrahau' iti* ॥

The *Rāmāyaṇa* quotation occurs at 2.2165* 5 post., read by many manuscripts within the *kaccitsarga*, 2.94.

Of the six *guṇas*, *sandhi* and *vigraha* are each twofold. When there is an occasion of war with the *vijigīṣu* the stronger enemy desires war; while the weaker enemy opts for *sandhi* – peace through negotiations, agreements etc. This is true for weaker and stronger enemies on both sides respectively (*ari* – *pārṣṇigrāha*).

Further, as far as the stronger enemy is concerned *vigraha* leads to *yāna* or *āsana*; while *sandhi* for the weaker enemy, and *dvaidhībhāva* for equals. Here the *Tattvavivṛti* quotes Kāmandaka: *sandhi* and *vigraha* are termed as *dvaiguṇya* (having two *guṇas* together). Then he explains *yāna* and *āsana* individually. Thereafter comes the explanation of *dvaidhībhāva* – march of the *vijigīṣu* towards the enemy is *yāna*, surrounding and forcing the enemy to stay within his fortress and thereby stopping the replenishment of his food and other supplies is *āsana*, while *dvaidhībhāva* is to wage war with a strong enemy and simultaneously surrendering to a stronger enemy merely by speech and not actually. This is supported by again quoting from Kāmandaka.

Thereafter the *Tattvavivṛti* explains *āśrayabhāva*. The king should serve the state, that is, take care of the state by assessing the enemies, creating goodwill with them by *sandhi* if it leads to his own *vṛddhi* (growth and progress). These are further referred to in the NP 8.80–86. Therefore the king should always resort to eightfold tasks: a) *kṛṣi* = agriculture, b) *vaṇik* = trade and commerce, c) *patha* = roadways, d) *durga* = fortifications, e) *setu* = building bridges, f) *kuñjarabandhana* = *bahubhakṣakatvāt teṣāṃ yathāvibhāgaṃ gramīṇadhanikaiḥ eva grāme grāme poṣaṇārthaṃ bandhanam* = allotting the elephants belonging to the army to different villages to maintain them so as to equally balance their cost, g) *khani* = mines from where gems are found, and h) *ākara* = mines from where precious metals like gold etc. can be found (*Tattvavivṛti* on NP 1.50).

These are not mentioned in the Bālakāṇḍa and Ayodhyākāṇḍa of the *Rāmāyaṇa*.

Upāyas

Upāyas are the means or techniques to know the strengths and weaknesses of oneself and the enemy. The *Nītiprakāśikā* says that the king must be able to judge perfectly the strengths and weaknesses of himself and of the enemy, whereby he must effectively use the policy of creating factions among the enemy through means such as *sāma*, *dāna*, *daṇḍa and bheda* – (conciliation, gift, sowing dissensions in the enemy camp and the use of force) the fourfold *upāyas* (*anumānavibhāgavit*) (NP 1.48b). Later, in the eighth *sarga* Vaiśaṃpāyana describes each of these at length (NP 8.73–79). They are seven in number – the four mentioned earlier and *mantra*, *auṣadhi* and *indrajāla* (negotiation, use of herbs and magic). All these have to be known through the espionage system before using the six *guṇas*.

These are again not referred to specifically in the Bālakāṇḍa and Ayodhyākāṇḍa.

Dūtas

The *Nītiprakāśikā* says that the *dūta* must be expert in all scriptures, clean and be able to read the enemy's mind (NP 8.21). However, in the *kaccitsarga* Rāma asks whether Bharata had appointed a *dūta* who is learned, diplomatic, truthful projecting things as they are and having impressing personality (2.94.29cd).

Cāra/cārakas

Spies have been of great importance in knowing the inner secrets of the administration as well as the inter-state relations. Much has been said about it elsewhere but in the *Nītiprakāśikā* there is no mention of it.

The *Rāmāyaṇa* depicts the espionage in its later *kāṇḍas*. However, in the *kaccitsarga* it is referred to from a theoretical point of view by Rāma. The spies were to find and report the ins and outs of all the eighteen heads of the departments (*tīrthas*) enumerated earlier (2.94.30). This helped the king to be vigilant over the sector of civil administration for effective control over the Elements – namely, ministers (*mantrin*), chief priest (*puro-*

hita), commander-in-chief (*senāpati*), crown prince (*yuvarāja*), lord chamberlain (*pratihārī*), palace manager (*antarvāsika*), head of the jails (*kārāgārādhikārī*), finance secretary (*arthasañcaya-kṛt*), chief secretary (*kāryaniyoktā*), judge (*prāḍvivāka*), army paymaster (*senādhināyaka*), civil paymaster (*karmāntika*), mayor (*nagarādhyakṣa*), chief justice (*dharmādhikṛt*), council secretary (*sabhya*), governor of the frontiers (*rāṣṭrāntapāla*), commander of the forts (*durgapāla*) and police commissioner (*daṇḍapāla*) (Rām 2.100.36; *Amṛtakataka* commentary of Śrī Mādhavayogī; Dharma 1989: 78).

Daily routine of the king

Several verses of the eighth chapter of the *Nītiprakāśikā* give a detailed narration of the daily routine of the king similar to the ones that we find in several of the treatises on *nīti*. These refer to his daily division of work including judiciary, counselling with his ministers, civil administrative tasks, constructing roadways, giving donations, and so on (NP 8.9–72). Here again is mentioned at length that *ṣāḍguṇya*, *kṣayasthānavṛddhi* and *śakti*s need to be discussed with the council of ministers (NP 8.80–91).

The *Rāmāyaṇa* refers to this in the *kaccitsarga*. There is reference to the daily routine of the king (2.94.12–13) such as daily checking the main roads of the city early in the morning (2.94.43), personally checking all the forts and fortifications of the kingdom so that they are always ready with their full stores of wealth, grains and food provisions, weapons, water resources, machineries (for war), artisans and archers (2.94.44) and giving away some wealth for worship of gods, ancestors, caring for the brahmins, welcoming the guests, warriors and friends (2.94.46).

Protocols

The *Nītiprakāśikā* refers to the protocol which the king follows in welcoming Vaiśaṃpāyana (NP 1.4–6) but not how the king is received officially.

The *Rāmāyaṇa* mentions this point more than once. Certain protocol was followed for the reception of the king formally and welcome etc. thereafter, on meeting another king on formal occasion like wedding etc. To illustrate, when on his way to meet Rāma, Bharata approached Bharadvaja's hermitage, he kept behind him his army and proceeded to meet the sage with the *purohita* (2.84.1–3). Formalities were exchanged and thereafter the sage offered hospitality to Bharata and his retinue overnight with meals. Here the entire arrangement was as per the laid down protocol (2.84.1–3, 85.37).

Succession

The point of the succession to the throne is not mentioned in the *Nītiprakāśikā*. But in the *Rāmāyaṇa* it forms the pivot on which the story evolved.

Succession to the throne was by the law of primogeniture or rather, it can be said that there was hereditary monarchy, that is, the eldest son was entitled to get the throne (1.1.19; 2.8.7, 73.7, 95.2, 102.29) – *jyeṣṭha* son was the *dāyāda*. The kingdom rightfully belonged to Rāma – *svakam* (2.102.29c) and *ikṣvākūṇāṃ hi sarveṣāṃ rājā bhavati pūrvajaḥ* I *pūrvaje nāvaraḥ putro jyeṣṭho rājye abhiṣicyate* II (2.102.30). Daśaratha had a soft spot for him (1.1.19) and moreover, he was born of the eldest queen (2.3.23). Hence the formal announcement and appointment of Rāma as the Crown Prince or as Sheldon Pollock calls it 'prince regency' (Pollock 1986: 12). However it had to undergo a formal procedure as described in Ayodhyākāṇḍa 2.2, namely, presentation of the proposal before the *pariṣad* and the *sabhā* for due support, proposal of alternative name/names, and then the approval of the members of the assembly. In the case of Rāma it was immediately and unanimously approved and endorsed. (Here it implies that the coronation of Rāma in this case would imply the retirement of the reigning king in near future. Moreover, this process would avoid the dispute of succession as well).

Another important aspect of the point arose in the case of Rāma, that is, when the eldest son's right was overlooked, the throne could be taken away in different ways suggested respectively by Lakṣmaṇa and Daśaratha – (1) to fight and kill the unjust king (2.20.23–32), (2) to imprison the ruling king who took such a decision (2.31.23) and (3) to win over the entire administration and subjects and found a new kingdom (2.31.23).

The *Rāmāyaṇa* mentions another tradition of selecting the crown prince which was probably geopolitical or as referred to later in the paper on the basis of the text itself. At the time of marriage with Kaikeyī, Daśaratha had promised her father to give the kingdom to her son; this is termed as *rājyaśulka* or the 'bride price' (2.99.3–8). Moreover she was granted a boon by him when she saved his life in the battle against Śambara (1.20). But Vālmīki solves this disastrous situation of fragmentation of the family and state by the code of submission to hierarchy.

Nature and aim of *rājadharma/rājanīti*

Both the *Nītiprakāśikā* and the *Rāmāyaṇa* project their views on this point. The very first verse of the eighth *sarga* of the *Nītiprakāśikā* lays down the purpose of *rājadharma*: *lokarañjanam eva* = satisfying the subjects, as is noted in the *Tattvavivṛti*: *janasantoṣakaraṇam.* Vaiśampāyana gives a realistic illustration here – just as a woman puts aside personal welfare and concentrates upon the welfare of her child that she is carrying, so is the king supposed to focus on the welfare of his subjects (NP 8.2).

In the *Rāmāyaṇa* Bharata poses a very significant interrogative question – how is *rājadharma* going to help him when he had deviated from his *dharma* by violating the tradition of the right to succession by primogeniture (2.95.1cd, 97.2).

This suggests an important point, namely, the king himself has to follow *dharma* for imposing it on his subjects; this is the fundamental concept of *rājadharma* / *rājanīti*. This is an eternal truth of *rājadharma*: *yadvṛttāḥ santi rājānas tadvṛttāḥ santi hi prajāḥ* = the people follow the behaviour of their king (2.101.9ab); for, the world always focuses on *kāma*, desire (2.101.9cd). Therefore the conduct of the kings must be truthful and beyond the censure of the people: *satyam evānṛśaṃsyaṃ ca rājavṛttaṃ sanātanam* (2.101.10ab). Besides this it requires looking after the subjects well and seeing to it that they are happy and progress, which is suggested in the words *prajāpālana* and *lokānurañjana* (1.50.20; 2.18.33, 98.57, 99.15). A minor but a very important point is projected by Rāma in the *kaccitsarga*: to achieve success is never to neglect the people who are weak and not to misbehave with them (2.94.31).

Further, the *Rāmāyaṇa* is aware that *rājadharma* was *ugra* – stern, a path of thorns on which one had to remain ever alert to safeguard one's interest. This was Mantharā's reproach to Kaikeyī (2.7.19). This implies the inborn shrewdness to understand the political situation and guard one's self-interest. Here the implication was to remind Kaikeyī of her son Bharata's right over the throne as promised to her father at the time of marriage (*rājyaśulka*), granting of boons after the war with Śambara and put legitimate claim right away. Mantharā rouses Kaikeyī to demand Bharata's coronation and Rāma's exile to the forest for fourteen years. Thereby Bharata would be able to establish himself as the king without obstacle, *niṣkaṇṭaka* (2.8.24), and would be able to gain support from his well-wishers as well as supporters in his favour; while danger from Rāma would be totally eradicated as a brother is one's natural enemy – *sahajaśatru* (2.8.24).

Rājadoṣas or the weaknesses of the king

These form the focus of discussion and comparison as Sītārāma very specifically observes the similarity between the *Nītiprakāśikā* and the *Rāmāyaṇa*. These weaknesses have to be avoided by the king so as to perform his *dharma* efficiently.

These are fourteen in number (*Tattvavivṛti* on NP 1.48–49): a) *nāstikya* = disregard for the other/next world (*paralokāpalāpa*), b) *anṛta* = falsehood, c) *krodha* = being short-tempered, d) *pramāda* = negligence, carelessness, e) *dīrghasūtratā* = procrastination of tasks to be done, f) *adarśana* = not appearing before the subjects (Tilaka in *kaccitsarga* = not visiting the wise), g) *jñānavatām alasya* = lethargy towards the wise and the learned, h) *pañcavṛttitā* = being under the control of the sensual pleasures yielded by the five sense organs, i) *ekacintanam arthānām* = considering purposes/means/goals to be achieved without a comprehensive outlook, (that is, with only one person – Tilaka in the *kaccitsarga*), j) *anarthajñaiḥ cintana* = counsel of people with wrong vision, (not initiating the tasks that must be undertaken along with the advisors – Tilaka in the *kaccitsarga*), k) *niścitānām anārambha* = not undertaking and commencing the tasks already decided for execution, l) *mantrasya apariraksaṇa* = not guarding the confidential discussions, counsel and decisions taken with the help of the advisory council, m) *maṅgalādyaprayoga* = putting aside the performance of auspicious rites like *pūjā* etc. of the propitiating the deities to be performed at dawn, and n) *pratyutthānaṃ sarvaśaḥ* = simultaneously marching towards all the surrounding enemies. The presence of these *doṣas* in the king is evidence of his lack of far-sightedness and maturity needed to win the love of the people. Therefore they must not be present in the king.

The *Tattvavivṛti* notes that these are mentioned in the *Rāmāyaṇa*. Therefore, in this particular context and otherwise in general, Oppert's introductory remark '... the *Nītiprakāśikā* coincides in many points with the *Rāmāyaṇa*, especially with its first two books ... ' in his edition of the *Nītiprakāśikā* may be recalled.

Now towards the end of his inquiry, Rāma emphasized that the king must be away from the fourteen *rājadoṣas*. These are the ones that reminded Sītārāma of the *Rāmāyaṇa* while commenting on NP 1.48–49. They are the same as enumerated there with two modifications and *anarthajñaiḥ mantraṇam* (Rām 2.94. 56–58b).

Probably the crux of the whole situation is summed up in the fact that Daśaratha was passionately fond of Kaikeyī – *vṛddho rājā taruṇī bhāryā prāṇebhyo 'pi garīyasī* – which Kaikeyī exploited; in fact this can be called one of the *vyasana*s of the kings. On the other hand there can be seen Daśaratha's desire to have Rāma as the crown prince, which as noted above was proposed and sanctioned by the assembly in haste and in the absence of Bharata, the heir to the throne by *rājyaśulka*.

In brief, about *rājanīti* of the *Rāmāyaṇa* John Brockington says,

> The king surrounded by his courtiers forms the natural focus of society throughout the epic, not just in the Ayodhyākāṇḍa where the narrative concentrates on the machinations of the court (Brockington 1984: 124).

He further observes,

> Nonetheless, the general impression is of a ruler intimately in contact with his people and responsive to their feelings, even though mechanisms for popular participation in government were rudimentary, if indeed they really existed at all (Brockington 1984: 124).

The above discussion is an attempt to compare the delineation of polity in the *Nītiprakāśikā* and the *Rāmāyaṇa* on the basis of the observation by Gustav Oppert, the first editor of the *Nītiprakāśikā*, which in turn is derived from the quotations from the *Rāmāyaṇa* in the *Tattvavivṛti*. Oppert's mention of the *Mahābhārata* reminds us of the dialogue between Nārada and Yudhiṣṭhira in the Sabhāparvan (Mbh 2.5.7–116), which despite being known as the *kaccitadhyāya* is not referred to by Sītārāma. P. V. Kane calls both of these *kaccitpraśna* chapters (Kane 1968: 399).

Here too each query of Nārada begins with the particle *kaccit.* Several verses found here have a striking similarity and at times are identical with those found in the *kaccitsarga* of the *Rāmāyaṇa.* To illustrate, again, Mbh 2.96–98, describing the *rājadoṣas*, are exactly the same as those in the *Rāmāyaṇa*, except for a word or two here or there – Mbh 2.5.96d has *kṣiptacittatām* while Rām 2.94.56d has *pañcavṛttitām*, Mbh 2.5.97b has *cintanam* instead of *mantraṇam* of Rām 2.94.57b, and Mbh 2.5.98b has *prasaṅgaṃ viṣayeṣu ca* whereas Rām 2.94.58b has *pratyutthānaṃ ca sarvaśaḥ.* E. W. Hopkins, while showing the parallel phraseology of the two epics, brings out the parallels between these cantos,

> But the most remarkable chapter of parallels in didacticism is to be seen in the *kaccit* sections of the two epics (Hopkins 1898: 147).

Towards the end of his paper, Hopkins draws conclusion on the basis of this and other parallel features of the two epics,

> I need scarcely point out how clearly the *kaccitsarga* shows the method by which an epic may be turned gradually into a text-book of morality. The *Rāmāyaṇa* is here entering on the path into which and along which the *Mahābhārata* was dragged, and its *kaccit* chapter furnishes by analogy a useful corrective to the superficial theory that the *Mahābhārata* was never an epic because it is now stuffed with maxims (Hopkins 1898: 151).

This may be supported by another but an extremely brief and formal query in the Bālakāṇḍa 51.7–9 containing Vasiṣṭha's questions pertaining to the proper administration to Viśvāmitra in his earlier life as a king. These verses also begin with the particle *kaccit* similarly translated by R. P. Goldman as 'I trust' (Goldman 1984: 222–223).

Here Sheldon Pollock's annotation may be cited,

> Such *kaccit-parvans* were simply a stock feature of ancient Indian epic, intended to inculcate kingly *dharma* ... and so the question of borrowing is in a sense misleading. Yet the *kaccit-parvan* of the *Ayodhyākāṇḍa* can lay claim to a special narrative propriety that is lacking in the case of the principal *Mahābhārata* version. ... For it seems to testify to Rāma's profound, encyclopedic knowledge of government, thereby validating his qualifications for the kingship and making his stern refusal to assume it not only the more regrettable and poignant, but incontestable ... (Pollock 1986: 495 – annotations on Rām 2.94).

John Brockington also expresses a similar opinion about the inter-relationship of the two epics while examining

> their stock of formulae deriving from the oral composition underlying both epics and often therefore assumed to be broadly similar (Brockington 1985: 77).

Thus we find parallels in the instructions on *nīti* in both the Epics. No one would readily identify the authors of *Nītiprakāśikā* and the *Mahābhārata* due to such similarities; for, basically the *Nītiprakāśikā* being a work on *dhanurveda*, the linguists too probably find it difficult to place its composition as being contemporary to the *Rāmāyaṇa* and *Mahābhārata*. However, it may safely be deduced that the trend of the depiction of polity and its aim in the later texts was idealistic, rooted in the concept of the ideal *Rāmarājya*. As far as *nīti* as both ethics/morality and as polity are concerned, Sheldon Pollock notes that we find in the epic poetry

> an extension of a private ethics into a public context, a grounding of political success on moral success, and an embodiment of these concerns, with self-conscious didactic purpose in epic poetry (Pollock 1986: 22).

It is because as he has observed in the beginning,

> at the very root, in the very heart of many of these epic narratives, can be found a political problem similar to the one with which the Ayodhyākāṇḍa confronts us (Pollock 1986: 9).

REFERENCES

Primary sources

NP = *Nītiprakāśika*. Ed. by Gustav Oppert. Madras: Government Press, 1882 (reprint: New Delhi: Kumar Brothers, 1970).

Nītiprakāśika. With *Tattvavivṛti* of Sītārāma. Ed. by T. Chandrasekharan. (Government Oriental Manuscripts Series, 24.) Madras: Government Press, 1953.

Rām = *Rāmāyaṇam*. With the Tilaka commentary of Rāma. Ed. by Vasudev Laxman Shastri Panashikar. Bombay: Nirnayasagar Press, 1930 (reprint: Delhi: Indological Book House, 1983).

Śrīmad-vālmīki-rāmāyaṇam. With the Amṛtakataka commentary of Mādhavayogī, vol. 3: *Ayodhyakanda* – II. Ed. by K. S. Varadacharya. Mysore: Oriental Research Institute, University of Mysore, 1965.

Vālmīki Rāmāyaṇa. Text as Constituted in its Critical Edition. Ed. by R. T. Vyas. Vadodara: Oriental Institute, 1992.

Secondary sources

BROCKINGTON J. L. 1984. *Righteous Rāma. The Evolution of an Epic*. New Delhi: Oxford University Press.

——— 1985. Fashions in formulae: Sanskrit Epic tradition: 3. *Proceedings of the Fifth World Sanskrit Conference, Varanasi*: 77–90. New Delhi: Rashtriya Sanskrit Sansthan. [Reprinted in: Greg Bailey & Mary Brockington (eds.), *Epic Threads. John Brockington on the Sanskrit Epics*. New Delhi: Oxford University Press, 2000, pp. 339–352.]

DANDEKAR, R. N. 1998. Ancient Indian polity: some notes. In: Rajendra Nanavati (ed.), *Purāṇa-itihāsa-vimarśaḥ* (Essays in Honour of Prof. S. G. Kantawala). New Delhi: Bharatiya Vidya Prakashan.

DHARMA, P. C. 1989. *The Rāmāyaṇa Polity*. 2nd edition. Bombay: Bharatiya Vidya Bhavan.

GOLDMAN, Robert P. (ed. & trans.) 1984. *The Rāmāyaṇa of Vālmīki. An Epic of Ancient India*, vol. 1: *Bālakāṇḍa*. Introduction by Robert P. Goldman. Annotation by Robert P. Goldman and Sally J. Sutherland. Princeton: Princeton University Press.

HOPKINS, E. Washburn 1898. Parallel features in the two Sanskrit Epics. *American Journal of Philology* 19.2: 138–151.

KANE, Pandurang Vaman 1968. The two Epics. In: P. V. Kane, *History of Dharmaśāstra*, vol. 1, part 1. 2nd edition, rev. and enlarged: 349–408. Poona: Bhandarkar Oriental Research Institute.

POLLOCK, Sheldon (trans.) 1986. *The Rāmāyaṇa of Vālmīki. An Epic of Ancient India*, vol. 2: *Ayodhyākāṇḍa*. Ed. by Robert P. Goldman. Introduction and annotation by Sheldon Pollock. Princeton: Princeton University Press.

——— 1995. Rāmāyaṇa and public discourse in medieval India. In: R. T. Vyas (ed.), *Silpa-samvit*: 141–158. Vadodara: Abhinav Publications.

The Locus of the *Ānandarāmāyaṇa*

VIDYUT AKLUJKAR

The medieval Sanskrit text of 12,252 verses, called *Ānandarāmāyaṇa* (ĀR) has been well-known for more than five centuries among many areas of India.[1] It was heavily drawn upon by Eknāth of Maharashtra (1528–1573) in the creation of his *Bhāvārtharāmāyaṇa* in Marathi, and also by Narahari (16th century) in Karnataka in his *Toraverāmāyaṇa* in Kannada. In Maharashtra, the temple priests used this text even in the 20th century for their daily recitation and story-telling sessions. I have seen on ĀR *pothi*s hand-written notes by the priests of Nasik to the effect, 'These cantos were lent to such and such priest to use at the *kathāpurāṇa* session at such and such temple.' I am told that in Tamilnādu it has had several thousands of followers, some of whom are now residents in the United States. There are manuscripts of it in Telugu script and editions and translations in Hindi and Marathi.[2] It is also popular among the

---

[1] *Yogavāsiṣṭha-* and *Adhyātmarāmāyaṇa* are mentioned in the list of many *Rāmāyaṇa*s in ĀR Manoharakāṇḍa 8.63 [61–70].

[2] There is no complete translation into English of the text. I have provided partial translations as needed in articles written on the text (Aklujkar 1991; 1995; 1997; 2000). I am working on a complete translation and an edition of the text. I have the first couple of pages and the last page of an early edition edited with critical comments and variant readings by Vasudev Balacharya Ainapure, published in 1902, from the Janata printing press of Gopal Narayan in Mumbai. Also an earlier printed edition published from Bombay is by Govardhandas Lakhmidas (Thakkar/Bhatiya), in *pothi* form. It was published in 1887 by the Jagadīśvara Printing press in Bombay. Some pages of earlier chapters are missing from it and a copy I obtained is uninterrupted

Rasik Sampradāya of North India since it anticipates the later trend of Rasika worship of Rāma.[3] The Rāmnāmis of Uttar Pradesh regard it as one of their sacred texts, since it states the importance of wearing a garment imprinted with the name of Rāma along with prescriptions of how and when to use such garments with *Rāmanāma* (ĀR Manohara 12.79–99).

With such a wide following among diverse people, we still do not know anything about the authorship of this text beyond the deliberate and anachronistic attribution of the entire work to the authorship of Muni Vālmīki. While analysing in an earlier paper the rationale of such a blatantly anachronistic attribution and the deliberate anonymity of its author, I had indicated my belief that the locus of ĀR could be Maharashtra (Aklujkar 2000: 86). My present paper puts forward some internal evidence to support the hypothesis that the composition of ĀR must have taken place in Maharashtra, more likely, in or around the river Godāvarī, near Nasik.[4] I shall argue that the composer was not just a resident of Maharashtra, but someone whose native language was Marathi, and who took great pride in his Marathi heritage. In passing, I shall also show how the Hindi translation of ĀR at times misses the linguistic peculiarities only apparent to a Marathi speaker.

The following points are presented to support my hypothesis:

---

only from Rājyakāṇḍa uttarārdha. A Marathi translation made by Pundit Vishnushastri Bapat was published by Dā. Sāṃ. Yande, in 1900. I have secured two pages of the *prastāvanā* or preface to this edition, but not the whole translation. A Hindi commentary called *Jyotsnā* written by Pundit Rāmtej Pandey is included in the edition published by Yugal Kishor Dwivedi, in 1977 by Pandit Pustakālaya, Varanasi. Another Hindi translation of all nine cantos, albeit in abridged form, is by Pundit Ramlagna Pandeya, published by Thakur Prasad Kailashnath Bookseller, Varanasi in 2001. Also see Gupta 1984.

[3] Bulcke 1970; Varnekar 1988; Simha 1957: 97, 99.

[4] I had already indicated this possibility in a footnote in my earlier article (Aklujkar 1995: 109, fn. 4).

1. The *Ānandarāmāyaṇa*'s choice of a pair of narrator–listener from Maharashtra

The composition of ĀR is in the form of a dialogue between two main pairs of narrator–listeners. Framing the entire work is the divine pair, Śiva and Pārvatī, who are seated on Mt. Kailāsa,[5] but the second pair introduced at the third *sarga* of the second canto is that of a mortal teacher Rāmadāsa and his disciple Viṣṇudāsa, who are situated in the central area of Godāvarī river in Maharashtra, in a holy place called Abjaka.[6] Abjakakṣetra is situated in the navel of Godāvarī, in Daṇḍakāraṇya, where a brahmin called Nṛsimha lived whose son Rāmadāsa is the narrator of most of the cantos of ĀR. That puts the locus of ĀR on two poetic planes, as it were, a divine one on Kailāsa and the mundane one in the region of Daṇḍakāraṇya, near Godāvarī, in Maharashtra. Elsewhere too, ĀR sings praises of Daṇḍakāraṇya by saying that it is the holiest of places and that no other place can equal its glory (ĀR Rājya [uttara] 18.106).[7] ĀR often refers to Godāvarī by its alternate name well-known in Maharashtra, as Gautamī, showing the river's connection with the sage Gautama.

The Hindi translation/commentary *Jyotsnā* of ĀR by Pandit Ramtej Pandey is mostly adequate, but often careless, full of omissions and sometimes it is simply inaccurate. It does not show awareness that Gautamī is another name of Godāvarī. For example, in the first canto, Sārakāṇḍa, when Sītā in Rāvaṇa's captivity is repenting her wrongful accusations of Lakṣmaṇa at the banks of Godāvarī, the words in ĀR (Sāra 9.108cd–109ab) are as follows:

---

5 In this, the ĀR seems to follow the model of the *Adhyātmarāmāyaṇa*.

6 ĀR Yātrā 3.1–12. The name *abjaka* means a lotus, literally, that which is born in water. The oldest name of Nasik, on the banks of Godāvarī, is Padmapura, because Brahmā, seated in the lotus, did penance there. It is highly likely that ĀR is here referring to Nasik by its older name.

7 *deśebhyaḥ sakalebhyaś ca supuṇyam daṇḍakā bhavet* I *daṇḍakena samo deśo na bhūto na bhaviṣyati* II 106 II

> *mayā virāgas saumitris trāsito* gautamītaṭe ‖ 108 ‖
> *prāyaścittaṃ karomy adya tasya tyaktvā svajīvitam* ।

The Hindi translation reads,

> *maine* gomatī nadī ke kināre *sumitrā ke nirdoṣa putra lakṣman ko jo dhamkāyā thā, usakā main āj prāṇ de kar prāyaścitta karūngī* [emphasis mine].

Here, an entirely different name, Gomatī, is substituted in the Hindi translation for the name Gautamī in the original text. Although the two names are similar and such an error of flipping the last two syllables is possible, we must realize that the Gomatī river, which is mentioned in *Vālmīkirāmāyaṇa*, is in North India, while Gautamī alias Godāvarī of ĀR is in central India.

At another place (ĀR Yātrā 3.2), the Sanskrit reads:

> *bhārate daṇḍakāraṇye godānābhau virājite* ।
> kṣetre 'bjake *nṛsiṃhākhyo munir agre bhaviṣyati* ‖ 2 ‖

The Hindi translation reads:

> *bhārata varṣa ke daṇḍakāraṇya mem godāvarī ke madhya pradeśīya* andhaka *kṣetra mem* ...

So sometimes Pandeyji translates Abjaka as Andhaka, for no apparent reason, and sometimes, he forgets that it is a place name, and wrongly translates as if it refers to a river.[8] In all this, the centrality given in ĀR to the river Gautamī, and *abjakakṣetra* alias present day Nasik is lost in Pandeyji's Hindi translation.

2. The *Ānandarāmāyaṇa* gives the most detailed descriptions of the holy places and rivers of Maharashtra in the context of the pilgrimages undertaken by Rāma

2(a) The second canto of ĀR, Yātrākāṇḍa, is full of descriptions of holy pilgrimages Rāma undertakes along with his family. The

[8] *godānābhau abjake 'tha snātvā natvā trivikramam* ‖ (ĀR Yātrā 7.78); '*godāvarī aur abjak nadī mem snāna kar ke*' (ĀR translation, p. 199).

last three *sargas*, 6–9 of Yātrākāṇḍa chart out his pilgrimages to the east, west and south. The southern pilgrimage in the seventh *sarga* is the most detailed one regarding the place names and rivers and temples. Along with many famous place names in South India, the following not so famous ones are mentioned from Maharashtra: Kolapura (Kolhapur), Karavīra (Satara), confluence of Kṛṣṇā and Veṇī (Veṇṇā), Ghaṭaprabhā, Mallāri, Karā river, Narasimha on Nirā river, Pāṇḍuraṅga on Candrabhāgā, Bhīmā and Chandalā, Mārtaṇḍa of Premapura, Nīladurgā, Devī of Tulajāpura, Māṇikyāmbā, Yogīśvarī of Ambāpura, Vaidyanātha (Vaijnath), Vañjarāsaṃgama, Nāgeśa, Pūrṇāsaṃgama, and Mudgala's *āśrama* on Godāvarī's north bank, Bāṇatīrtha, Sindhuphenasaṃgama, Abjakakṣetra, Trivikrama, Ambikā, Caṇḍikā, Ātmatīrtha, Vijnāneśvara, Mahālakṣmī on Vaḍavāsaṃgama, Pratiṣṭhāna or Paithan, Vṛddhāilasaṃgama, Nṛsiṃha on Śivanandāsaṃgama, Pravarāsaṃgama, Siddheśvara, Nivāsa (or Newase), Nūpura (Neūr), Puṇyastaṃbha (Puṇtāmbe) on the bank of Godāvarī, Kadrūsaṃgama, Vinatāsaṃgama, and lastly, Janasthāna, and Tryambaka also near Godāvarī.

These names show thorough familiarity with the region of Maharashtra, and a fondness for the places and temples around the Godāvạrī. Not just that, but the pilgrimages of Rāma to the east and west in the two other *sargas* either begin or end near the Godāvarī as well (ĀR Yātrā 6.136, 8.8–9), so ĀR mentions some more place names on the banks of the Godāvarī or Gautamī. Thus, ĀR 2.6.136 mentions Saptagodāvarībheda, and the mountain bearing Rāma's name while ĀR 2.8.8–9 mention Tryambaka near Nasik, Saptaśṛṅgagiri (which is to the east of the present day Nasik), and the mountain where Rāma had earlier slept with Sītā.

2(b) Explanations and folk etymologies are supplied foı many minor place names all around Pañcavaṭī, connected with some incidents such as the chase of the golden deer and abduction of Sītā. For example, in ĀR Sāra 7.120–132, it is said that

> Where the deer with quick moving eyes fell down to the arrow shot by Rāma, there arose the town called 'Quickness' Cāpalya (Cāphaḷ in present day Marathi.). Even today one can see the river all around Pañcavaṭī where Lakṣmaṇa had drawn a line with his arrow. Even today one can see the foot of Rāvaṇa on a boulder where he had stood in front of the hut in Pañcavaṭī. The cave where Rāma had hidden Sītā while fighting with Khara is still seen there with the name of Sītāgumphā. The mountain where Rāma had slept with Sītā is called Rāmaśayyāgiri, and even today, the grass blades on that mountain look flat as if someone has slept there.

This mountain is indeed called in the present day Marathi, appropriately, Ramshej Dongar or Rāma's Bed mountain and is shown by the locals to be to the north of the present day Nasik. To wrap up the above explanations, ĀR says that even today, one can visualize Rāma and Lakṣmaṇa in the vicinity of Pañcavaṭī, if only one has the vision of wisdom. The ones lacking such a vision see only the rocks and boulders (ĀR Sāra 7.129).

3. Prominence given to the holy places near Godāvarī even at the expense of the otherwise famous North Indian holy sites

For example, in ĀR Yātrā 6, there is a story of Rāma's *piṇḍadāna* or funeral rites for king Daśaratha on the *pretaśilā* in Gayā. Daśaratha's hand is not seen accepting the *piṇḍa* as it should, so Rāma is puzzled. When asked, Lakṣmaṇa says, that earlier, when they had offered the *piṇḍa*s on the banks of Godāvarī, their father had accepted them very well. What's the matter now? Finally they find out that Daśaratha has already been sated by the 108 *piṇḍa*s or balls of sand made by Sītā to create an image of Durgā, which Daśaratha had consumed by snatching them as soon as she put them on the ground. When Rāma asks, why did you make such haste to grab the sand-balls made by Sītā, Daśaratha answers, 'Well, you know, here in Gayā, there are so many obstructions at the time of these offerings (ĀR Yātrā 6.113cd–114ab).[9] So I made haste while they were there.'

At another place, the narrator Rāmadāsa tells a story of his own father, Nṛsimha, who had become old but had no son. He goes to the holy site of his family deity, Mohinī of Nivāsa (or Newase) on the banks of Pravarā. There he is told to go to Ayodhyā, in order to bathe there at the *Rāmatīrtha* in the month of Caitra in spring so as to obtain a son. As he is quite old, he finds it difficult to make the long journey, and is extremely worried and harried, when Rāma himself appears to him and tells him not to go to Ayodhyā. Rāma instructs him to go to the holy site of Pratiṣṭhāna, on the northern bank of Godāvarī, and bathe there. Rāma assures him that the outcome will be the same. Nṛsiṃha obeys the command of Rāma, and obtains the son, Rāmadāsa, the narrator of ĀR. From then onwards, the pond where he bathed is called Rāma's pond (*Rāmahrada*, or *Rām ḍoh* in Marathi).

4. Inclusion of Marathi myths which appear odd or lacklustre in Sanskrit

There are at least two such tales which I have treated in separate articles elsewhere.[10]

4(a) In ĀR Sara 7.138–150, we find a story where Pārvatī goes to test Rāma's divinity by disguising herself as Sītā when Rāma is wandering in the forests looking for his lost Sītā. Rāma sees through her disguise, and says to her, 'I know who you are. I know you are not Sītā. Why are you here instead of on Mount Kailāsa?' Then ĀR concludes that since Rāma repeatedly used

---

9 *sa prāhātra gayāyām tu bahu vighnaḥ rāghava* ll 113 ll *bhavanti śrāddhasamaye kṛtā tasmāt tvarā mayā* l

10 'Who Was Vālmīki?: Gender and Retribution in Ānanada-rāmāyaṇa', and 'Is She or Is She Not Sītā? The Identity of Pārvatī and Sītā in Ānanda-rāmāyaṇa'. Both unpublished.

the words '*tvaṃ kā, tvaṃ kim*' in the above speech, a holy place of Devī with the name 'Tvaṃkā' became famous in Daṇḍakāraṇya (ĀR Sāra 7.148cd–149ab).[11] And yet, there is no Goddess called Tvaṃkā anywhere to correspond to this story. We know that this is a local legend which reads better in Marathi, since it explains the Marathi name 'Tukāī' of the Goddess situated near Nasik in Daṇḍakāraṇya. The Goddess Tukāī at this place is also called Rāmavaradāyinī, or Giver of boon to Rāma. Eknāth relates the same story in his *Bhāvārtharāmāyaṇa* (Araṇyakāṇḍa 20.64–66), where his words are

> *śrīrāma mhaṇe tū kāī, teṃci nāma lokāṃcyā ṭhāyīṃ* I *adyāpi rāhilemseṃ pāhī* I *mhaṇatī tukāī jagadaṃbā* II 64 II *tukāī nāmeṃ nāmābhidhāna* I *teṃ jagadaṃbeceṃ nijasthāna* I *adyāpi pūjitī jana* I *ati pāvana daṇḍakāraṇyīṃ* II 65 II *śrīrāmavaradāyinī mātā* I *tukāī mhaṇatī rāmavākyārthā* I *puḍhe kāya vartalī kathā* I *sāvadhāna śrotāṃ parisāvī* II 66 II

In Eknāth's Marathi account the words used by Rāma '*Tuṃ kāī?*' and the famous name 'Tukāī' of the topical goddess known to people for several centuries fit each other much better than the forced etymology given in Sanskrit to explain a non-existent name such as Tvaṃkā. It is an obvious case of Sanskritization of a local myth.

4(b) In Rājyakāṇḍa 14, while describing the former births of Sage Vālmīki, we are told that he was a *kirāta*, a forest-dwelling hunter who waylaid travellers to support his family. One day, he came upon the seven sages and wanted to rob them when they told him to go home and ask if his family was willing to be party to his sins thus acquired. He did so, and was disillusioned, and fell at the feet of the seven sages. In order to help him achieve a better goal of deliverance, they advised him to practice chanting an inverted mantra, '*marā marā*' instead of the proper name of Rāma, and left. He sat still, and went on chanting as they told

---

[11] *tvaṃ kā, tvaṃ kim iti proktaṃ rāghaveṇa punaḥ punaḥ* II 148 II *yā tvaṃ sā daṇḍake jātā tvaṃ kā nāṃnāṃbikā vane* I

him, and finally emerged from an anthill (*valmīka*) that grew all around him, and thereupon became known as the sage out of the anthill or Vālmīki. The story is found with some variations in *Skandapurāṇa*, etc. and finally in the *Adhyātmarāmāyaṇa*. The ĀR story is taken straight from the *Adhyātmarāmāyaṇa*, with a few additional verses inserted to embellish the *Adhyātmarāmāyaṇa* account (Ayodhyākāṇḍa 6.42–88). It appears to be a local legend popular in Marathi[12], since the inverted words *marā marā* have got a specific meaning in Marathi which makes them more appropriate to be in the mouth of a hunter who was willing to kill for money. In Marathi, the word *marā* is a verbal form in imperative mood, second person plural of the verb *marṇe*, to die. So in Marathi, the repetition of these words becomes meaningful as it seems that a highway robber was saying 'Die, die' to his victims when his practice was converted into a holy chanting by the grace of his compassionate benefactors and this became the story of Vālmīki.

5. Some words and phrases used in the *Ānandarāmāyaṇa* appear to be common Marathi vocabulary

Immediately recognizable for native Marathi speakers like myself, such words remain opaque to others. I shall elaborate on one example in detail.

Rājyakāṇḍa 19.83 describes Rāma's leisure activities with Sītā as playing with dice or playing chess. The words ĀR uses are:

sāribhiḥ sītayā krīḍāṃ tathā buddhibalena hi ।
nānākṛtrimasainyaiś cākarod anyair api prabhuḥ ॥ 83 ॥

Chess was a popular game which most likely originated in India and right from the epic times it is mentioned in many an-

[12] A summary of its conclusion appears in Eknāth's *Bhāvārtharāmāyaṇa* Bālakāṇḍa 15.68–70.

cient works under different names, including *caturaṅga*[13], *aṣṭāpada*[14], and *ayānayīna*. Only in Maharashtra, it is known by this specific word, *buddhibala*, continuously since at least 16th century until now.[15] In Hindi, they use the word *śatranj*, derived from the Pharsi word *śatranj* itself related to the older Sanskrit word, *caturaṅga*. Marathi uses the word *buddhibaḷa* for chess because it focuses on the most essential aspect of the game, intelligent pre-planning or, strength of intellect.

Again, at this point, the Hindi translator Pandeyji does not realize that *buddhibalena* is a reference to the game of chess. He translates it literally, as *tab rām sītā ke sāth buddhibal se kuch der tak causar ādi ke khel khelte the*, meaning that Rāma played with Sītā for a while, using the strength of intellect, games such as *chausar* etc. However, *chausar* is not chess; it is a board game similar to Parcheesi played with dice. Pandeyji has here left the pregnant word *nānākṛtrimasainyaiś* untranslated, and he has missed the significance of both the words *buddhibalena* and *tathā* on the line in question.

Another context for the occurrence of the word *buddhibala* in the sense of chess is in ĀR Vilāsa 6.11cd–12. Here the verses are:

---

[13] The name focuses on its fourfold armies, the elephant-riders, the cavalry, the camel-riders and the foot soldiers.

[14] The name focuses on its game-board divided in eight lines and eight houses. It is mentioned by this name in Rām 1.5.16, in describing *Ayodhyāṣṭāpadakarā* and the commentator, Tilaka tells us that *aṣṭāpada* is *dyūtaphalaka* or the board of chess. *Bharatīya Saṃskṛti kośa* vol. 6 lists many other references.

[15] The earliest reference I can locate is from Eknath's *Bhavartharāmāyaṇa*, where in five verses he describes the game accurately and uses it as a metaphor for the game of living in the world for a living liberated soul. *Bhavartharāmāyaṇa* Bālakāṇḍa 11.54–58. Monier Williams lists *sinhāsan dvātriṃśikā* as another context, but I have not been able to verify that.

jevī buddhibaḷe vāḍe koḍe mithyā rājā mithyā pradhāna hattī ghoḍe | *heṃ jāṇonī kheḷa kheḷatī puḍhe* | *teṃvī muktākaḍe bhavabhāva* || 54 || *soṃkaṭiyāṃ jīvabhāva nāhīṃ* | *tarī mārileṃ mhaṇatī pāhī* | *teṃvīṃ mukta dehīṃca videhī* | *bhay nāhīṃ maraṇāce* || 55 ||

> *hastyaśvoṣṭramantrirājadūtaiḥ kṛtrimanirmitaiḥ* || 11 ||
> *hemaratnahastidantasaṃbhūtair aticitritaiḥ* |
> *krīḍāṃ buddhibalenaiva cakāra sītayā sukhaṃ* || 12 ||

Here there is again a clear reference to various game pieces of chess as they are called in India, that is, the Elephant (bishop), the Horse (knight), the Camel, the Minister (queen) and the King's messenger (pawn). Nevertheless, the Hindi translator Pandeyji has missed the point that it is the game of chess, called *buddhibala* that is being described here. He takes these to be different toys with which Rāma and Sītā play. His translation simply reads

> *vahāṃnakalī hāthī, ghoḍe ūṃṭ, mantrī, aur rājdūt ādi ke khilaune rakkhe hue the* | *unke sāth rām tathā sītā ne baḍī der tak khelvāḍ kiyā* | *unmeṃ bahut se khilaune suvarṇ, hāthīdāṃt, evaṃ ratno ke bane hue the, aur, unpar baḍhiyā raṅgāī kī huī thī* |

Note that in this translation, Pandeyji leaves out the troublesome words *buddhibalenaiva* completely untouched. As I said before, this is a clear case of the ĀR composer's Marathi–Sanskrit idiom, transparent to another Marathi speaker but elusive to others.

6. References to some social conventions common in Maharashtra

6(a) Sītā when separated from Rāma, talks to the crow and asks him to verify the homecoming of Rāma by flying away (ĀR Rājya 5.10–11). This superstition linking a crow's sign to the desired guest's homecoming is very much in line with the superstition prevalent in Maharashtra, then and now. A famous expression of it is found in the song attributed to the famous Marathi Sant Dnyāndev, who lived in the 13th century. The song reads,

> That crow is cawing on the other side, Mai, it is giving me a sign. Fly away, O crow, fly away and I shall decorate your feet with gold, when the desired guest, the Lord of Paṇḍharī comes to my house.

paila to ge kāū kokatāhe । *śakuna ge māye sāṅgatāhe* ॥ *uḍa uḍa re kāū* । *tuze soneni maḍhavin pāū* । *pāhuṇe paṇḍharīrāū* । *gharā kaiṃ yetī* ॥

6(b) Sītā when separated from Rāma, also does rituals similar to the ones performed by women and men in present day Maharashtra, such as *dev pāṇyānt ghālṇe* or *dev koṇḍṇe*. She immerses the images of Gaṇeśa, Māruti and other gods in water in a *sthaṇḍila*, and by shutting the doors of the sanctum pours water over them (ĀR Rājya 5.7). It is believed that by doing these rituals, you bring about the desired outcome. Marathi literature abounds in references to this custom.[16]

6(c) The greeting 'Rām Rām' used by the *śūdra*s in Maharashtra until the 20th century is explained in ĀR by citing a story. The chanting of Rāma's name when greeting others is taught by Rāma as his blessing to the *śūdra*s so as to lead them to their own deliverance in *kaliyuga* (ĀR Rājya 10.111–115).

7. Some distinctive attitudes of the composer of the *Ānandarāmāyaṇa* which increase the possibility of his being a Marathi language speaker, and a Maharashtriyan

(Here I am aware that I am entering an impressionistic thicket which may turn out to be deadwood. I am, however, listing these, at present as additional albeit abstract points strengthening my overall argument.)

7(a) ĀR's extremely obvious liberalism in attempting to show the ultimate oneness of divinity.
Granted, it is a trait of philosophical discussions in Hinduism since the time of Upaniṣads and is found in various measures also in several other texts. However, it comes as a bit of a surprise in ĀR, since on the one hand, it is definitely a Vaiṣṇava text, exalting Rāma above all other forms of Divinity, including

[16] Vijay Tendulkar's 1973 play 'Ghashiram Kotwal' has a reference to it in Act 2, p. 57.

Viṣṇu's other incarnations, but on the other hand, it takes great pains to point out at every opportunity, in no uncertain words, that essentially, there is no difference whatsoever between different forms of divinity. ĀR uses all available methods to underscore this ultimate unity of Gods. There are descriptions of Rāma's daily practices where he worships Śiva, there are incidents where Rāma and Śiva pay visits to each other, give precious gifts, and spend a jolly time together. There are deliberate explanations given by Rāma to others regarding the unity of him and Śiva, or him and Kṛṣṇa. There are also heated debates and quarrels by the devotees of Rāma and Kṛṣṇa, where they hurl insults at each other's favourite Gods, only to clear the air with revelations of the ultimate non-duality or Advaita. Since I have already written three papers on this topic, I shall not elaborate on it here, except to point out that this unequivocal liberalism coupled with worship of Rāma/Viṣṇu seems to be also a product of the Marathi/Maharashtrian brand of Bhakti, founded by Dnyāndev in the 13th century and expanded by Nāmdev in the 14–15th centuries. Bhakti in Maharashtra always had a strong influence of Śaiva philosophy on it, via the Sants' connection with the Nāth panth.

7(b) Another trait of ĀR which makes me strongly suspect the composer's Maharashtrian heritage is his admission that along with the texts in Sanskrit, *Rāmāyaṇas* and other Bhakti texts written in vernacular languages are also respectable. While acknowledging respectability and worth of vernacular language compositions, ĀR also has the honesty to condemn bad poetry even if it is composed in Sanskrit. This unequivocal message regarding the real worth of poetry in Devavāṇī, the language of Gods and other languages, was also a strong trait of the Bhakti movement in Maharashtra. Dnyāndev and others have made similar statements to defend poetry composed in vernacular languages. The uniqueness of the composer of ĀR is that he defends other languages while himself composing this voluminous, fabulous epic in Sanskrit. Thereby the composer of ĀR shows not just his pride in his own vernacular, which I believe to be Marathi, but he also exhibits his mastery over the language of prestige, Sanskrit.

ĀR Manohara, 8.83–88ab puts forth his message thus:

ataḥ śrīrāmabhaktaiś ca kavitvais tatstutiḥ kṛtā ।
sā 'pi mānyā sadā śravyā paṭhanīyā budhaiḥ muhuḥ ॥ 83 ॥
bhāratāc ca śatāṃśena phaladātrī smṛtātra hi ।
nindanti ye bhaktakṛtāṃ kavitāṃ te kharāḥ smṛtāḥ ॥ 84 ॥
tattadbhāṣākṛtaṃ kāvyaṃ rāmavarṇanasaṃyutam ।
bhāratasya sahasrāṃśaṃ śrotur vaktuḥ phalaṃ smṛtam ॥ 85 ॥
nirmātus tu bhavet puṇyaṃ sādhuśabdaśatāṃśataḥ ।
gīrvāṇīkavitākartā so 'pi vyāsāṃśa īritaḥ ॥ 86 ॥
svasvabhāṣākavitvānāṃ kartāras te kavīśavarāḥ ।
gīrvāṇīkavitā cāpi padānvayasamanvitā ॥ 87 ॥
arthapramāṇasahitā saiva mānyā na cetarā ।

7(c) The unusually feminist attitude of ĀR in spite of incorporating a few traditionally misogynist passages.

There are many indications of the feminist attitude of the composer of ĀR strewn throughout the text. Here for the first time one finds *Sītākavaca*, and Sītā's 108 names.[17] There is also the story of the killing of Mūlakāsura by Sītā in Rājyakāṇḍa 6. ĀR stipulates that one is to worship Rāma always along with Sītā, never should one worship the image of Rāma devoid of Sītā, by himself, alone. There are eloquent defences of Sītā's purity by her sons, Lava and Kuśa. Rāma behaves with utmost tenderness towards Sītā in his daily life in ĀR. He even brings her back from the earth, when she disappears in the earth to escape the repeated fire-tests. His monogamy is described in most believable details. Sītā's character is developed in the most flattering manner. Her intelligence, her steadfast love for Rāma, her generosity toward her people, her strength of character under adversity, her

[17] ĀR Manohara 14. Both these are attributed to Ṛṣi Agasti.

exquisite manners when dealing with elders, peers and young ones are described in great detail.

All women of the royal family including the youngest step-mother of Rāma, Kaikeyī, are shown to have reached the highest level of deliverance, by pondering upon a simple incident and thereby achieving self-realization. In addition, there are many stories to celebrate intelligent and morally superior women, and their victories over their not so intelligent nor moral husbands.

The feminism of the composer of ĀR stands out as being ahead of his times and it also lends support to my supposition, as it is a forerunner of many later day social reformers in Maharashtra who were feminist males, and were ahead of many other areas in India.

7(d) A refreshing sense of humour which is not afraid to poke fun at his own revered, loved and cherished deity.

I have written from time to time on the humour in ĀR. There are many original incidents in ĀR that are not found in other works on *Rāmakathā*. There is the incident when Rāma is afraid of laughter and puts a ban on laughter in his kingdom, or the time when he explains to Sītā why she has no daughter, nor more children, or when he punishes a monk by granting him the headship of a Maṭha, etc.

8. Many minor socio-linguistic peculiarities point to a Marathi context

8(a) For example, Hanumān is mentioned throughout ĀR mostly by his patronym, as Māruti, rather than any other name. This is the name more commonly used in Maharashtra than any of his other names.

8(b) ĀR uses the word *hāṭaka*, or *haṭṭa* consistently for the market rather than other words such as *paṇya*. This word *hāṭa* is still used in rural Maharashtra for market.

8(c) There are several references throughout ĀR to *drākṣāvaḷḷī* or grape-vines which are not found anywhere in *Vālmīkirāmāyaṇa*. Rāma and Sītā often relax under a pergola adorned with grape-vines. Sītā, when she is awaiting Rāma's return also sits under the grape-pergola (ĀR Rājya 5.2). In Rājyakāṇḍa 22.3–4, a friend called Bhūrikīrti sends thousands of baskets full of grapes to Rāma as a gift.

Grapes from Nasik and nearby area are famous even in the twenty-first century in Maharashtra and, even today, the baskets full of grapes are sent as gifts among friends.

* * *

Examples can be accumulated further. However, on the strength of what I have said so far, I submit that it is highly likely that the composer of ĀR was a Marathi language speaker, a Maharastrian with a great sense of humour, proud of his intellectual and cultural heritage, and unique in creating an unusual, highly influential literary text of Rāma worship.

REFERENCES

Primary sources

Adhyātmarāmāyaṇa. Ed. with a Hindi translation by Munilal. Gorakhpur: Gita Press, 1967.

Ānandarāmāyaṇa. *Ānandarāmāyaṇam*. Ed. by Pandit Yugalkishor Dvivedi with the commentary 'Jyotsnā' by Pandeya Ramtej Shastri. Vārāṇasī: Pandita Pustakalaya, 1977.

Bālarāmāyaṇa. Rājaśekhara, *Bālarāmāyaṇa*. Ed. by Pundit Govinda Deva Shastri. Benaras: Medical Hall Press, 1869.

Bhāvārtharāmāyaṇa. Ekanātha, *Bhāvārtha rāmāyaṇa by Śrī santa Ekanātha*. Ed. by Narahari Viṣṇuśāstrī Paṇaśīkar. 2 vols. Pune: Yashwant Prakashan, 1968.

Gargasaṃhitā. Composed by Garga Mahāmuni. Ed. by Pandey Ramtej Shastri. Varanasi: Pandit Pustakalaya, 1971.

Hanumannāṭaka. Hanumannāṭakaṃ śrīmadhanumadviracitam vibhāsanskṛta-hindī-vyākhyopetam. Ed. by Pandit Jagadish Mishra Kavyatīrtha. (Haridas Sanskrit Series, 271.) Varanasi: Chowkhamba Sanskrit Series Office, 1967.

Kālidāsagranthāvalī. Complete Works of Kālidāsa. Critically ed. by Revāprasāda Dvivedī. Varanasi: Banaras Hindu University, 1976.

Mukteśvarakṛtarāmāyaṇa. Ed. by Bhā Śrī Paranjape. Nagpur: Suvicāra Prakashan Mandal, 1969.

Padmapurāṇa. Padmapurāṇam. ṣaṣṭham uttarakhaṇḍam, pūrvārddham, caturtho bhāgaḥ. (Gurumandala Granthamālā, 18.) Calcutta: Mansukhray Mor, 1959.

Uttararāmacarita. Bhavabhūti, *The Uttara-rāma-carita of Bhavabhūti.* Ed. with the commentary of Vīrarāghava and English trans. by M. R. Kale. 4th edition. Bombay: Gopal Narayan and Co, 1934.

Vālmīkirāmāyaṇa.

——— *The Vālmīki Rāmāyaṇa.* Edition based on the North-eastern recension. Ed. by Gaspare Gorresio. 5 vols. Parigi: Stamperia Reale, 1843 (Reprint, ed. by V. Raghavan: Madras: Indian Heritage Trust, 1981–82).

——— *The Vālmīki Rāmāyaṇa According to Southern Recension.* Ed. by T. R. Krishnacharya. 2 vols. (Sri Garibdas Oriental Series, 3.) Kumbakonam: Nirnaya Sagar Press, 1905 (Reprint: Delhi: Sri Satguru Publications, 1982).

——— *Śrīmadvālmīkirāmāyaṇaṃ śrīmadgovindarājīyarāmānujīyataniślokīmāheśvaratīrthīyākhyavyākhyācatuṣṭayasamanvitā.* Bombay: Lakshmi Venkateshwar Press, 1935.

——— *The Vālmīki-Rāmāyaṇa.* Critical Edition. Ed by Govindlal Hargovind Bhatt *et al.* 7 vols. Baroda: Oriental Institute, 1960–75.

——— *The Rāmāyaṇa of Vālmīki.* With the commentary Tilaka by Rāma. Ed. by Wāsudev Laxman Śāstrī Paṇaśīkar. Delhi: Indological Book House, 1983.

——— *The Vālmīkiya Rāmāyaṇā.* With Hindi Translation. Gorakhpur: Gītā Press, 1990.

——— *Vālmīki Rāmāyaṇam. Text as Constituted in its Critical Edition.* Vadodara: Oriental Institute, 1992.

Secondary sources

AKLUJKAR, V. 1991. Battle as banquet: a metaphor in Sūradāsa. *Journal of the American Oriental Society* 111: 353–361.

——— 1995. Rāmāvatāra recycled. *Annals of the Bhandarkar Oriental Research Institute* 76: 107–118.

——— 1997. Why build a bridge over the ocean: propriety in the Ānanda Rāmāyaṇa. Read at the Annual South Asia Conference at Madison, Wisconsin, Oct. 17, 1997. (Forthcoming)

——— 2000. Crying dogs and laughing trees in Rāma's kingdom: self-reflexivity in *Ānanda Rāmāyaṇa*. In: Paula Richman (ed.), *Questioning Ramayanas. A South Asian Tradition*: 83–103. New Delhi: Oxford University Press.

BULCKE, Camille. 1971. *Rām-kathā. Utpatti aur vikās*. Rev. edition. Allahabad: Allahabad University.

GUPTA, Aruna. 1984. *Ānanda-rāmāyaṇa. A Cultural Study*. Delhi, Eastern Book Linkers.

SIMHA, Bhagavati Prasad 1957. *Rāmabhakti mem Rasika Sampradāya*. Balrampur: Avadh Sahitya Mandir.

VARNEKAR, S. B. 1988. *Saṃskṛta vāṅmaya kośa*. 2 vols. Kalakattā: Bhāratīya Bhāṣā Pariṣada.

Contributors

VIDYUT AKLUJKAR
Centre for India and South Asia Research
Institute of Asian Research
University of British Columbia

NICHOLAS J. ALLEN
Institute of Social and Cultural Anthropology
University of Oxford

HORST BRINKHAUS
Abteilung für Indologie
Seminar für Orientalistik
Christian-Albrechts-Universität zu Kiel

JOHN BROCKINGTON
Emeritus Professor of Sanskrit
Asian Studies
University of Edinburgh

MARY BROCKINGTON
Asian Studies
University of Edinburgh
Research Fellow, International Association of Sanskrit Studies

SIMON BRODBECK
School of Religious and Theological Studies
Cardiff University

ANTONELLA COSI
Asian Studies
University of Edinburgh

DANIELLE FELLER
Section de langues et civilisations orientales
Faculté des Lettres
Universite; de Lausanne

JAMES L. FITZGERALD
Department of Classics
Brown University

SALLY J. SUTHERLAND GOLDMAN
Department of South and Southeast Asian Studies
University of California at Berkeley

JAMES M. HEGARTY
School of Religious and Theological Studies
Cardiff University

ALF HILTEBEITEL
Religion Department
George Washington University

PAOLO MAGNONE
Istituto di Glottologia
Università Cattolica di Milano

ANGELIKA MALINAR
Universität Zürich
Indogermanisches Seminar
Abteilung für Indologie

WENDY J. PHILLIPS-RODRIGUEZ
Faculty of Oriental Studies
University of Cambridge

ADHEESH SATHAYE
Department of Asian Studies
University of British Columbia

SVEN SELLMER
Department of Oriental Studies
Adam Mickiewicz University, Poznań

URMI S. SHAH
Department of Sanskrit
St Xavier's College, Ahmedabad

YAROSLAV VASSILKOV
Peter the Great Museum of Anthropology and Ethnography
St Petersburg

Index of Epic Passages Cited

Harivaṃśa (Hv)
(References to the Poona Critical Edition, except † = Bombay edition)

[Vālmīki-]*Rāmāyaṇa* (Rām)
(References to the Baroda Critical Edition, except † = Nirnayasagar edition)

BOOK 1: Bālakāṇḍa

General Index